STRUCTURE AND BONDING is intended to the publication of papers dealing with problems in all fields of modern inorganic chemistry, chemical physics and biochemistry, where the general subject are problems of chemical structure and bonding forces.

STRUCTURE AND BONDING is issued at indefinite intervals, according to the material received. With the acceptance for publication of a manuscript, copyright of all countries is vested exclusively in the publisher. Only papers not previously appearing elsewhere should be submitted. Likewise, the author guarantees against subsequent publication elsewhere. The text should be clear and concise as possible, the manuscript written on one side of the paper only. Illustrations should be limited to those actually necessary.

Manuscripts will be accepted by the editors:

Dr. *C. K. Jørgensen*   Cyanamid European Research Institute
                            91, Route de la Capite
                            CH-1223 Cologny GE

Professor *J. B. Neilands* University of California, Biochemistry Department
                            Berkeley, California / USA

Professor *R. S. Nyholm*   University College London, Department of Chemistry
                            Gower Street
                            London WC 1 / Great Britain

Dr. *Dirk Reinen*   Anorganisch-Chemisches Institut der Universität Bonn
                            D-5300 Bonn, Meckenheimer Allee 168

Professor   Wadham College, Inorganic Chemistry Laboratory
R. J. P. Williams   Oxford / Great Britain

SPRINGER-VERLAG   SPRINGER VERLAG
                                        NEW YORK INC.

D-6900 Heidelberg 1   D-1000 Berlin 31
P. O. Box 1780   Heidelberger Platz 3   175, Fifth Avenue
Telephone 4 91 01   Telephone 83 03 01   New York, N. Y. 10010
Telex 04-61 723   Telex 01-83 319   Telephone 673-2660

STRUCTURE AND BONDING

Volume 2

Editors: C. K. Jørgensen, Cologny · J. B. Neilands, Berkeley · R. S. Nyholm, London · D. Reinen, Bonn · R. J. P. Williams, Oxford

With 79 Figures

Springer-Verlag
Berlin Heidelberg GmbH 1967

ISBN 978-3-540-03989-1   ISBN 978-3-540-35558-8 (eBook)
DOI 10.1007/978-3-540-35558-8

Title-No. 4526

Contents

STRUCTURE AND BONDING

Volume 2

Editors: C. K. Jørgensen, Cologny · J. B. Neilands,
Berkeley · R. S. Nyholm, London · D. Reinen,
Bonn · R. J. P. Williams, Oxford

With 79 Figures

Springer-Verlag Berlin Heidelberg GmbH 1967

ISBN 978-3-540-03989-1   ISBN 978-3-540-35558-8 (eBook)
DOI 10.1007/978-3-540-35558-8

Title-No. 4526

Contents

The Physics of Hemoglobin[1]

Prof. Dr. M. Weissbluth

Biophysics Laboratory, Stanford University, Stanford, California, USA

Table of Contents

[1] This research was supported by the National Science Foundation under Grant NSF GB 3994 and by the Office of Naval Research under Contract Nonr 225 (87).

I. Introduction

Hemoglobin has long been an object of intensive study because of its central importance in respiration. Quite naturally, the initial emphasis has been on what may be termed the more obvious chemical properties. More recently, the application of a variety of physical methods has led to considerably deeper insights and it may be said that hemoglobin is probably the best understood biological molecule from the standpoint of physics. It therefore serves as an excellent example of research in a rapidly expanding area which lies at the borders of chemistry, physics, and biology or what is now called biophysics.

On the experimental side there exist absorption spectra (*10, 13, 39*), measurements of magnetic susceptibility (*46, 52*), electron spin resonance data (*4, 14, 15, 28*), and Mössbauer spectra (*3, 16, 41, 42*). In addition, through the work of *Perutz* and his collaborators (*43, 48*) on the x-ray diffraction patterns of hemoglobin, and the companion work of *Kendrew* (*31, 32, 33*) on myoglobin, we now have a detailed description of the three dimensional structure of both molecules. Most of the theoretical effort has been devoted to the interpretation of magnetic properties and has led to an understanding of the low-lying electronic states of hemoglobin. This work is mainly that of *Griffith* (*18, 19, 20, 21, 22, 23, 24*) and *Kotani* (*35, 36, 37, 38*) who employed ligand field theory to provide a coherent theoretical framework encompassing most of the known experimental information. Additional insight, particularly in regard to spectra, has been provided by molecular orbital calculations (*45, 50, 59*).

We shall, in the first instance, summarize the experimental information on the physical properties of hemoglobin and then proceed to the main task which is to develop, in a reasonably leisurely fashion, the theoretical background for the interpretation of the experimental data.

The discussion will be confined almost exclusively to hemoglobin; nevertheless, because of the close relationship between hemoglobin and myoglobin most of the conclusions apply equally well to myoglobin.

II. Some Properties of Hemoglobin

A. Chemistry

Hemoglobin is the main constituent of red blood cells. It serves as the major vehicle for the transport of oxygen from the lungs to the tissues and participates in the return transport of carbon dioxide.

For the reader who needs some assistance with the chemical terminology we provide a brief description of the pertinent chemical terms and structures.

Pyrrole: A 5 -membered ring shown in Fig. 1.

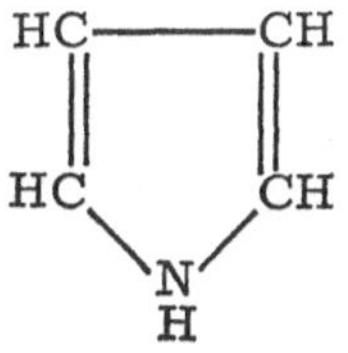

Fig. 1. Pyrrole.

Porphyrin: Four pyrrole rings linked together by $-$ CH $=$ bridges. There can be eight side chains on the ring in the positions marked *a* to *h* (see Fig. 2 on page 4).

Protoporphyrin: A porphyrin with the following side chains: Propionic acid ($-$ CH_2 $-$ CH_2 $-$ COOH) in positions *a, h*; methyl ($-$ CH_3) in positions *b, c, e, g*; and vinyl ($-$ CH $=$ CH_2) in positions *d, f*. The structure is shown in Fig. 3 on page 4.

Heme: When the two hydrogens on the center nitrogens in protoporphyrin are replaced by one iron atom the resulting compound is heme, also known as ferrous protoporphyrin (see Fig. 4 on page 5). The iron is coordinated to the four pyrrole nitrogens, the entire structure being approximately planar. Two other coordination positions, labelled 5 and 6, are available in directions perpendicular to the heme plane (Fig. 4b). In heme the iron atom is in the divalent or ferrous (Fe^{2+}) state.

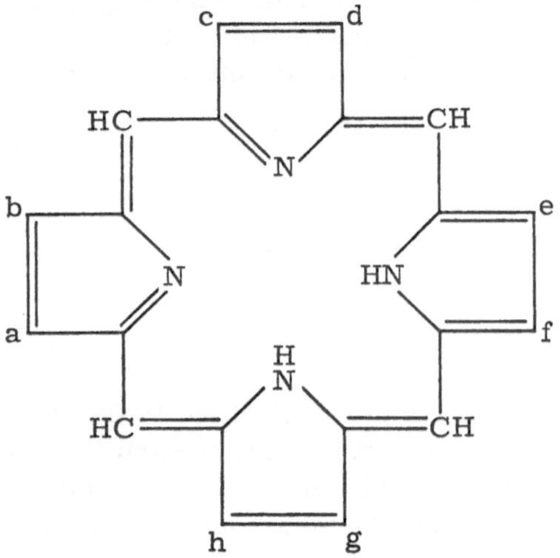

Fig. 2. Porphyrin skeleton.

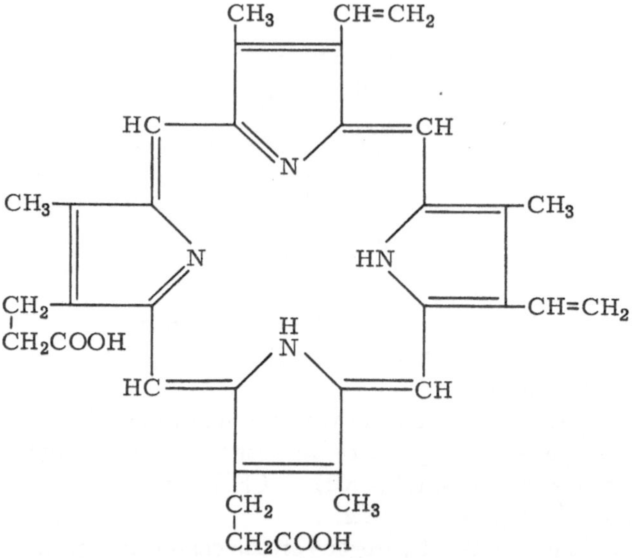

Fig. 3. Protoporphyrin.

Amino acid: Amino acids found in proteins have the general structure shown in Fig. 5a on page 5. R represents any one of approximately 20 molecular groups which distinguish one amino acid from another. A particular amino acid – histidine – is shown in Fig. 5b on page 5.

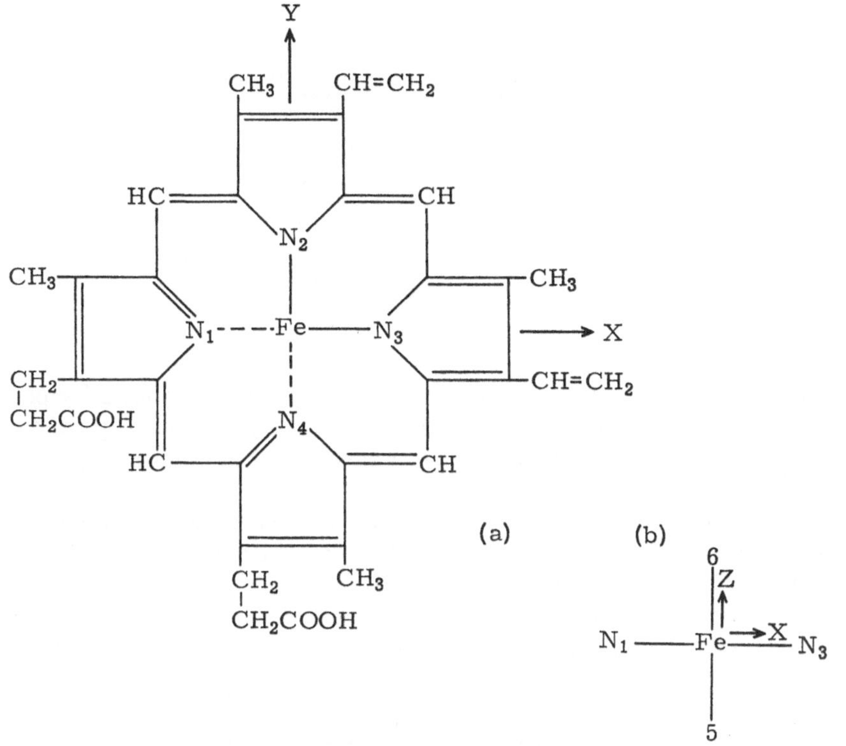

Fig. 4a and b. The heme complex with the numbering system for (a) in-plane ligands, and (b) out-of-plane ligands.

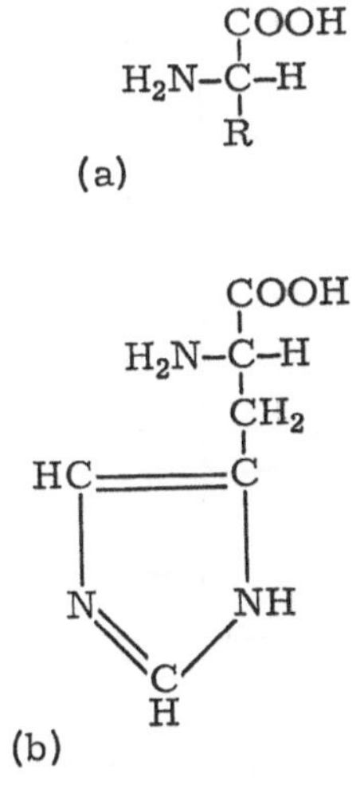

Fig. 5a and b. (a) General form of an amino acid. (b) The amino acid histidine; the five-membered, planar, ring structure is imidazole.

5

Peptide Bond: Amino acids may bind to one another by eliminating a molecule of water between the —COOH of one and the —NH$_2$ of another. The formation of a dipeptide consisting of two amino acid residues is shown in Fig. 6; the peptide bond is —CO—NH.

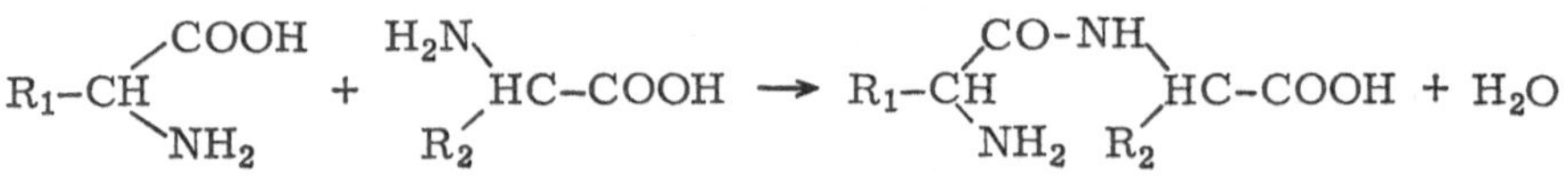

$$R_1-\overset{\displaystyle COOH}{\underset{\displaystyle NH_2}{CH}} \quad + \quad \overset{\displaystyle H_2N}{\underset{\displaystyle R_2}{HC-COOH}} \quad \rightarrow \quad R_1-\overset{\displaystyle CO-NH}{\underset{\displaystyle NH_2}{CH}} \quad \underset{\displaystyle R_2}{HC-COOH} \quad + \quad H_2O$$

Fig. 6. Formation of a dipeptide.

Polypeptides: A large number of amino acid residues joined by peptide bonds. Depending on external conditions polypeptides may exist in a helical configuration (Fig. 7) or in a randomly coiled form (see Fig. 8 on page 7).

Hemoglobin: The hemoglobin molecule has a molecular weight of about 67,000. There are some 10,000 atoms of which 4 are iron — the rest are C, H, O, N, S. Each of the four iron atoms is bound into a heme structure. In a direction perpendicular to the plane of the heme, the iron atom is attached to a polypeptide chain through the imidazole ring belonging

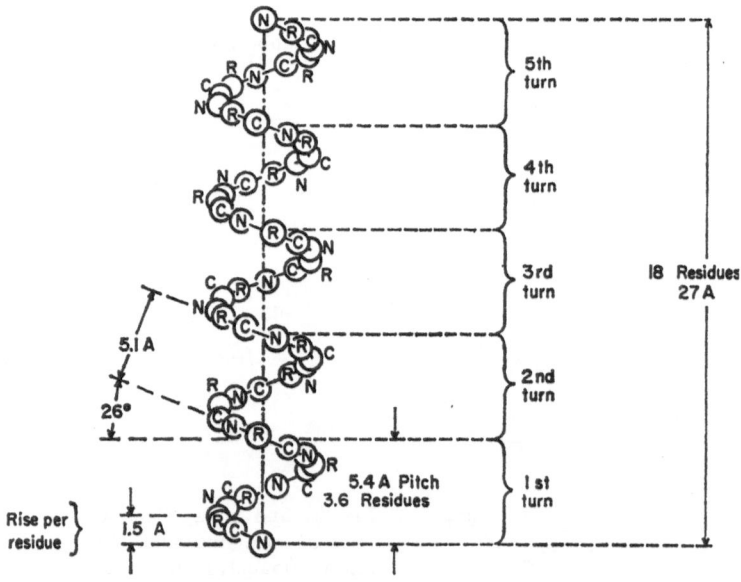

Fig. 7. Protein in the form of an α-helix (3.6 amino acid residues per turn) *(9)*.

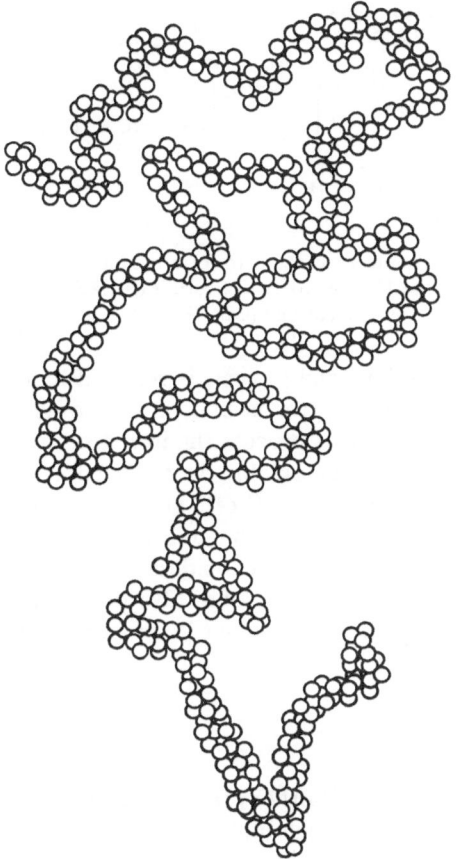

Fig. 8. Model of a protein in the configuration of a random coil.

to histidine (see Fig. 9 on page 8); the nitrogen of imidazole is designated as the fifth ligand. In the opposite direction, the sixth ligand may be any one of several small molecules or ions, e. g. H_2O, O_2, CO, F^-, OH^-, CN^-, etc. From the standpoint of biology the O_2 ligand is the most important. Hemoglobin thus consists of four hemes which are the functional units or active sites in the binding of oxygen, and four polypeptide chains; the latter taken together make up the protein part or globin of hemoglobin. The polypeptide chains are identical in pairs, labeled α and β, that is, there are two α-chains and two β-chains. Altogether, there are a total of 574 amino acid residues in the 4 chains.

The process of oxygen attachment is reversible; in the lungs where the partial pressure of oxygen is high, an O_2 molecule will bind to the iron atom forming oxyhemoglobin while in the tissues where the partial

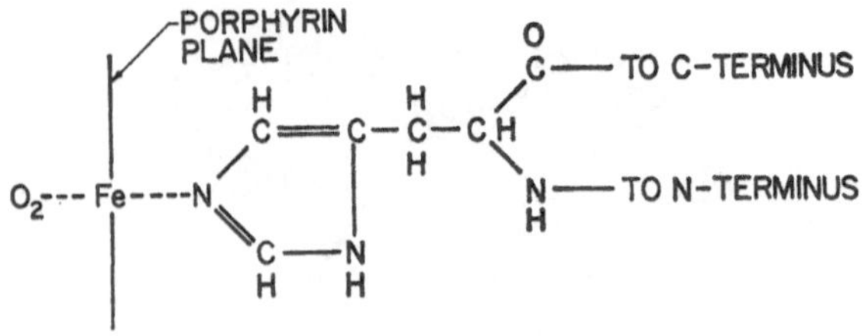

Fig. 9. Attachment of heme to protein to form hemoglobin. The histidine is linked to two amino acids, one leading to the N-terminus (NH$_2$ end) of the polypeptide chain and the other to the C-terminus (COOH end).

pressure of oxygen is low, the oxygen is released and an H$_2$O molecule attaches to the iron. The latter is known as deoxygenated or reduced hemoglobin. Throughout this process the iron atom remains in the divalent (Fe^{2+}) state. When hemoglobin is removed from the red cells it readily oxidizes to the trivalent (Fe^{3+}) state. It is important to recognize that the binding of oxygen to heme-iron in hemoglobin requires the presence of the globin — if the heme is detached from the globin it loses its ability to bind oxygen. Hemoglobin also participates in the transfer of carbon dioxide, but the latter is not bound to the heme.

B. Structure

In contrast to practically all other proteins, the three-dimensional structures of hemoglobin and myoglobin are known in great detail from the x-ray diffraction work of *Perutz, Kendrew* and their collaborators (*31, 33, 43, 47*). In hemoglobin each chain winds itself into a complicated spatial structure with the heme tucked into a pocket. The four chains taken together form an approximately tetrahedral array with overall dimensions of 50 Å x 55 Å x 69 Å. An idealized drawing of the two β-chains is shown in Fig. 10. The arrangement of the four heme groups relative to one another is shown in Fig. 11, and their spacing is given in Table 1. The heme itself has a somewhat distorted octahedral or tetragonal shape.

It has been found more recently (*33*) that in myoglobin, which may be pictured quite accurately as one fourth of a hemoglobin molecule, as well as in several iron porphyrins, the iron atom is displaced by approximately 1/4 Å from the porphyrin plane. In metmyoglobin the displacement is away from the histidine. It is conjectured that a similar displacement of the iron atom occurs in hemoglobin although the presently attained resolution in structural analysis is insufficient to bear this out.

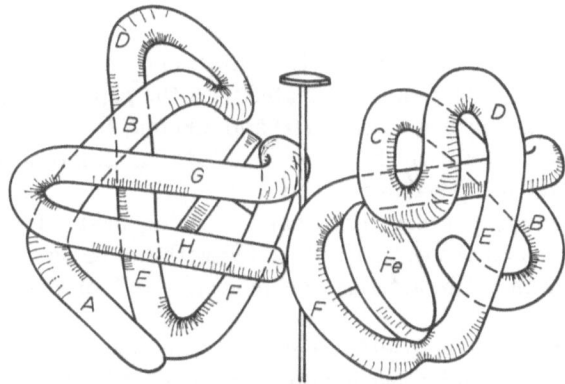

Fig. 10. The two β-chains in horse oxyhemoglobin. The α-chains form a similar pair (*43*).

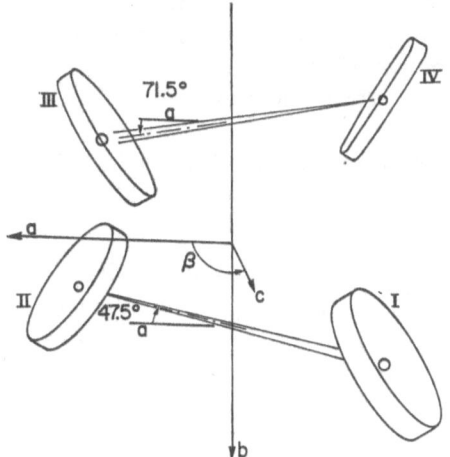

Fig. 11. Arrangement of heme groups in hemoglobin. The small circles are Fe atoms (*48*).

Table 1. *Spacing of hemes in hemoglobin* (*43*)

	Distance between hemes (Å)*	
	Horse Oxyhemoglobin	Human Reduced Hemoglobin
Fe_1–Fe_2	33.4	40.3
Fe_3–Fe_4	36.0	35.0
Fe_1–Fe_3	25.2	25.0
Fe_1–Fe_4	30.4	37.4

* See Fig. 11

9

C. Absorption Spectra

All proteins absorb strongly in the ultraviolet. At wavelengths of 190–200 mμ the absorption is associated with the peptide bond, at somewhat longer wavelengths between 260 and 280 mμ, the absorption is due to the aromatic group which occurs in the three amino acids tyrosine, tryptophan and phenyl-alanine. These absorptions are characteristic of all proteins which contain the aromatic amino acids. In hemoglobin there are additional absorptions associated with the heme group and it is these bands which characterize the molecule and distinguish it from other proteins.

One feature common to all hemoglobin absorption spectra (Fig. 12, 13) is the occurrence of an intense band in the region 400–420 mμ. The millimolar extinction coefficient ε_{mM} lies in the range of 120 to 140. This band is known as the Soret band and is due to an electronic transition in the prophyrin structure which is common to all hemoglobins. To discuss other features of the spectra, it is necessary to distinguish between the ferric and ferrous compounds and these classes must further be subdivided into compounds in which the iron is either in a state of high spin or low spin.

Ferric (Fe^{3+}) Hemoglobins (Fig. 12): In high spin compounds like ferrihemoglobin fluoride the Soret band is located at 405–410 mμ. In the

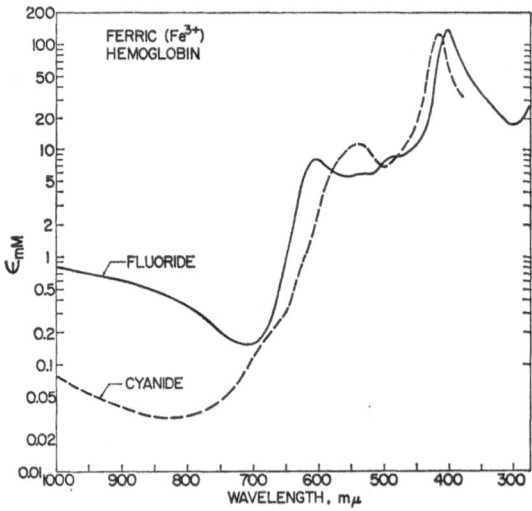

Fig. 12. Absorption spectra for ferrihemoglobin fluoride (high spin) and ferrihemoglobin cyanide (low spin).

visible there is an absorption band at about 500 mµ ($\varepsilon_{mM} \approx 9$) and a second band at 610–630 mµ ($\varepsilon_{mM} \approx 4$). In the infrared there is a broad band beginning at 850 mµ with ε_{mM} close to 1. Low spin compounds like ferrihemoglobin cyanide have a Soret band at 420 mµ. The band in the visible is at 545 mµ ($\varepsilon_{mM} \sim 10$); in the infrared there is a broad region of low absorption ($\varepsilon_{mM} < 0.1$) beginning at about 900 mµ. The spectra of low spin derivatives of a series of heme proteins e. g., the cyanide derivatives of hemoglobin, myoglobin and peroxidase are quite similar; this also holds for the high spin derivatives. The hydroxides, however, show no such regularities and it is thought that they are mixtures of low and high spin forms in thermal equilibrium (*13*).

Ferrous (Fe^{2+}) Hemoglobins (Fig. 13): In addition to the Soret band it is common for ferrous derivatives to have two absorption bands in the range 525–580 mµ with approximately the same intensity ($\varepsilon_{mM} = 13 - 15$).

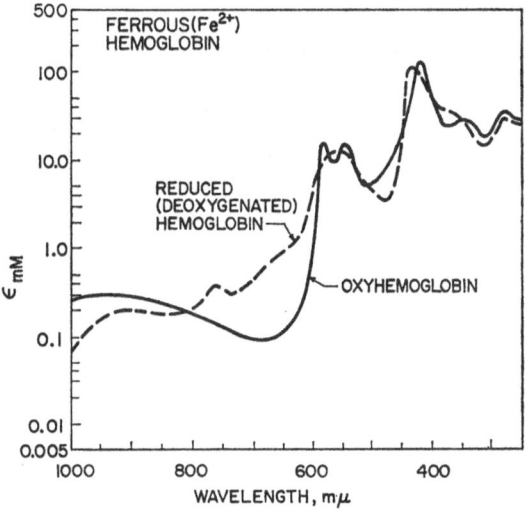

Fig. 13. Absorption spectra for (deoxygenated) hemoglobin (high spin) and oxyhemoglobin (low spin).

These are often designated by α (longer wavelength) and β (shorter wavelength). Thus in the low spin oxyhemoglobin there is the Soret band at 412–415 mµ, and α-band at 577 mµ, and a β-band at 541 mµ. Carboxyhemoglobin is very similar; however, in the infrared, carboxyhemoglobin is essentially transparent whereas oxyhemoglobin has a broad band centered at 920 mµ. The high spin reduced (deoxygenated) hemoglobin has a Soret band at 430 mµ, a rather weak α band at 590 mµ and a distinct β band at 555 mµ.

11

It may be mentioned that metal porphyrins have spectra which consist of a Soret band located in the vicinity of 400 mμ and two other bands, generally designated as the α- and β-bands situated at about 650 and 550 mμ respectively. The precise locations of the bands will vary according to the particular compound (7, 25). Metal-free porphyrins have a Soret band and four other rather weak absorption bands.

D. Magnetic Properties

Hemoglobin has paramagnetic properties in addition to the diamagnetism normally present in all proteins. The paramagnetism is associated with the iron atoms in the heme portions of the molecule. Measurements of the static susceptibility (13, 46, 52) are summarized in Table 2. Electron

Table 2. *Some properties of hemoglobin and derivatives.*

Compounds	No. 6 Ligand	Valence State of Iron	μ Bohr Magnetons	S	ESR*
Ferrihemoglobin hydroxide Methemoglobin hydroxide Alkaline methemoglobin	OH^-	3+	4.5—4.7	1/2, 5/2	
Ferrihemoglobin Methemoglobin Acid methemoglobin	H_2O	3+	5.6—5.8	5/2	$g_\| = 2$ $g_\perp = 6$
Ferrihemoglobin fluoride	F^-	3+	5.8—5.9	5/2	$g_\| = 2$ $g_\| = 6$
Ferrihemoglobin azide	N_3^-	3+	2,4—2.8	1/2	$g_x = 1.72$ $g_y = 2.22$ $g_z = 2.80$
Ferrihemoglobin cyanide	CN^-	3+	2.3—2.5	1/2	
Oxyhemoglobin	O_2	2+	0	0	Impossible
Reduced hemoglobin (deoxygenated)	H_2O	2+	5.2—5.5	2	None Observed
Carboxy hemoglobin	CO	2+	0	0	Impossible
**Hemin, Ferriporphyrin Chloride	Cl^-	3+			
**Hematin, Ferriporphyrin	H_2O OH^-	3+			

* $g_\| = g_z$, $g_\perp = g_x$ or g_y. The z-axis is perpendicular to the porphyrin plane which contains the x- and y-axes.

** In hemin and hematin there is no protein; the fifth ligand is H_2O in place of N of imidazole as in the hemoglobins.

spin resonance (ESR) has been observed in the ferric complexes both for low (S = 1/2) and high spin (S = 5/2) derivatives (*4, 5. 14, 15*). The data are given in Table 2. It is seen that in low spin hemoglobin derivatives such as the azide there are three distinct g-values while the high spin derivatives like the fluoride are characterized by two g-values, $g_{\parallel} = 2$ and $g_{\perp} = 6$. Here parallel and perpendicular refer to the fourfold symmetry axis or z-axis which is perpendicular to the porphyrin plane; the latter is taken as the xy plane. We note that the description of a spin resonance experiment by means of a g-value implies a direct proportionality between microwave frequency and magnetic field. The hemoglobin experiments have been carried to frequencies of 50,000 Mc/s (*29*).

If the magnetic field is oriented in an arbitrary direction θ relative to the z-axis, the g-value is given by

$$g_{\theta}^{2} = g_{\parallel}^{2} \cos^2\theta + g_{\perp}^{2} \sin^2\theta \tag{1}$$

On the basis of this expression, *Bennett* and *Ingram* (*4*), working with myoglobin, and *Ingram, Gibson* and *Perutz* (*28*) working with hemoglobin were able to determine the orientation of the heme planes relative to the crystallographic axes. With the magnetic field along one of the axes in hemoglobin, four separate resonances were observed. This immediately indicated that the four heme planes are not parallel. Each resonance determined a value of θ which gave the orientation of a heme plane relative to the crystallographic axis along which the magnetic field was oriented. The structural information derived from electron spin resonance was subsequently verified, and extended, by x-ray diffraction methods.

In ferrous derivatives, low spin compounds like oxyhemoglobin have S = O; hence they have no paramagnetism and are entirely diamagnetic. Electron spin resonance in high spin (S = 2) ferrous compounds has not been observed. In the case of ferrihemoglobin hydroxide, data on magnetic susceptibility suggested the possibility of a state with S = 3/2, but as has already been mentioned, closer investigation (*13*) revealed these compounds to be thermal mixtures of S = 1/2 and S = 5/2.

E. Mössbauer Resonance

The presence of the stable isotope Fe^{57} to an abundance of 2.2% in all natural compounds of iron makes it possible to study hemoglobin by the methods of Mössbauer spectroscopy (*1*). In such experiments one exposes a sample of hemoglobin to a beam of highly monochromatic

γ-rays emanating from a radioactive source which contains Fe^{57} nuclei in an excited state (as a daughter product of Co^{57}). By imparting small velocities to the source, relative to the absorber, the γ-rays may be shifted in energy as a result of the Doppler effect. At certain velocities, equivalent to certain γ-ray energies, the Fe^{57} nuclei in hemoglobin will absorb the γ-rays resonantly, that is, the absorption of a γ-ray photon will cause a nucleus of Fe^{57} to make a transition from its ground state to an excited state. Such absorptions are recognized by a decrease in the intensity of the transmitted γ-rays. A plot of transmitted γ-ray intensity as a function of Doppler velocity constitutes a Mössbauer absorption spectrum.

The nucleus is sensitive to its surroundings by virtue of being bound to other nuclei and through various electric and magnetic hyperfine interactions which manifest themselves in certain characteristic features of the absorption spectra. For this reason, Mössbauer spectra may vary from one iron compound to another or for the same compound under various conditions of temperature, aggregation, external magnetic fields, etc. Two parameters are often used in describing Mössbauer spectra. One is the isomer shift which may be taken as the position of the center of gravity of the absorption spectrum with respect to zero velocity. The measured value of the isomer shift depends on the material in which the radioactive source is embedded; however, the relative shift for two materials whose spectra are obtained with the same source is independent of the matrix material. For future reference, we give the empirical relationships among the isomer shifts (δ) for platinum, palladium and stainless steel

$$\delta \text{ (Pt)} = \delta \text{ (SS)} - 0.043 \text{ cm/sec}$$

$$\delta \text{ (Pd)} = \delta \text{ (SS)} - 0.025 \text{ cm/sec} \qquad (2)$$

If the Mössbauer spectrum contains two lines, their separation maesured in units of relative velocity between source and absorber (Doppler velocity) defines another parameter known as the quadrupole splitting (Δ).

Mössbauer spectra in hemoglobin and related compounds have been obtained by *Gonser* and *Grant* (16); *Bearden* et al. (3); *Maling* and *Weissbluth* (41, 42). Spectra in red cells kept at 5 ° K were obtained by *Gonser* and *Grant* (16). Fig. (14a) shows a spectrum obtained with red cells that had been saturated with gaseous oxygen. The spectrum therefore represents (ferrous) oxyhemoglobin which, of course, has O_2 as the sixth ligand. Upon replacing the O_2 ligand with H_2O, as was done by saturating the red cells with N_2 or CO_2, the spectrum in Fig. (14b) resulted. The latter may be taken to represent deoxygenated or reduced (ferrous) hemoglobin.

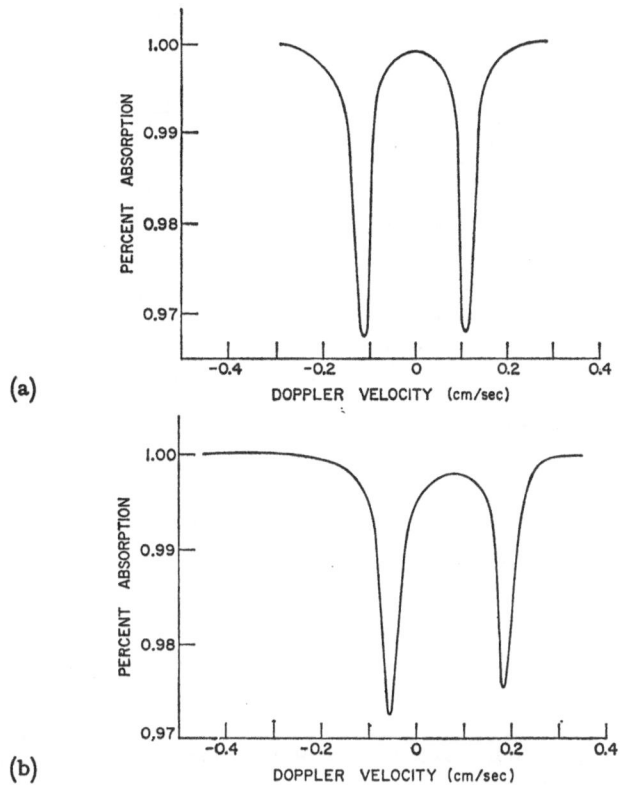

(a)

(b)

Fig. 14a and b. Mössbauer spectra of human red cells (a) saturated with O_2, (b) saturated with N_2. The source (Co^{57} in platinum) was at room temperature, the absorbers at 5 °K (*16*).

The most prominent feature of both spectra is that they show a quadrupole splitting of 0.22 to 0.24 cm/sec. However, the isomer shift in oxyhemoglobin is practically zero (Co^{57} in Pt) while in reduced hemoglobin it is + 0.059 cm/sec. In addition, the doublet in oxyhemoglobin is symmetric, by which it is meant that the two components are of the same intensity, whereas in reduced hemoglobin the doublet is decidely asymmetric. Carboxyhemoglobin, with CO as the sixth ligand, shows essentially no quadrupole splitting.

In high spin (ferric) methemoglobin (*42*) the Mössbauer spectra taken in a dry powder sample at various temperatures are shown in Fig. (15). At room temperature there is a symmetric doublet with a quadrupole splitting of 0.20 cm/sec and an isomer shift which is zero when referred to Co^{57} in Pd and − 0.018 cm/sec referred to Co^{57} in Pt. As the temperature

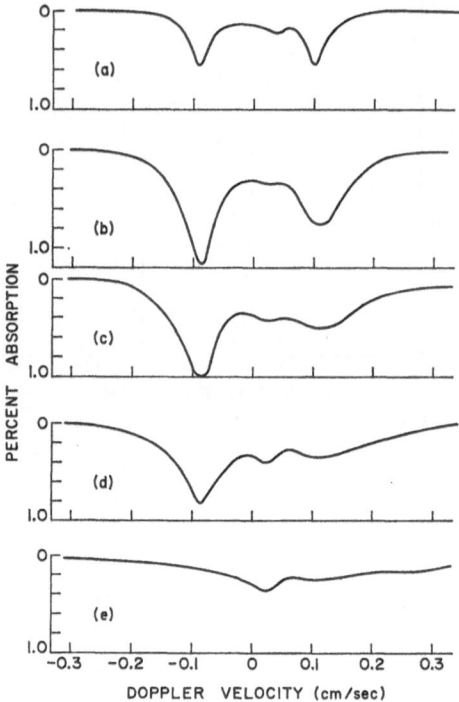

Fig. 15a-e. Mössbauer spectra of human methemoglobin as a dry powder at (a) 297 °K, (b) 88 °K, (c) 80 °K, (d) 50 °K, (e) 19 °K. The source (Co⁵⁷ in palladium) was at room temperature (*42*).

is reduced to the boiling point of liquid nitrogen there is a general increase in absorption intensity but the doublet becomes highly asymmetric. Further reduction in temperature, approaching that of liquid helium, results in a broad resonance of low intensity.[2]

III. Electronic States of Iron

A. Free Ions

A neutral iron atom has 26 electrons of which 18 reside in closed shells and the remaining 8 have the orbital configuration $(3d)^6 (4s)^2$. In hemoglobin iron occurs either in the trivalent (Fe^{3+}, ferric) or divalent (Fe^{2+}, ferrous) state. In the former case the orbital configuration outside

---

[2] An extensive experimental and theoretical investigation of the Mössbauer effect in hemoglobin compounds has recently been reported by *G. Lang* and *W. Marshall*: Proc. Phys. Soc. *87*, 3 (1966).

of closed shells is $(3d)^5$ and in the latter it is $(3d)^6$. These configurations taken together with the environment produced by the ligands form the basis for the theory of magnetic and optical properties of hemoglobin.

The 3d shell consists of five orbitals which can accommodate a maximum of 10 electrons provided their spins are paired (antiparallel) as required by the exclusion principle. When there are fewer than 10 electrons, various arrangements are possible. Specifically, with 5 electrons there may be one electron in each d-orbital; the spins may all be aligned parallel to one another resulting in a total spin $S = 5/2$ as shown in Fig. 16a. Other possible arrangements are shown in Fig. 16b and c leading to $S = \frac{3}{2}$ and $\frac{1}{2}$. With 6 electrons similar considerations lead to $S = 2, 1, 0$ as shown in Fig. 16d, e, f. Hund's rule favors the high spin state as the state of lowest energy so that the lowest states for $(3d)^5$ and $(3d)^6$ are ^6S and ^5D respectively, consistent with spectroscopic observations.

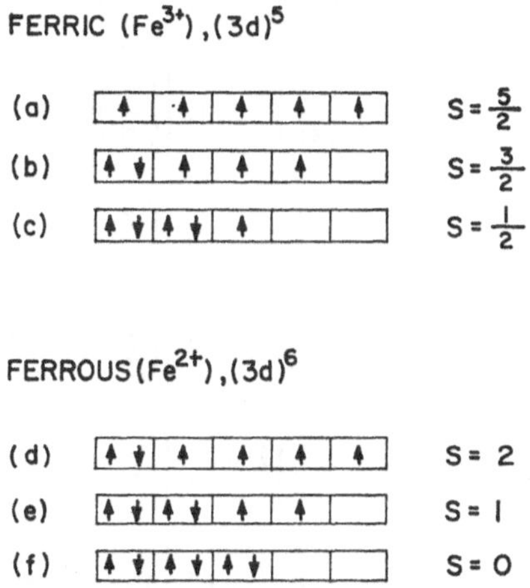

FERRIC (Fe^{3+}), $(3d)^5$

FERROUS (Fe^{2+}), $(3d)^6$

Fig. 16a-f. Possible spin alignments in Fe^{3+} and Fe^{2+}.

The paramagnetic properties of hemoglobin have their origin in the various possible spin alignments. It would be supposed on the basis of what has been said so far that there are five classes of compounds to consider namely those with $S = 1/2, 1, 3/2, 2, 5/2$. The sixth with $S = 0$ would, of course, exhibit no paramagnetism since this particular configuration has no net magnetic moment, but it must be remembered that diamagnetic effects will be present under all circumstances. However, there are no known derivates of hemoglobin, either ferric or ferrous, in which the iron resides in the state with intermediate spin. Moreover, electron spin resonance has not been observed in hemoglobin compounds with $S = 2$. We are left then, for the consideration of magnetic properties, with compounds whose ground states have $S = 1/2, 2, 5/2$, and of these, electron spin resonance data is confined to those with $S = 1/2$ and $5/2$, or the ferric compounds. The nucleus of iron has four stable isotopes: Fe^{54}, Fe^{56}, Fe^{57} and Fe^{58} with abundances of 5.84%, 91.68%, 2.17% and 0.31% respectively. Only Fe^{57} has a non-vanishing nuclear spin in its ground state. Hence hyperfine interactions will contribute negligibly to electron spin resonance. Nevertheless, the presence of Fe^{57}, despite its low abundance, makes it possible to obtain Mössbauer resonance spectra with natural hemoglobin.

We come now to a more detailed consideration of d-orbitals. For the case of a free ion the five d-orbitals are degenerate; they consist of products of a radial function $R(r)$ and the spherical harmonics Y_2^m. They are given by:

$$d_2 = R(r)\, Y_2^2 = \left(\frac{5}{4\pi}\right)^{\frac{1}{2}} R(r) \left[\left(\frac{3}{8}\right)^{\frac{1}{2}} \frac{(x+iy)^2}{r^2}\right]$$

$$= \left(\frac{5}{4\pi}\right)^{\frac{1}{2}} R(r) \left[\left(\frac{3}{8}\right)^{\frac{1}{2}} \sin^2\theta\, e^{2i\varphi}\right]$$

$$d_1 = R(r)\, Y_2^1 = -\left(\frac{5}{4\pi}\right)^{\frac{1}{2}} R(r) \left[\left(\frac{3}{2}\right)^{\frac{1}{2}} \frac{z\,(x+iy)}{r^2}\right]$$

$$= -\left(\frac{5}{4\pi}\right)^{\frac{1}{2}} R(r) \left[\left(\frac{3}{2}\right)^{\frac{1}{2}} \sin\theta\cos\theta\, e^{i\varphi}\right]$$

$$d_0 = R(r)\, Y_2^0 = \left(\frac{5}{4\pi}\right)^{\frac{1}{2}} R(r) \left[\frac{1}{2}\frac{3z^2-r^2}{r^2}\right]$$

$$= \left(\frac{5}{4\pi}\right)^{\frac{1}{2}} R(r) \left[\frac{1}{2}\,(3\cos^2\theta - 1)\right]$$

$$d_{-1} = R(r)\, Y_2^{-1} = \left(\frac{5}{4\pi}\right)^{\frac{1}{2}} R(r) \left[\left(\frac{3}{2}\right)^{\frac{1}{2}} \frac{z\,(x-iy)}{r^2}\right]$$

$$= \left(\frac{5}{4\pi}\right)^{\frac{1}{2}} R(r) \left[\left(\frac{3}{2}\right)^{\frac{1}{2}} \sin\theta\cos\theta\, e^{-i\varphi}\right]$$

$$d_{-2} = R(r)\, Y_2^{-2} = \left(\frac{5}{4\pi}\right)^{\frac{1}{2}} R(r) \left[\left(\frac{3}{8}\right)^{\frac{1}{2}} \frac{(x-iy)^2}{r^2}\right]$$

$$= \left(\frac{5}{4\pi}\right)^{\frac{1}{2}} R(r) \left[\left(\frac{3}{8}\right)^{\frac{1}{2}} \sin^2\theta\, e^{-2i\varphi}\right] \qquad (3)$$

in which the pase convention employed is that of *Condon* and *Shortley*. It will generally be convenient to omit the common factor $(\frac{5}{4\pi})^{\frac{1}{2}} R(r)/r^2$ in discussing d-orbitals. The functions $d_2 \ldots d_{-2}$ are eigenfunctions of a spherically symmetric Hamiltonian; hence they must be basis functions for a representation of the three dimensional rotation group, R_3. Since the angular parts of $d_2 \ldots d_{-2}$ are the spherical harmonics Y_2^m, the d-orbitals belong to the representation $D^{(2)}$.

When the d-orbitals are combined with the spin functions α and β, the total degeneracy is 10. In a useful notation we may write

$$
\begin{aligned}
d_2 \; \alpha &= |2^+ > &\quad d_2 \; \beta &= |2^- > \\
d_1 \; \alpha &= |1^+ > &\quad d_1 \; \beta &= |1^- > \\
d_0 \; \alpha &= |0^+ > &\quad d_0 \; \beta &= |0^- > \\
d_{-1} \; \alpha &= |-1^+ > &\quad d_{-1} \; \beta &= |-1^- > \\
d_{-2} \; \alpha &= |-2^+ > &\quad d_{-2} \; \beta &= |-2^- >
\end{aligned}
\tag{4}
$$

This set is often described as an $|n \; \ell \; m_\ell \; m_s >$ basis set where the quantum members n, ℓ have been suppressed since n = 3, $\ell = 2$ in all cases of the present discussion; the spin functions α and β are indicated by + and − respectively.

When electrostatic interactions among electrons are taken into account a particular configuration gives rise to a number of terms. The $(3d)^5$ configuration leads to the following terms:

> Doublets: S P D D D F F G G H I
>
> Quartets: P D F G
>
> Sextets: S

Of these terms we shall be particularly interested in ^6S and ^4P.

The energies of the various terms are expressed either by Racah coefficients A, B and C, or by the Slater-Condon parameters F_0, F_2 and F_4 to which the former set are related by

$$
\begin{aligned}
A &= F_0 - 49\,F_4 \\
B &= F_2 - 5\,F_4 \\
C &= 35\,F_4
\end{aligned}
\tag{5}
$$

For ^6S and ^4P the energies are

$$
\begin{aligned}
E\,(^6S) &= 10A - 35B \\
E\,(^4P) &= 10A - 28B + 7C
\end{aligned}
\tag{6}
$$

and

$$\Delta E = E\ (^4P) - E\ (^6S) = 7\ (B + C) \tag{7}$$

For Fe^{3+} ions, spectroscopie data yields (36)

$$B = 1133\ \text{cm}^{-1} \qquad C = 3883\ \text{cm}^{-1} \tag{8}$$

The degeneracy of each term is $(2S + 1)\ (2L + 1)$. Each component may be expressed as a 5-electron wave function. Thus the component of ^4G with $M_L = 4$ and $M_s = 3/2$ may be written

$$\left|\ ^4G\ 4\ \frac{3}{2}\ \right> = |\ 2^+\ 2^-\ 1^+\ 0^+\ -1^+\ > \tag{9}$$

where the right hand ket represents a normalized, 5-electron antisymmetrized wave function called a microstate because it specifies the m_ℓ and m_s values of the individual electrons:

$$|\ 2^+2^-1^+0^+-1^+ > = \frac{1}{\sqrt{5!}} \begin{vmatrix} d_2(1)\alpha(1) & d_2(1)\beta(1) & d_1(1)\alpha(1) & d_0(1)\alpha(1) & d_{-1}(1)\alpha(1) \\ d_2(2)\alpha(2) & d_2(2)\beta(2) & d_1(2)\alpha(2) & d_0(2)\alpha(2) & d_{-1}(2)\alpha(2) \\ d_2(3)\alpha(3) & d_2(3)\beta(3) & d_1(3)\alpha(3) & d_0(3)\alpha(3) & d_{-1}(3)\alpha(3) \\ d_2(4)\alpha(4) & d_2(4)\beta(4) & d_1(4)\alpha(4) & d_0(4)\alpha(4) & d_{-1}(4)\alpha(4) \\ d_2(5)\alpha(5) & d_2(5)\beta(5) & d_1(5)\alpha(5) & d_0(5)\alpha(5) & d_{-1}(5)\alpha(5) \end{vmatrix} \tag{10}$$

This is known as a Slater determinant or a determinantal wave function.

Since

$$M_L = \sum_{i=1}^{5} m_{\ell i}, \quad M_S = \sum_{i=1}^{5} m_{si} \tag{11}$$

the ket written in Eq. (9) is the only one possible. It therefore serves as a useful starting point to generate other wave functions by means of the shift operators defined by

$$L_\pm |^{2S+1}\ L\ M_L\ M_s > = [L(L + 1) - M_L(M_L \pm 1)]^\frac{1}{2}\ |^{2S+1}\ L\ M_L \pm 1\ M_s > \tag{12}$$

The operators $L_\pm$ are sums of the one-electron operators $\ell_\pm$. Still other functions may be generated by the shift operators $S_\pm$ which are defined in a manner analogous to Eq. (12).

The method will be illustrated by operating with L_- on $|\ ^4G\ 4\ 3/2 >$. Application of Eq. (12) gives

$$L_- \mid {}^4G \; 4 \; 3/2 > \; = 2 \; \sqrt{2} \mid {}^4G \; 3 \; 3/2 > \tag{13}$$

Also

$$L_- \mid 2^+ \; 2^- \; 1^+ \; 0^+ \; -1^+ > \; = \sum_{i=1}^{5} \ell_-^{(i)} \mid 2^+ \; 2^- \; 1^+ \; 0^+ \; -1^+ > \tag{14}$$

We find

$$\ell_-^{(1)} \mid 2^+ \; 2^- \; 1^+ \; 0^+ \; -1^+ > \; = [\ell \, (\ell + 1) - m_\ell^{(1)} \, (m_\ell^{(1)} - 1)]^{\frac{1}{2}} \mid 1^+ \; 2^- \; 1^+ \; 0^+ \; -1^+ >$$

The right hand ket has two electrons with $m_\ell = 1$ and $m_s = \frac{1}{2}$; they correspond to two identical columns in the Slater determinant which, accordingly, must vanish. Alternatively we may say that the right hand ket vanishes because it violates the exclusion principle. Thus

$$\ell_-^{(1)} \mid 2^+ \; 2^- \; 1^+ \; 0^+ \; -1^+ > \; = 0 \tag{15}$$

Continuing in this fashion

$$\ell_-^{(2)} \mid 2^+ \; 2^- \; 1^+ \; 0^+ \; -1^+ > \; = 2 \mid 2^+ \; 1^- \; 1^+ \; 0^+ \; -1^+ >$$

$$\ell_-^{(5)} \mid 2^+ \; 2^- \; 1^+ \; 0^+ \; -1^+ > \; = 2 \mid 2^+ \; 2^- \; 1^+ \; 0^+ \; -2^+ > \tag{16}$$

We now have the expansion of $\mid {}^4G \; 3 \; \frac{3}{2} >$ in terms of the microstates. After normalization

$$\mid {}^4G \; 3 \; \tfrac{3}{2} > \; = \tfrac{1}{\sqrt{2}} \; [\mid 2^+ \; 1^- \; 1^+ \; 0^+ \; -1^+ > \; + \mid 2^+ \; 2^- \; 1^+ \; 0^+ \; -2^+ >] \tag{17}$$

It is now a simple matter to find $\mid {}^4F \; 3 \; \frac{3}{2} >$ for it must be orthogonal to $\mid {}^4G \; 3 \frac{3}{2} > :$

$$\mid {}^4F \; 3 \; \tfrac{3}{2} > \; = \tfrac{1}{\sqrt{2}} \; [\mid 2^+ \; 1^- \; 1^+ \; 0^+ \; -1^+ > \; - \mid 2^+ \; 2^- \; 1^+ \; 0^+ \; -2^+ >] \tag{18}$$

By this process, with an increasing amount of labor, we may generate any component of any term expressed as an expansion in microstates.

After two more steps we arrive at an expansion of $| \ ^4P \ 1 \ \frac{3}{2} >$,

$$| \ ^4P \ 1 \ \tfrac{3}{2} > = (\tfrac{1}{5})^{\frac{1}{2}} \ | \ 2^+ \ 2^- \ 0^+ \ -1^+ \ -2^+ > + (\tfrac{3}{10})^{\frac{1}{2}} \ | \ 2^+ \ 1^+ \ 1^- \ -1^+ \ -2^+ >$$

$$+ (\tfrac{3}{10})^{\frac{1}{2}} \ | \ 2^+ \ 1^+ \ 0^+ \ 0^- \ -2^+ > + (\tfrac{1}{5})^{\frac{1}{2}} \ |2^+ \ 1^+ \ 0^+ \ -1^+ \ -1^- > \quad (19)$$

Slater (53) gives expansions for $M_L = 0$ and $M_S = 0$ or 1/2. With these as a starting point it is possible to generate other functions by means of the shift operators $L_{\pm}$ and $S_{\pm}$.

For the ^6S state we have

$$| \ ^6S \ 0 \ \tfrac{5}{2} > = |2^+ \ 1^+ \ 0^+ \ -1^+ \ -2^+ > \quad (20)$$

B. Spin-Orbit Coupling in Fe^{3+}

The ground state for the $(3d)^5$ configuration is ^6S. Since $L = 0$ there can be no spin-orbit coupling within this term. However, there may be spin-orbit interaction with excited states. In the Russell-Saunders coupling scheme the selection rules on the non-vanishing matrix elements are

$$\Delta L = 0, \pm 1$$
$$\Delta S = 0, \pm 1$$
$$\Delta J = 0$$
$$\Delta M_J = 0.$$

From $\Delta L = 0, \pm 1$ we conclude that ^6S may have non-vanishing matrix elements with other S states and P states. In $(3d)^5$ there are no other S states and there is only one P state, namely ^4P; the selection rule on the spin quantum number, S, is satisfied. Since ^6S has only one value of J, namely, 5/2, the selection rule on J restricts the spin-orbit interaction to $^6S_{5/2}$ with $^4P_{5/2}$. Finally we must connect states of the same M_J, but the matrix element is independent of the particular value of M_J chosen. Thus, the interaction between $^6S_{5/2}$ and $^4P_{5/2}$ will be described by the matrix element

$$< \ ^6S_{5/2} \ |\mathscr{H}_s| \ ^4P_{5/2} > \quad (21)$$

where $\mathscr{H}_s$ is the spin-orbit coupling Hamiltonian given by

$$\mathscr{H}_s = \sum_{i=1}^{5} \xi(r_i)\,\vec{\ell}_i \cdot \vec{s}_i. \tag{22}$$

It will also be of interest to examine the spin-orbit interaction within the ⁴P term, that is, matrix elements of the form

$$< {}^4P_J |\mathscr{H}_s| {}^4P_J >. \tag{23}$$

Since $\mathscr{H}_s$ is a sum of one-electron operators, matrix elements such as (21) and (23) may be evaluated directly after the terms have been expanded into microstates of the form of Eq. (19) and (20). Although the process is straightforward it is generally quite tedious. Much more elegant and powerful methods have been developed by *Racah*. These methods make full use of the Wigner-Eckart theorem to evaluate matrix elements of operators written in the form of irreducible tensors. Descriptions are to be found in *Slater* (*53*) and *Judd* (*30*). We shall apply these methods to evaluate (21) and (23).

Judd (*30*) gives the following expression for the matrix element of the spin-orbit coupling:

$$< \ell^n\ W\ \xi\ S\ L\ J\ M_J\ |\mathscr{H}_s| \ell^n\ W'\ \xi'\ S'\ L'\ J'\ M'_J >$$

$$= \delta(J,\,J')\,\delta(M_J,\,M'_J)\,(-1)^{S'+L+J}$$

$$\times \left[\frac{\ell\,(\ell+1)\,(2\ell+1)}{6}\right]^{\frac{1}{2}} \begin{Bmatrix} S & S' & 1 \\ L' & L & J \end{Bmatrix} \tag{24}$$

$$\times (\ell^n\ W\ \xi\ S\ L\|W^{(11)}\|\ell^n\ W'\ \xi'\ S'\ L')$$

in which ℓ^n is the electron configuration, e. g. d^5; $S\ L\ J\ M_J$ and $S'\ L'\ J'\ M'_J$ are the quantum numbers of spin, orbital angular momentum, total angular momentum and projection of the total angular momentum for the initial and final states respectively; $\mathscr{H}_s$ is the spin-orbit coupling operator, Eq. (22); $\begin{Bmatrix} S & S' & 1 \\ L' & L & J \end{Bmatrix}$ is a 6-j symbol (*51*). The reduced matrix element is given by

$$(\ell^n \ W \ \xi \ S \ L \| W^{(11)} \| \ell^n \ W' \ \xi' \ S' \ L')$$

$$= 3n \ \{[S] \ [S'] \ [L] \ [L']\}^{\frac{1}{2}} \ (-1)^{\frac{1}{2} \ + \ell + S + L}$$

$$\times \sum_{\bar{\theta}} \ (\theta \ \{|\bar{\theta}) \ (\theta' \ \{|\bar{\theta}) \ (-1)^{\bar{S} + \bar{L}} \ \begin{Bmatrix} S & 1 & S' \\ \frac{1}{2} & \bar{S} & \frac{1}{2} \end{Bmatrix} \begin{Bmatrix} L & 1 & L' \\ \ell & \bar{L} & \ell \end{Bmatrix} \tag{25}$$

in which [S], [S'], etc. stand for $2S + 1$, $2S' + 1$, etc. W, ξ, W', ξ' are group theoretical symbols which distinguish terms of the same kind arising from a given configuration ℓ^n. This situation does not arise in the present discussion and we shall not need to refer to these symbols. $(\theta\{|\bar{\theta})$ and $(\theta'\{|\bar{\theta})$ are coefficients of fractional parentage. They are tabulated in various places e. g. *Judd (30)*, *Slater (53)*, *Nielson* and *Koster (44)*. $\bar{S}$ and $\bar{L}$ are quantum numbers of spin and orbital angular momentum of those states, arising from the configuration ℓ^{n-1}, which are considered to be the parents of the state described by the quantum numbers S, L, the latter arising from the configuration ℓ^n. $\begin{Bmatrix} S & 1 & S' \\ \frac{1}{2} & \bar{S} & \frac{1}{2} \end{Bmatrix}$ and $\begin{Bmatrix} L & 1 & L' \\ \ell & \bar{L} & \ell \end{Bmatrix}$ are 6-j symbols. Eq. (24) gives the spin-orbit coupling in units of ζ, known as the one electron spin-orbit coupling constant, and defined by

$$\zeta = \hbar^2 \int_0^{\infty} [R(r)]^2 \ \xi(r) \ r^2 \ dr \tag{26}$$

in which $R(r)$ is the radial function associated with the one-electron orbitals. Most often, ζ is derived from experimental data; for Fe^{3+}, $\zeta \cong 400 \ cm^{-1}$.

We shall now apply this formalism to compute the matrix element

$$M = \ < d^5 \ {}^6S_{\frac{5}{2}} \ | \ \mathscr{H}_s \ | \ d^5 \ {}^4P_{\frac{5}{2}} \ >$$

for which

$$\ell = 2 \quad S = 5/2 \quad S' = 3/2 \quad L = 0 \quad L' = 1 \quad J = 5/2 \quad (-1)^{S'+L+J} = 1$$

$$\left[\frac{\ell \ (\ell + 1) \ (2\ell + 1)}{6} \right]^{\frac{1}{2}} = \sqrt{5}$$

From tables such as *Rotenberg* et al. (*51*)

$$\begin{Bmatrix} S & S' & 1 \\ L' & L & J \end{Bmatrix} = \begin{Bmatrix} \frac{5}{2} & \frac{3}{2} & 1 \\ 1 & 0 & \frac{5}{2} \end{Bmatrix} = -\frac{\sqrt{2}}{6}$$

Therefore

$$M = -\frac{(10)^{\frac{1}{2}}}{6}\,(d^5\ ^6S\|W^{(11)}\|d^5\ ^4P) \tag{27}$$

Referring to Eq. (25),

$$n = 5$$

$$3n\,\{[S]\,[S']\,[L]\,[L']\}^{\frac{1}{2}} = 90(2)^{\frac{1}{2}}$$

$$(-1)^{\frac{1}{2}+\ell+S+L} = -1$$

The ^6S state has only one parent state, $d^4\ ^5D$. Hence the sum over $\bar\theta$ reduces to one term and we have (*53*)

$$(\theta\ \{\ |\bar\theta) = 1$$

$$(\theta'\ \{\ |\bar\theta) = -1/2$$

$$\bar S = 2$$

$$\bar L = 2$$

$$(-1)^{\bar S+L} = 1$$

$$\begin{Bmatrix} S & 1 & S' \\ \frac{1}{2} & \bar S & \frac{1}{2} \end{Bmatrix} = \begin{Bmatrix} \frac{5}{2} & 1 & \frac{3}{2} \\ \frac{1}{2} & 2 & \frac{1}{2} \end{Bmatrix} = -\left(\tfrac{1}{15}\right)^{\frac{1}{2}}$$

$$\begin{Bmatrix} L & 1 & L' \\ \ell & \bar L & \ell \end{Bmatrix} = \begin{Bmatrix} 0 & 1 & 1 \\ 2 & 2 & 2 \end{Bmatrix} = -\left(\tfrac{1}{15}\right)^{\frac{1}{2}}$$

giving

$$(d^5\ ^6S\|W^{(11)}\|d^5\ ^4P) = 90\,\sqrt{2}\ \cdot -1 \cdot 1 \cdot -\tfrac{1}{2}\ \cdot -\frac{1}{\sqrt{15}}\ \cdot -\frac{1}{\sqrt{15}} = 3\,\sqrt{2} \tag{28}$$

and

$$M = -\frac{\sqrt{10}}{6}\cdot 3\,\sqrt{2}\,\zeta = -\sqrt{5}\,\zeta \tag{29}$$

The same expression, Eq. (24), may be used to calculate the spin-orbit coupling within the ^4P term. The reduced matrix element now becomes

$$(d^5\ ^4P\|W^{(11)}\|d^5\ ^4P) = -3n[S]\,[L]$$

$$\times \sum_{\bar\theta} (\theta\{\ |\bar\theta)^2\,(-1)^{\bar S+L} \begin{Bmatrix} S & 1 & S \\ \frac{1}{2} & \bar S & \frac{1}{2} \end{Bmatrix}\begin{Bmatrix} L & 1 & L \\ 2 & \bar L & 2 \end{Bmatrix} \tag{30}$$

We list below the coefficients of fractional parentage and the values of the 6-j symbols

	$(\theta\{\,|\bar{\theta})^2$	$(-1)^{\bar{S}+\bar{L}}$	$\begin{Bmatrix} S & 1 & S \\ \frac{1}{2} & S & \frac{1}{2} \end{Bmatrix}$	$\begin{Bmatrix} L & 1 & L \\ 2 & \bar{L} & 2 \end{Bmatrix}$
^5D	$\frac{1}{4}$	1	$\frac{\sqrt{10}}{20}$	$\frac{\sqrt{5}}{30}$
$^3F_1$	$\frac{14}{75}$	1	$\frac{\sqrt{10}}{12}$	$\frac{\sqrt{5}}{15}$
$^3F_2$	$\frac{14}{75}$	1	$\frac{\sqrt{10}}{12}$	$\frac{\sqrt{5}}{15}$
^3D	$\frac{7}{60}$	-1	$\frac{\sqrt{10}}{12}$	$\frac{\sqrt{5}}{30}$
$^3P_1$	$\frac{16}{75}$	1	$\frac{\sqrt{10}}{12}$	$-\frac{\sqrt{5}}{10}$
$^3P_2$	$\frac{7}{150}$	1	$\frac{\sqrt{10}}{12}$	$-\frac{\sqrt{5}}{10}$

Upon performing the summation we find that the reduced matrix element vanishes and there is no splitting by spin-orbit coupling within the ^4P term itself. This may seem to contradict the Lande' interval rule which predicts a splitting into three levels with energies given by

$$E(^4P_{5/2}) = 3/2\,\lambda$$
$$E(^4P_{3/2}) = -\lambda$$
$$E(^4P_{1/2}) = -5/2\,\lambda. \tag{31}$$

where λ is the coefficient appearing in the Hamiltonian appropriate for spin-orbit coupling within a single term:

$$\mathscr{H}_s = \lambda\,\vec{L}\cdot\vec{S} \tag{32}$$

However, when λ is evaluated in terms of the one-electron spin-orbit coupling constant ζ, it is found that $\lambda = 0$, consistent with the result obtained above.

Equivalent expressions for spin-orbit coupling are given by *Slater* (53). For a configuration ℓ^n, *Slater* gives

$$< \alpha\,S\,L\,J|\mathscr{H}_s|\alpha'\,S'\,L'\,J' >$$
$$= (-1)^{S+L'-J}\,[\ell\,(\ell+1)\,(2\ell+1)]^{\frac{1}{2}}$$
$$\times\;(\alpha\,S\,L\|V''\|\alpha'\,S'\,L')\,W(S\,L\,S'\,L';\,J\,1) \tag{33}$$

where α represents additional quantum numbers that may be required to specify a state and W(S L S' L'; J 1) is known as Racah's W-coefficient. It is related to the 6-j symbol by

$$\begin{Bmatrix} S & L & J \\ L' & S' & 1 \end{Bmatrix} = (-1)^{S+L+S'+L'} \ W(S \ L \ S' \ L'; \ J \ 1) \qquad (34)$$

The matrix element may then be written

$$< \alpha \ S \ L \ J \ |\mathscr{H}_s| \ \alpha' \ S' \ L' \ J > = (-1)^{L+S'+J}$$

$$\times [\ell(\ell+1)(2\ell+1)]^{\frac{1}{2}} (\alpha \ S \ L \| V'' \| \alpha' \ S' \ L') \begin{Bmatrix} S & L & J \\ L' & S' & 1 \end{Bmatrix}$$

$$(35)$$

in units of ζ. The reduced matrix elements are tabulated by *Slater* (53); for the configuration d^5

$$(^6S\|V''\|^4P) = \sqrt{3}$$

For

$$S = \tfrac{5}{2}, \ L = 0, \ J = \tfrac{5}{2}$$

$$S' = \tfrac{3}{2}, \ L' = 1$$

the 6-j symbol has the value $-\frac{\sqrt{2}}{6}$. Hence

$$< \ ^6S_{5/2} \ |\mathscr{H}_s| ^4P_{5/2} > = \sqrt{30} \cdot \sqrt{3} \cdot \tfrac{-\sqrt{2}}{6} \ \zeta$$

$$= -\sqrt{5} \ \zeta$$

From *Slater's* tables we may verify that the reduced matrix element

$$(^4P\|V''|^4P) = 0$$

showing again that the ^4P term is not split by spin orbit coupling.

In summary, the matrix of $\mathscr{H}_s$ within the set of terms consisting of ^6S and ^4P is

$\mathscr{H}_s$	^6S	^4P
^6S	0	$-\sqrt{5}\,\zeta$
^4P	$-\sqrt{5}\,\zeta$	0

Using Eqs. (6) and (7) we have, to second order in ζ,

$$E(^6S) = 10A - 35B - \frac{5\zeta^2}{7(B+C)} \qquad (36)$$

27

$$|^6S >' = |^6S > - \frac{\sqrt{5}\,\zeta}{7(B + C)}\,|^4P > \tag{37}$$

where $|^6S >'$ is the ground state corrected for spin-orbit coupling.

IV. Cubic Symmetry

A. Ligand Field Potential

In preparation for the discussion of magnetic properties we present, in this chapter and the next, some of the pertinent formalism of ligand field theory. This approach concentrates attention on the central ion which, in the present case, is an open shell ion, that is an ion with a partially filled set of orbitals. The ligands are presumed to be charged, or dipolar with a charged end facing the central ion. It is therefore possible to describe the effect of the ligands by means of an electrostatic potential which exerts a perturbing influence on the orbitals of the central ion. The altered properties of the complex relative to the free ion are seen as consequences of the changes in energy and occupation of the orbitals brought about by the perturbation. A rather different approach, based on the method of molecular orbitals, will be discussed in chap. IX.

In hemoglobin, the iron atom is coordinated to six ligands of which four are nitrogens belonging to prophyrin, the fifth is nitrogen belonging to the imidazole ring of histidine and the sixth is a ligand which varies from one hemoglobin derivative to another. The six ligands form an approximately octahedral arrangement, that is, one ligand is located at each of the positions $x = \pm\,a, y = \pm\,a, z = \pm\,a$. Ligands disposed in this fashion are also said to produce a cubic environment for the iron atom or alternatively, the symmetry elements of the environment are those associated with the group O_h. The effect of a lower symmetry arrangement will be discussed in chap. V.

We shall now examine some of the properties of the cubic group O_h. It is a group which consists of the elements of the group O together with the elements obtained by multiplying each element of O by the inversion operator, i. Alternatively

$$O_h = O \times C_i \tag{38}$$

which states that O_h is a direct product of the group O and the group C_i, the latter consisting of two elements: identity and inversion. It is therefore less cumbersome to describe O and subsequently to take account of inversion separately. The character table for the group O is given in Table 3.

Table 3. *Character table for the group O.*

O		E	8 C$_3$	3 C$_2$	6 C$_2'$	6 C$_4$
Γ_1	A$_1$	1	1	1	1	1
Γ_2	A$_2$	1	1	1	-1	-1
Γ_3	E	2	-1	2	0	0
Γ_4	T$_1$	3	0	-1	-1	1
Γ_5	T$_2$	3	0	-1	1	-1

Definitions (refer to Fig. 17).

E: identity operation,

C$_3$: a rotation of $+\,120°$ or $-\,120°$ about any one of four body diagonals such as AB,

C$_2$: a rotation of $180°$ about any one of the there coordinate axes,

C$_2'$: a rotation of $180°$ about any one of six axes bisecting opposite sides such as CD,

C$_4$: a rotation of $+\,90°$ or $-\,90°$ about any one of the three coordinate axes.

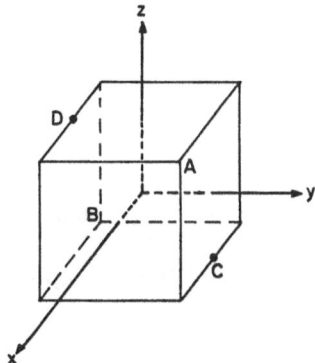

Fig. 17. Coordinate system for the cubic group O.

A ligand field potential which is invariant under O$_h$ and which yields non-vanishing matrix elements with d-orbitals is given by

$$V_c = C\,[Y_4^0 + (\tfrac{5}{14})^{\frac{1}{2}}\,(Y_4^4 + Y_4^{-4})] \tag{39}$$

or, in cartesian coordinates

$$V_c = D[x^4 + y^4 + z^4 - \tfrac{3}{5}\,r^4] \tag{40}$$

The constants C and D depend on the disposition of the ligands and their charges. For singly charged ligands situated at $x = \pm\,a$, $y = \pm\,a$, $z = \pm\,a$

$$D = \frac{35\,e^2}{4a^5} \tag{41}$$

where e is the electronic charge.

B. Basis Functions

It has been stated earlier that the d-orbitals (Eq. 3) are eigenfunctions of a spherically symmetric Hamiltonian and that they are basis functions for the irreducible representation $D^{(2)}$ of the three dimensional rotation group. When the ligand field potential, V_c, is added to an initially spherically symmetric Hamiltonian, the total Hamiltonian is no longer invariant under all three dimensional rotations. It is invariant only under the operations of O_h and the orbitals $d_2 \ldots d_{-2}$ are no longer eigenfunctions of such a Hamiltonian. To analyze the situation with respect to the new Hamiltonian we note that $D^{(2)}$ is also a representation in O_h; however, with respect to O_h, the representation $D^{(2)}$ is reducible. The reduction is easily obtained by means of a fundamental theorem of group theory which states that a reducible representation Γ may be expanded in terms of irreducible representations Γ_i according to

$$\Gamma = \sum_i a_i \, \Gamma_i \tag{42}$$

$$a_i = \frac{1}{h} \sum_R \chi_i(R)\, \chi(R) \tag{43}$$

where h is the order of the group, $\chi_i(R)$ are characters for the operations R in the representation Γ_i, $\chi(R)$ are the characters of Γ, and a_i are integers which indicate the number of times a representation Γ_i appears in the decomposition of Γ.

The character of a representation $D^{(j)}$ is given by

$$\chi(\varphi) = \frac{\sin (j + 1/2)\, \varphi}{\sin 1/2\, \varphi} \tag{44}$$

For $j = 2$ we have

Operation	φ	$\chi(\varphi)$
E	0	5
C_3	$2\pi/3$	-1
C_2	π	1
C_2'	π	1
C_4	$\pi/2$	-1

The decomposition of $D^{(2)}$ into representations of the group O is now obtained directly from Eq. (42) with the aid of the character table (Table 3).

It is readily found that

$$a_1 = a_2 = a_4 = 0$$

$$a_3 = a_5 = 1 \tag{45}$$

so that

$$D^{(2)} = e + t_2 \tag{46}$$

where lower case letters have been used to describe the representations rather than upper case as in Table 3. The distinction is unimportant; it is customary to use lower case letters when referring to one-electron orbitals and upper case when referring to terms. We see, then, that $D^{(2)}$, an irreducible representation of the three dimensional rotation group, has been decomposed into the representations e and t_2 in the group O.

The parity of $D^{(2)}$ may be deduced from the fact that the d-orbitals, which are basis functions for $D^{(2)}$ contain the spherical harmonics Y_2^m which do not change sign upon reflection in the origin. Therefore the representation $D^{(2)}$ is of even parity. This property is carried over into the cubic group so that we may now say that $D^{(2)}$ has been reduced to the representations e_g and t_{2g} in the group O_h, where the suffix g indicates even parity.

It is now necessary to construct linear combinations of the d-orbitals which transform according to the representations e_g and t_{2g}. Generally speaking, the construction of basis functions may be quite tedious, apart from a number of simple cases where it may be done practically by inspection (as for example, the one-dimensional representations). Basis sets for the common situations are tabulated in various places e. g., *Koster* et al. (*34*), *Ballhausen* (*2*), *Griffith* (*21*). It will be sufficient for our purpose to give the basis sets for e_g and t_{2g} and to demonstrate that they satisfy the necessary requirements. Since our discussion is confined to systems of d-electrons, all states will be of even parity or g-states. To simplify the notation we shall henceforth suppress the parity index, unless specifically needed.

$$| e\theta > = | 0 > = d_{z2} = d_0 = \tfrac{1}{2}(3z^2 - r^2)$$

$$| c\varepsilon > = \tfrac{1}{\sqrt{2}} [| 2 > + | -2 >] = d_{x^2-y^2} \tag{47}$$

$$= \tfrac{1}{\sqrt{2}} [d_2 + d_{-2}] = \tfrac{\sqrt{3}}{2}(x^2 - y^2)$$

$$| t_2\xi > = \tfrac{i}{\sqrt{2}} [| 1 > + | -1 >] = d_{yz}$$

$$= \tfrac{i}{\sqrt{2}} [d_1 + d_{-1}] = \sqrt{3}\, yz$$

$$| t_2\eta > = -\tfrac{1}{\sqrt{2}} [| 1 > - | -1 >] = d_{zx} \tag{48}$$

$$= -\tfrac{1}{\sqrt{2}} [d_1 - d_{-1}] = \sqrt{3}\, zx$$

$$|t_2\zeta> = \frac{1}{i\sqrt{2}} [|2> - |{-}2>] = d_{xy}$$

$$= \frac{1}{i\sqrt{2}} [d_2 - d_{-2}] = \sqrt{3}\,xy$$

The wave functions $|e0>$, $|e\varepsilon>$, belong to e and $|t_2\xi>$, $|t_2\eta>$, $|t_2\zeta>$ belong to t_2. These orbitals are shown in Fig. 18.

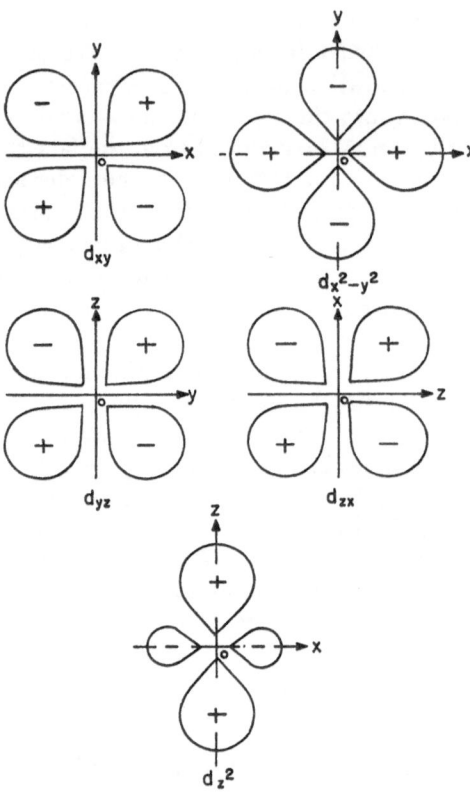

Fig. 18. The d-orbitals (Eqs. (47) and (48)).

We shall illustrate the behavior of the basis functions under the operations of the group O. For convenience take the z-axis along one of the 4-fold symmetry axes as shown in Fig. 17. The six operations C_4 are $\pm C_4^x$, $\pm C_4^y$, $\pm C_4^z$. For illustrative purposes we take C_4^z. This operation may be described by

$$x \rightarrow y$$
$$y \rightarrow -x$$
$$z \rightarrow z \tag{49}$$

or by

$$C_4^z \begin{pmatrix} x \\ y \\ z \end{pmatrix} = \begin{pmatrix} y \\ -x \\ z \end{pmatrix} \tag{50}$$

with

$$C_4^z = \begin{pmatrix} 0 & 1 & 0 \\ -1 & 0 & 0 \\ 0 & 0 & 1 \end{pmatrix} \tag{51}$$

If these replacements are made in the d-functions (Eq. 3) we get

$$C_4^z |2> = -|2>$$
$$C_4^z |1> = -i|1>$$
$$C_4^z |0> = |0>$$
$$C_4^z |-1> = i|-1>$$
$$C_4^z |-2> = -|-2> \tag{52}$$

and, with Eqs. (47, 48),

$$C_4^z |e\theta> = |e\theta>$$
$$C_4^z |e\varepsilon> = -|e\varepsilon>$$
$$C_4^z |t_2\xi> = -|t_2\eta>$$
$$C_4^z |t_2\eta> = |t_2\xi>$$
$$C_4^z |t_2\zeta> = -|t_2\zeta> \tag{53}$$

We note the important fact that under C_4^z there is no mixing of e-orbitals with t_2-orbitals. The above relations may then be expressed in matrix form

$$C_4^z \begin{pmatrix} e\theta \\ e\varepsilon \end{pmatrix} = \begin{pmatrix} 1 & 0 \\ 0 & -1 \end{pmatrix} \begin{pmatrix} e\theta \\ e\varepsilon \end{pmatrix} = \begin{pmatrix} e\theta \\ -e\varepsilon \end{pmatrix} \tag{54}$$

$$C_4^z \begin{pmatrix} t_2\xi \\ t_2\eta \\ t_2\zeta \end{pmatrix} = \begin{pmatrix} 0 & -1 & 0 \\ 1 & 0 & 0 \\ 0 & 0 & -1 \end{pmatrix} \begin{pmatrix} t_2\xi \\ t_2\eta \\ t_2\zeta \end{pmatrix} = \begin{pmatrix} -t_2\eta \\ t_2\xi \\ -t_2\zeta \end{pmatrix} \tag{55}$$

Thus the matrix

$$\begin{pmatrix} 1 & 0 \\ 0 & -1 \end{pmatrix}$$

is the e-representation of C_4^z in the group O. The character is zero, consistent with Table 3. Similarly, the matrix

$$\begin{pmatrix} 0 & 1 & 0 \\ -1 & 0 & 0 \\ 0 & 0 & -1 \end{pmatrix}$$

is the t_2-representation of C_4^z in the group O; the character is -1 consistent with Table 3.

The same procedure may be repeated with all the elements of the group O to verify that (a) the set of e-orbitals does not mix with the set of t_2-orbitals; each set mixes only within itself, and (b) the set of e-orbitals and the set of t_2-orbitals each generates matrices which are irreducible representations corresponding to e and t_2 respectively of the group O.

Certain linear combinations of the t_2-orbitals are often useful:

$$|t_2 1> = -\tfrac{i}{\sqrt{2}} [|t_2\xi> + i|t_2\eta>]$$

$$= |-1>$$

$$|t_2 0> = i|t_2\zeta> = |t_2\zeta_1>$$

$$= \tfrac{1}{\sqrt{2}} [|2> - |-2>] \equiv |\zeta_1> \tag{56}$$

$$|t_2 -1> = \tfrac{i}{\sqrt{2}} [|t_2\xi> - i|t_2\eta>]$$

$$= -|1>$$

The converse relations are

$$|t_2\xi> = \tfrac{i}{\sqrt{2}} [|t_2 1> - |t_2 -1>]$$

$$|t_2\zeta> = -i|t_2 0> \tag{57}$$

$$|t_2\eta> = \tfrac{1}{\sqrt{2}} [|t_2 1> + |t_2 -1>]$$

It may be well to note that if the z-axis in Fig. 17 is oriented in another direction, say along a 3-fold axis, the e- and t_2-orbitals will be quite different linear combinations of d-orbitals. In another notation the t_2-orbitals are called dϵ, and the e-orbitals are dγ.

C. Energetic Considerations

It is seen from Fig. 18 that the $d_{x^2-y^2}$ orbital lies along the x- and y-axes. This means that an electron in a $d_{x^2-y^2}$ orbital will have a maximum probability density along the coordinate axes, while an electron in a d_{xy}-orbital will have maximum probability density along directions which lie at 45° to the x- and y-axes. If the ligands are situated on the axes and are negatively charged the electrostatic repulsion with an electron in a $d_{x^2-y^2}$ orbital will be greater than with an electron in a d_{xy}-orbital. Therefore the energy of the d_{xy} orbital is depressed relative to a $d_{x^2-y^2}$-orbital, as shown in Fig. 19. Similar arguments apply to the d_{yz}-

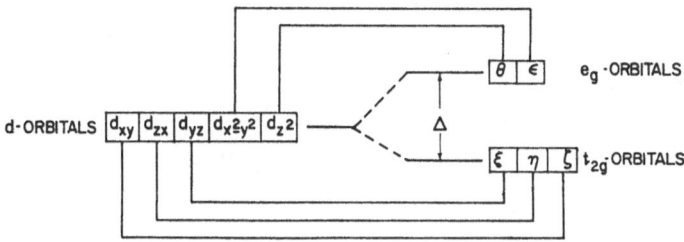

Fig. 19. The splitting of d-orbitals in a cubic field.

and d_{zx}-orbitals which lie in the yz and zx planes respectively. For the d_{z^2}-orbital, simple pictorial arguments may not be entirely convincing; calculation shows that d_{z^2} is degenerate with $d_{x^2-y^2}$ in a cubic field. We see, then, that states possessing electron distributions which point strongly towards negatively charged regions in the surroundings (e. g., the ligands of a complex) will have a relatively high energy because of the mutual repulsion of negative charges. Others, with electron distributions avoiding the negative charges, will consequently have a lower energy.

The effect of the cubic ligand field potential, V_c, has been to remove the five-fold degeneracy of the d-orbitals in the free ion and to separate the orbitals into a three-fold degenerate set belonging to the t_2-representation of O and a two-fold degenerate pair belonging to the e-representation. This is shown in Fig. 19. Moreover, if the six ligands are electronegative, as they are in heme compounds, the t_2-levels will lie lower than the e-levels. The separation between the two sets of levels is designated by Δ (or 10 Dq), which is taken as a measure of the strength of the ligand field. Clearly Δ must be related to the ligand field potential as given by Eq. (39) or (40). Thus for six-fold coordination

$$\Delta = \tfrac{4}{21} D < r^4 > \tag{58}$$

where D is given by Eq. (41). Although the computation of Δ has been accomplished in a few instances (56), more often one treats Δ as an empirical parameter to be obtained from experimental data such as absorption spectra. It is important to recognize that the value of Δ depends on the ligands. Thus Δ for ferrihemoglobin cyanide is considerably greater than for ferrihemoglobin fluoride. As will be shown below, the high value of Δ in the cyanide derivative is responsible for the low spin ($S = 1/2$), whereas the high spin ($S = 5/2$) in the fluoride derivative is a consequence of the low value of Δ.

Since the t_2-orbitals lie lower in energy than the e-orbitals by an amount Δ, occupation of the t_2-orbitals is favored. On the other hand there are two effects which work in the opposite direction. One is that electrons in the same spatial orbitals tend to have a higher electrostatic repulsion than electrons in separate orbitals. The second arises from an exchange energy which favors states with high spin. But these, according to the Pauli principle, arise from states in which the electrons are distributed in separate orbitals. The last two effects taken together are often called the pairing energy. Hence the relative magnitude of Δ, the orbital separation due to the cubic crystal field, and the pairing energy will determine the distribution of electrons among the t_2- and e-orbitals.

Two limiting cases may be distinguished. When Δ is much smaller than the pairing energy, the electrons tend to distribute themselves so as to achieve maximum spin. For Fe^{3+}, the configuration $t_2^3 \, e^2$ with $S = \frac{5}{2}$ has the lowest energy and for Fe^{2+} it is the configuration $t_2^4 \, e^2$ with $S = 2$ that lies lowest (Fig. 20). Conversely when Δ is much larger than the pairing energy the electrons tend to fill the t_2-orbitals; for Fe^{3+} and Fe^{2+} the lowest energy configurations are t_2^5 ($S = \frac{1}{2}$) and t_2^6 ($S = 0$) respectively. It is seen then, that to a large extent, the value of Δ determines the magnetic properties of the complex. In particular, a strong ligand field may produce a diamagnetic ferrous complex whereas other ferrous complexes possessing weaker ligand fields will, as a consequence, be paramagnetic.

The t_2-orbitals are three-fold degenerate; hence they may contain a maximum of six electrons. Similarly the e-orbitals are two-fold degenerate and may therefore accommodate a maximum of four electrons. With this information we may list all the possible electronic configurations without regard as to their energies:

$$(3d)^5, \; Fe^{3+}: \; t_2^5, \; t_2^4 e, \; t_2^3 e^2, \; t_2^2 e^3, \; t_2 e^4$$

$$(3d)^6, \; Fe^{2+}: \; t_2^6, \; t_2^5 e, \; t_2^4 e^2, \; t_2^3 e^3, \; t_2^2 e^4$$

These configurations together with the spin alignments corresponding to the maximum value of S are shown in Fig. 20. Each electron configuration

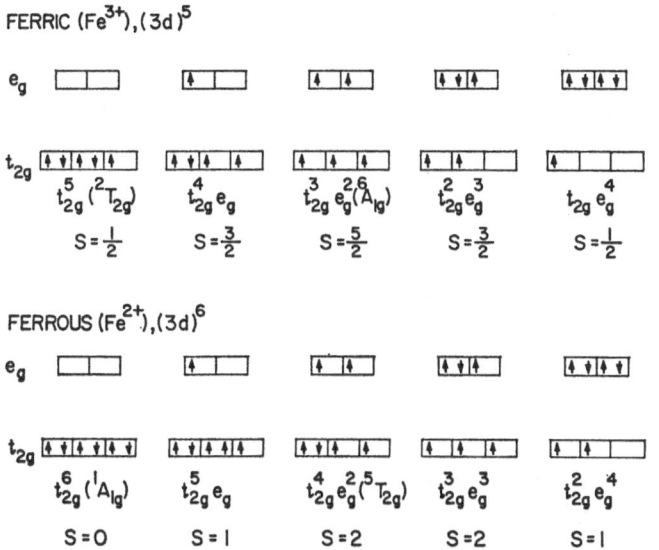

Fig. 20. Occupation of t_{2g}- and e_g-orbitals in ferric and ferrous systems.

listed above gives rise to a set of terms consistent with the Pauli principle. The situation is precisely analogous to that in the free ion. For example, the configuration t_2^5 leaves a single hole in the t_2-shell; therefore the only possible term from t_2^5 is $^2T_2$ which symbolizes the fact that the total spin $S = 1/2$ and the spatial part of the wave function transforms according to the T_2-representation of O. With respect to O_h the term would be written as $^2T_{2g}$; however, we shall continue to suppress the parity index since we are dealing exclusively with d-electrons. Apart from a few other simple situations it is not possible to obtain the terms arising from a particular configuration merely by inspection. Thus the configuration $t_2^3 e^2$ belongs to the product representation

$$t_2 \times t_2 \times t_2 \times e \times e$$

which may be expanded by Eqs. 42 and 43 into irreducible representations of O. This gives all the possible terms including some which violate the Pauli principle and must therefore be eliminated. The complete list of terms (satisfying the Pauli principle) is given in Table 4.

Tanabe and *Sugano* (57) computed the energies of the terms as a function of the cubic splitting Δ and the Racah parameters A, B, and C. The energies of the lowest states for Fe^{3+} are given by

$$E(t_2^3 e^2, \, ^6A_1) = 10A - 35B$$
$$E(t_2^5, \, ^2T_2) = 10A - 20B + 10C - 2\Delta$$
$$E(t_2^4 e, \, ^4T_1) = 10A - 25B + 6C - \Delta \quad\quad (59)$$

Table 4. *Terms arising from configurations in a cubic field.*

Ferric (Fe^{3+}), $(3d)^5$

Electron Configurations	Terms		
	$S = 1/2$	$S = 3/2$	$S = 5/2$
t_2^5	T_2		
$t_2^4 e$	$A_1 \, A_2 \, E \, E \, T_1 \, T_1 \, T_2 \, T_2$	$T_1 \, T_2$	
$t_2^3 e^2$	$A_1 \, A_1 \, A_2 \, E \, E \, E$	$A_1 \, A_2 \, E \, E \, T_1 \, T_2$	A_1
	$T_1 T_1 T_1 T_1 \; T_2 T_2 T_2 T_2$		
$t_2^2 e^3$	$A_1 \, A_2 \, E \, E \, T_1 \, T_1 \, T_2 \, T_2$	$T_1 \, T_2$	
$t_2 e^4$	T_2		

Ferrous (Fe^{2+}), $(3d)^6$

Electron Configurations	Terms		
	$S = 0$	$S = 1$	$S = 2$
t_2	A_1		
$t_2^5 e$	$T_1 \, T_2$	$T_1 \, T_2$	
$t_2^4 e^2$	$A_1 \, A_1 \, A_2 \, E \, E \, E \, T_1$	$A_2 \, E \, T_1 \, T_1 \, T_1 \, T_2 \, T_2$	T_2
	$T_2 \, T_2 \, T_2$		
$t_2^3 e^3$	$A_1 \, A_2 \, E \, T_1 \, T_1 \, T_2 \, T_2$	$A_1 \, A_2 \, E \, E \, T_1 \, T_1 \, T_2 \, T_2$	E
$t_2^2 e^4$	$A_1 \, E \, T_2$	T_1	

in which the electron configuration which gives rise to the particular term is indicated. To obtain relative energies we may set $E(t_2^3 e^2, \, {}^6A_1) = 0$ whence

$$E(t_2^5, \, {}^2T_2) = 15B + 10C - 2\Delta$$

$$E(t_2^4 e, \, {}^4T_1) = 10B + 6C - \Delta \tag{60}$$

Using the values of B and C given in Eqs. (8) we may plot the energies of Eq. (59) as a function of Δ (Fig. 21). It is seen that ${}^2T_2$ crosses ${}^6A_1$ when $\Delta = \frac{15}{2} B + 5C = 27{,}900$ cm^{-1}; for larger values of Δ, ${}^2T_2$ remains below ${}^6A_1$. Similarly ${}^4T_1$ crosses ${}^6A_1$ at $\Delta = 10B + 6C = 34{,}630$ cm^{-1}. However, ${}^4T_1$ can never become the ground state since, for all values of Δ, either ${}^6A_1$ or ${}^2T_2$ lies below ${}^4T_1$.

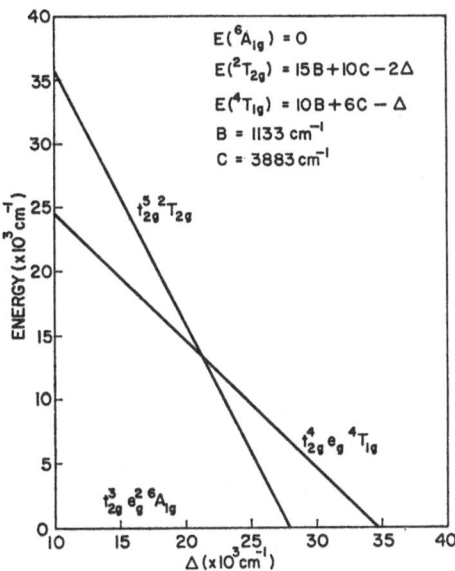

Fig. 21. Energies of $^2T_{2g}$ and $^4T_{1g}$ relative to $^6A_{1g}$ as a function of the cubic splitting parameter Δ.

In the discussion of terms in low symmetry fields it is useful to adopt a ket notation of the form

$$|^{2S+1}h\, M\, \theta >$$

in which S is the spin quantum number, h is a representation in O, M is the spin projection quantum number and θ is a component of h. For example, the $^6A_1$ term is spatially non-degenerate (an orbital singlet) but has a six-fold spin degeneracy. The six wave functions or kets are designated by

$$|^6A_1\, \tfrac{1}{2}\, a_1 >,\ |^6A_1\, \tfrac{3}{2}\, a_1 >,\ |^6A_1\, \tfrac{5}{2}\, a_1 >$$

$$|^6A_1 - \tfrac{1}{2}\, a_1 >,\ |^6A_1 - \tfrac{3}{2}\, a_1 >,\ |^6A_1 - \tfrac{5}{2}\, a_1 >$$

Since A_1 has only one component, the index a_1 which labels the components of A_1 is redundant and may be omitted if so desired.

The $^4T_1$ state has a 12-fold degeneracy: a 3-fold orbital degeneracy and a 4-fold spin degeneracy. The wave functions will be written

$$|^4T_1 \tfrac{3}{2} \, x>, \; |^4T_1 \tfrac{1}{2} \, x>, \; |^4T_1 - \tfrac{1}{2} \, x>, \; |^4T_1 - \tfrac{3}{2} \, x>$$

$$|^4T_1 \tfrac{3}{2} \, y>, \; |^4T_1 \tfrac{1}{2} \, y>, \; |^4T_1 - \tfrac{1}{2} \, y>, \; |^4T_1 - \tfrac{3}{2} \, y>$$

$$|^4T_1 \tfrac{3}{2} \, z>, \; |^4T_1 \tfrac{1}{2} \, z>, \; |^4T_1 - \tfrac{1}{2} \, z>, \; |^4T_1 - \tfrac{3}{2} \, z>$$

in which x, y, and z represent real components of the representation T_1. Alternatively, the three components of T_1 may be written in complex form; thus for $M = \tfrac{3}{2}$ we would have

$$|^4T_1 \tfrac{3}{2} \, 1>, \; |^4T_1 \tfrac{3}{2} \, 0>, \; |^4T_1 \tfrac{3}{2} - 1>$$

in which 1, 0 and -1 designate the complex components of T_1. The notation both for the real and complex components of T_1 is designed to be reminiscent of atomic p-functions to which the T_1 representation in the cubic group is closely related.

The real and complex components of T_1 are related by

$$|T_1 \, x> = \tfrac{i}{\sqrt{2}} \, [|T_1 \, 1> - |T_1 - 1>]$$

$$|T_1 \, y> = \tfrac{1}{\sqrt{2}} \, [|T_1 \, 1> + |T_1 - 1>]$$

$$|T_1 \, z> = -i \, |T_1 \, 0> \tag{61}$$

or by the inverse relations

$$|T_1 \, 1> \quad = -\tfrac{i}{\sqrt{2}} \, [\, |T_1 \, x> + i \, |T_1 \, y>|$$

$$|T_1 \, 0> \quad = i \, |T_1 \, z>$$

$$|T_1 - 1> = \tfrac{i}{\sqrt{2}} \, [\, |T_1 \, x> - i \, |T_1 \, y>] \tag{62}$$

D. Classification in O*

The discussion of spin-orbit coupling is facilitated by the classification of states according to the representations of the spinor or double groups which arise whenever it is necessary to deal with a system possessing half-integral angular momenta. We shall discuss this classification scheme with respect to the spinor group O* which is related to the group O in the manner shown by the character table (Table 5). It may be noted that O* contains all the irreducible representations contained in O plus three additional representations: E', E" and U' (or Γ_6, Γ_7 and Γ_8) which have the dimensions 2, 2 and 4 respectively. Discussions of these groups are given by *Griffith* (*21*), *Ballhausen* (*2*), and *Koster* et al. (*34*).

Table 5. *Character table for the group O**.

O*		E	R	$4C_3$ $4C_3^2R$	$4C_3R$ $4C_3^2$	$3C_2$ $3C_2R$	$6C_2'$ $6C_2'R$	$3C_4$ $3C_4^3R$	$3C_4R$ $3C_4^3$
Γ_1	A_1	1	1	1	1	1	1	1	1
Γ_2	A_2	1	1	1	1	1	-1	-1	-1
Γ_3	E	2	2	-1	-1	2	0	0	0
Γ_4	T_1	3	3	0	0	-1	-1	1	1
Γ_5	T_2	3	3	0	0	-1	1	-1	-1
Γ_6	E'	2	-2	1	-1	0	0	$\sqrt{2}$	$-\sqrt{2}$
Γ_7	E"	2	-2	1	-1	0	0	$-\sqrt{2}$	$\sqrt{2}$
Γ_8	U'	4	-4	-1	1	0	0	0	0

Definitions
R: a rotation through an angle of 2π about an arbitrary axis
Other operations are defined as in Table 3.

Consider first the $^6A_1$ term. Since $S = \frac{5}{2}$ the spin part of the wave function belongs to the $D^{(5/2)}$ representation of the three-dimensional rotation group. By means of Eqs. (42 and 43) $D^{(5/2)}$ may be decomposed into irreducible representations of O* with the result that

$$D^{(5/2)} = E" + U' \tag{63}$$

When the spin and orbital parts of the wave function are coupled, two product representations are obrained:

$$A_1 \times E" = E"$$

and

$$A_1 \times U' = U'$$

indicating that the $^6A_1$ term, which has a 6-fold spin degeneracy, develops into two levels with degeneracies of 2 and 4 respectively under the influence of spin-orbit coupling.

The basis functions in O* can be written as certain linear combinations of the basis functions in O. In analogy with the Clebsch-Gordan coefficients or the 3-j symbols there exist tables of coupling coefficients for the finite point groups (27, 34). A convenient notation for a ket in O* is

$$|^{2S+1} h\ t\tau >$$

in which S is the spin quantum number, h is a representation in O, t is a representation in O* and τ is a component of t.

If the two components of E" are denoted by $\alpha"$ and $\beta"$ and the four components of U' by $\varkappa$, λ, μ, ν, the basis functions in O* are given by

$$|^6A_1\ E"\ \alpha" > = \tfrac{1}{\sqrt{6}}\ |^6A_1\ \tfrac{5}{2}\ a_1 > - (\tfrac{5}{6})^{\frac{1}{2}}\ |^6A_1 - \tfrac{3}{2}\ a_1 >$$

$$|^6A_1\ E"\ \beta" > = \tfrac{1}{\sqrt{6}}\ |^6A_1 - \tfrac{5}{2}\ a_1 > - (\tfrac{5}{6})^{\frac{1}{2}}\ |^6A_1\ \tfrac{3}{2}\ a_1 >$$

41

$$|{}^6A_1\, U'\, \varkappa > \; = -\tfrac{1}{\sqrt{6}}\, |{}^6A_1\, \tfrac{3}{2}\, a_1 > - (\tfrac{5}{6})^{\frac{1}{2}}\, |{}^6A_1 - \tfrac{5}{2}\, a_1 >$$

$$|{}^6A_1\, U'\, \lambda > \; = |{}^6A_1\, \tfrac{1}{2}\, a_1 >$$

$$|{}^6A_1\, U'\, \mu > \; = -\, |{}^6A_1 - \tfrac{1}{2}\, a_1 >$$

$$|{}^6A_1\, U'\, \nu > \; = (\tfrac{5}{6})^{\frac{1}{2}}\, |{}^6A_1\, \tfrac{5}{2}\, a_1 > + \tfrac{1}{\sqrt{6}}\, |{}^6A_1 - \tfrac{3}{2}\, a_1 > \qquad (64)$$

An analogous procedure may be followed for the ${}^4T_1$ term. Here we have $S = \tfrac{3}{2}$ and

$$D^{(3/2)} = U'$$

The product representation expands according to

$$U' \times T_1 = E' + E'' + 2U'$$

and we have two 2-dimensional and two 4-dimensional representations. The basis functions in O^* are now the following:

$$|{}^4T_1\, E'\, \alpha' > \; = \tfrac{1}{\sqrt{2}}\, |{}^4T_1\, \tfrac{3}{2} -1 > - \tfrac{1}{\sqrt{3}}\, |{}^4T_1\, \tfrac{1}{2}\, 0 > + \tfrac{1}{\sqrt{6}}\, |{}^4T_1 - \tfrac{1}{2}\, 1 >$$

$$|{}^4T_1\, E'\, \beta' > \; = \tfrac{1}{\sqrt{6}}\, |{}^4T_1\, \tfrac{1}{2} -1 > - \tfrac{1}{\sqrt{3}}\, |{}^4T_1 - \tfrac{1}{2}\, 0 > + \tfrac{1}{\sqrt{2}}\, |{}^4T_1 - \tfrac{3}{2}\, 1 >$$

$$|{}^4T_1\, E''\, \alpha'' > \; = \tfrac{1}{\sqrt{6}}\, |{}^4T_1\, \tfrac{3}{2}\, 1 > - \tfrac{1}{\sqrt{2}}\, |{}^4T_1 - \tfrac{1}{2} -1 > - \tfrac{1}{\sqrt{3}}\, |{}^4T_1 - \tfrac{3}{2}\, 0 >$$

$$|{}^4T_1\, E''\, \beta'' > \; = -\tfrac{1}{\sqrt{3}}\, |{}^4T_1\, \tfrac{3}{2}\, 0 > - \tfrac{1}{\sqrt{2}}\, |{}^4T_1\, \tfrac{1}{2}\, 1 > + \tfrac{1}{\sqrt{6}}\, |{}^4T_1 - \tfrac{3}{2} -1 > \quad (65)$$

Since there are two U' representations an additional label is required to distinguish between them. Such a label, often called a J-value (27) is again suggested by the p-isomorphism. In the present case the two J-values are 3/2 and 5/2. We have, then, for the U' basis sets

$$|{}^4T_1\tfrac{3}{2}\, U'\, \varkappa > \; = (\tfrac{3}{5})^{\frac{1}{2}}\, |{}^4T_1\, \tfrac{3}{2}\, 0 > - (\tfrac{2}{5})^{\frac{1}{2}}\, |{}^4T_1\, \tfrac{1}{2}\, 1 >$$

$$|{}^4T_1\tfrac{3}{2}\, U'\, \lambda > \; = (\tfrac{2}{5})^{\frac{1}{2}}\, |{}^4T_1\, \tfrac{3}{2} -1 > + \tfrac{1}{\sqrt{15}}\, |{}^4T_1\, \tfrac{1}{2}\, 0 > - 2\,(\tfrac{2}{15})^{\frac{1}{2}}\, |{}^4T_1 - \tfrac{1}{2}\, 1 >$$

$$|{}^4T_1\tfrac{3}{2}\, U'\, \mu > \; = (\tfrac{2}{15})^{\frac{1}{2}}\, |{}^4T_1\, \tfrac{1}{2} -1 > - \tfrac{1}{\sqrt{15}}\, |{}^4T_1 - \tfrac{1}{2}\, 0 > - (\tfrac{2}{5})^{\frac{1}{2}}\, |{}^4T_1 - \tfrac{3}{2}\, 1 >$$

$$|{}^4T_1\tfrac{3}{2}\, U'\, \nu > \; = (\tfrac{2}{5})^{\frac{1}{2}}\, |{}^4T_1 - \tfrac{1}{2} -1 > - (\tfrac{3}{5})^{\frac{1}{2}}\, |{}^4T_1 - \tfrac{3}{2}\, 0 >$$

$$|{}^4T_1\tfrac{5}{2}\, U'\, \varkappa > \; = -\tfrac{1}{\sqrt{15}}\, |{}^4T_1\, \tfrac{3}{2}\, 0 > - \tfrac{1}{\sqrt{10}}\, |{}^4T_1\, \tfrac{1}{2}\, 1 > - (\tfrac{5}{6})^{\frac{1}{2}}\, |{}^4T_1 - \tfrac{3}{2} -1 >$$

$$|{}^4T_1\tfrac{5}{2}\, U'\, \lambda > \; = \tfrac{1}{\sqrt{10}}\, |{}^4T_1\, \tfrac{3}{2} -1 > + (\tfrac{3}{5})^{\frac{1}{2}}\, |{}^4T_1\, \tfrac{1}{2}\, 0 > + (\tfrac{3}{10})^{\frac{1}{2}}\, |{}^4T_1 - \tfrac{1}{2}\, 1 >$$

$$|{}^4T_1\tfrac{5}{2}\, U'\, \mu > \; = -(\tfrac{3}{10})^{\frac{1}{2}}\, |{}^4T_1\, \tfrac{1}{2} -1 > - (\tfrac{3}{5})^{\frac{1}{2}}\, |{}^4T_1 - \tfrac{1}{2}\, 0 > - (\tfrac{1}{10})^{\frac{1}{2}}\, |{}^4T_1 - \tfrac{3}{2}\, 1 >$$

$$|^4T_1 \tfrac{5}{2}\ U'\ \nu> \ = (\tfrac{5}{6})^{\frac12}\ |^4T_1\ \tfrac{3}{2}\ 1> + (\tfrac{1}{10})^{\frac12}\ |^4T_1 - \tfrac{1}{2} - 1> + (\tfrac{1}{15})^{\frac12}|^4T_1 - \tfrac{3}{2}\ 0> \tag{66}$$

As a final example of a set of basis functions belonging to representations in O* we consider a single d-electron. When a cubic field is applied, the electron may be found either in an e-orbital (Eqs. 47) or in a t_2-orbital (Eqs. 48). As we have seen, the latter is energetically favored and we shall therefore confine our attention to it. The term arising from a single electron in a t_2-orbital is clearly $^2T_2$. Since $S = \tfrac12$, the spin part of the wave function belongs to the $D^{(1/2)}$ representation of the three dimensional rotation group. The decomposition into representations in O* is simply

$$D^{(1/2)} = E'$$

and the product representation is then $E' \times T_2$ which expands into

$$E' \times T_2 = E'' + U'.$$

As previously, the basis functions in O* are now obtained from a table of coupling coefficients. They are written in three equivalent forms: the first is an expansion in $|^2T_2\ M\ \theta>$ where M is a component of spin and θ is a complex component of the T_2 representation, the second is in terms of the t_2-orbitals with complex components as defined by Eqs. (56) and the third is based on the d-orbitals (Eqs. 4).

$$
\begin{aligned}
|^2T_2\ E''\ \alpha''> &= \tfrac{1}{\sqrt3}\ |^2T_2\ \tfrac12\ 0> - (\tfrac23)^{\frac12}\ |^2T_2 - \tfrac12\ 1>\\
&= \tfrac{1}{\sqrt3}\ |t_2\ 0^+> - (\tfrac23)^{\frac12}\ |t_2\ 1^->\\
&= \tfrac{1}{\sqrt6}\ [|2^+> - |-2^+>] - (\tfrac23)^{\frac12}\ |-1^->\\
|^2T_2\ E''\ \beta''> &= (\tfrac23)^{\frac12}\ |^2T_2\ \tfrac12 - 1> - \tfrac{1}{\sqrt3}\ |^2T_2 - \tfrac12\ 0>\\
&= (\tfrac23)^{\frac12}\ |t_2 - 1^+> - \tfrac{1}{\sqrt3}\ |t_2\ 0^->\\
&= -(\tfrac23)^{\frac12}|1^+> - \tfrac{1}{\sqrt6}\ [|2^-> - |-2^->]\\
|^2T_2\ U'\ \varkappa> &= -\tfrac{1}{\sqrt3}\ |^2T_2\ \tfrac12 - 1> - (\tfrac23)^{\frac12}\ |^2T_2 - \tfrac12\ 0>\\
&= \tfrac{1}{\sqrt3}\ |t_2 - 1^+> - (\tfrac23)^{\frac12}\ |t_2\ 0^->\\
&= \tfrac{1}{\sqrt3}\ |1^+> - \tfrac{1}{\sqrt3}\ [|2^-> - |-2^->]\\
|^2T_2\ U'\ \lambda> &= |^2T_2 - \tfrac12 - 1>\\
&= |t_2 - 1^->\\
&= -|1^->
\end{aligned}
$$

$$|^2T_2\, U'\, \mu> \ = |^2T_2\, \tfrac{1}{2}\, 1>$$

$$= |t_2\, 1^+>$$

$$= |-1^+>$$

$$|^2T_2\, U'\, \nu> \ = -(\tfrac{2}{3})^{\frac{1}{2}}\, |^2T_2\, \tfrac{1}{2}\, 0> -\tfrac{1}{\sqrt{3}}\, |^2T_2 -\tfrac{1}{2}\, 1>$$

$$= -(\tfrac{2}{3})^{\frac{1}{2}}\, |t_2\, 0^+> -\tfrac{1}{\sqrt{3}}\, |t_2\, 1^->$$

$$= -\tfrac{1}{\sqrt{3}}\, [\,|2^+> - |-2^+>\,] -\tfrac{1}{\sqrt{3}}\, |-1^->. \tag{67}$$

V. Tetragonal Symmetry

A. Ligand Field Potential and Basis Functions

Next in importance to the cubic term in the ligand field potential is a term having tetragonal (axial) symmetry. For example, the ligands may be located at $x = \pm a$, $y = \pm a$, $z = \pm b$, with $a \neq b$. Such an environment, which departs from cubic symmetry by a distortion along the z-axis, is described by a potential, V_t, of the form (40)

$$V_t = B_2^0 (3z^2 - r^2) \tag{68}$$

in which certain, smaller, fourth order terms have been neglected. The discussion that follows parallels that given previously for fields possessing cubic symmetry.

The potential V_t is invariant under the operations of the group D_{4h} whose characters are listed in Table 6. However, since d-electrons generare representations of even parity only, we may refer to the simpler group D_4 whose characters are also contained in Table 6. Moreover, since an irreducible representation in O is generally a reducible representation in D_4, it is possible, with the help of the character table and Eqs. (42 and 46), to decompose the e- and t_2-representations in O into irreducible representations in D_4. The results are

$$e = a_1 + b_1$$

$$t_2 = b_2 + e. \tag{69}$$

We conclude therefore that the general effect of an axial field is to split levels which are degenerate in a field of cubic symmetry. In particular, the 2-fold degenerate e-orbitals in O are separated into two non-degenerate orbitals labeled a_1 and b_1 in D_4; at the same time the three-fold degenerage t_2-orbitals in O are split into one b_2 (non-degenerate) and one set of e (two-fold degenerate) orbitals. A diagram illustrating these relationships is given in Fig. 22. The basis functions in D_4, expressed in

Table 6. *Character table for the hroup* D_{4h}.

D_{4h}	E	$2C_4$	C_2	$2C_2'$	$2C_2''$	i	$2S_4$	σ_h	$2\sigma_v$	$2\sigma_d$
A_{1g}	1	1	1	1	1	1	1	1	1	1
A_{2g}	1	1	1	−1	−1	1	1	1	−1	−1
B_{1g}	1	−1	1	1	−1	1	−1	1	1	−1
B_{2g}	1	−1	1	−1	1	1	−1	1	−1	1
E_g	2	0	−2	0	0	2	0	−2	0	0
A_{1u}	1	1	1	1	1	−1	−1	−1	−1	−1
A_{2u}	1	1	1	−1	−1	−1	−1	−1	1	1
B_{1u}	1	−1	1	1	−1	−1	1	−1	−1	1
B_{2u}	1	−1	1	−1	1	−1	1	−1	1	−1
E_u	2	0	−2	0	0	−2	0	2	0	0

Definitions (based on the z-axis along the 4-fold symmetry axis)
E: identity operation
C_2: a rotation of 180° about the z-axis
C_4: a rotation of ±90° about the z-axis
C_2': a rotation of 180° about the x- or y-axis
C_2'': a rotation of 180° about an axis inclined at 45° to the x- and y-axis
σ_h: reflection in the xy-plane
i: inversion in the origin $(\sigma_h C_2)$
S_4: combined rotation of ±90° about the z-axis with reflection in the xy-plane
$(\sigma_h C_4)$
σ_v: reflection in the xz- or yz-plane $(\sigma_h C_2')$
σ_d: reflection in a plane inclined at 45° to both the xz- and yz-plane $(\sigma_h C_2'')$
The group D_4 contains the elements E, $2C_4$, C_2, $2C_2'$, $2C_2''$.

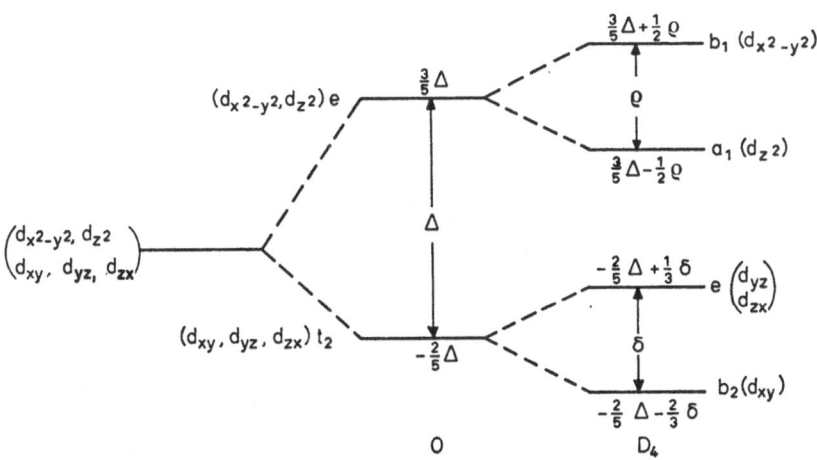

Fig. 22. Splitting of a d-state in a tetragonal (D_4) field. The negatively charged
ligands are at x = ± a, y = ± a, z = ± b with b > a.

various popular notations, are the following.

$$|a_1 > \; = \; |e\,\theta > \; = \; |0 > \; = d_{z^2} = d_0$$

$$= \tfrac{1}{2}\,(3z^2 - r^2)$$

$$|b_1 > \; = \; |e\,\varepsilon > \; = \; \tfrac{1}{\sqrt{2}}\,[|2 > + |-2 >]$$

$$= \tfrac{1}{\sqrt{2}}\,[d_2 + d_{-2}] = \tfrac{\sqrt{3}}{2}\,(x^2 - y^2) = d_{x^2-y^2}$$

$$|b_2 > \; = \; |t_2\,\zeta > \; = \; \tfrac{1}{i\sqrt{2}}\,[|2 > - |-2 >]$$

$$= \tfrac{1}{i\sqrt{2}}\,[d_2 - d_{-2}] = \sqrt{3}\; xy = d_{xy}$$

$$= \tfrac{1}{i}\,|\zeta_1 >$$

$$|ex > \; = \; |t_2\,\xi > \; = \; \tfrac{i}{\sqrt{2}}\,[|1 > + |-1 >]$$

$$= \tfrac{i}{\sqrt{2}}\,[d_2 + d_{-2}] = \sqrt{3}\; yz = d_{yz}$$

$$|ey > \; = \; - |t_2\,\eta > \; = \; \tfrac{1}{\sqrt{2}}\,[|1 > - |-1 >]$$

$$= - \tfrac{1}{\sqrt{2}}\,(d_1 - d_{-1}) = \sqrt{3}\; zx = d_{zx} \tag{70}$$

We may define

$$|e\,1 > \; = \; - \tfrac{i}{\sqrt{2}}\,[|ex > + i\,|ey >] = |1 >$$

$$|e-1 > \; = \; - \tfrac{i}{\sqrt{2}}\,[|ex > - i\,|ey >] = |-1 > \tag{71}$$

with the converse relations

$$|ex > \; = \; \tfrac{i}{\sqrt{2}}\,[|e\,1 > + |e-1 >]$$

$$|ey > \; = \; \tfrac{1}{\sqrt{2}}\,[|e\,1 > - |e-1 >] \tag{72}$$

When the symmetry is reduced from cubic to tetragonal, terms in O such as $^4T_1$, are decomposed into terms which are labeled according to irreducible representations in D_4. The multiplicity of a term remains unaffected by the reduction in symmetry. Applying the same method which leads to Eqs. (69) we get

$$^4T_1 = {}^4A_2 + {}^4E. \tag{73}$$

B. Classification in D_4^*

Corresponding to the group D_4 there is the double group D_4^* which is related to D_4 in the same manner as O^* is to O. As may be seen from the character table (Table 7), the irreducible representations of D_4^* are A_1, A_2, B_1, B_2, E, E', and E". Thus D_4^* contains all the irreducible representations found in D_4 plus two additional 2-dimensional representations, namely E' and E".

Table 7. *Character table for the group D_4^*.*

D_4^*		E	R	C_2 $C_2 R$	C_4 $C_4^3 R$	$C_4 R$ C_4^3	$2C_2'$ $2C_2' R$	$2C_2''$ $2C_2'' R$
Γ_1	A_1	1	1	1	1	1	1	1
Γ_2	A_2	1	1	1	1	1	−1	−1
Γ_3	B_1	1	1	1	−1	−1	1	−1
Γ_4	B_2	1	1	1	−1	−1	−1	1
Γ_5	E	2	2	−2	0	0	0	0
Γ_6	E'	2	−2	0	$\sqrt{2}$	$-\sqrt{2}$	0	0
Γ_7	E"	2	−2	0	$-\sqrt{2}$	$\sqrt{2}$	0	0

Basis functions for the irreducible representations of D_4^* may be obtained directly from basis functions in O*. A table showing the correspondence between the two groups is given in Table 8 in which the signs that have been chosen constitute one of several phase combinations which are consistent with the transformation properties.

Table 8. *Correspondence between basis functions in O* and basis functions in D_4^*.*

O*	D_4^*
A_1	A_1
A_2	B_1
$E\theta$	A_1
$E\varepsilon$	B_1
$T_1 x$	E x
$T_1 y$	E y
$T_1 z$	A_2
$T_2 \xi$	E x
$T_2 \eta$	− E y
$T_2 \zeta$	B_2
$E'\alpha'$	$E'\alpha'$
$E'\beta'$	$E'\beta'$
$E''\alpha''$	$E''\alpha''$
$E''\beta''$	$E''\beta''$
$U'\varkappa$	$E''\beta''$
$U'\lambda$	$-E'\alpha'$
$U'\mu$	$E'\beta'$
$U'\nu$	$-E''\alpha''$

M. Weissbluth

The parity of a representation remains unchanged so that a g (or u)-state in O* goes over into a g (or u)-state in D_4^*. With the aid of Table 8 we may immediately write the basis functions in D_4^* in terms of those in O* which in turn have already been expressed as expansions of basis functions in O as in Eqs. (64, 65, 66, 67).

As an example we consider a single electron confined to a t_2-orbital. Relative to the group O the single electron configuration gives rise to a six-fold degenerate term $^2T_2$. Relative to O*, which corresponds to the application of spin-orbit coupling, the $^2T_2$ term splits into two levels labeled E″ and U′ whose wave functions are given by Eqs. (67). When the symmetry is lowered from cubic to tetragonal the E″-level is unaffected while the U′-level splits into an E′-and an E″-level. Using Table 8 the basis functions in D_4^* are the following:

$$|^2T_2 \tfrac{1}{2} E'' \alpha'' > \; = |^2T_2 E'' \alpha'' >$$
$$|^2T_2 \tfrac{1}{2} E'' \beta'' > \; = |^2T_2 E'' \beta'' >$$
$$|^2T_2 \tfrac{3}{2} E'' \alpha'' > \; = - |^2T_2 U' \nu >$$
$$|^2T_2 \tfrac{3}{2} E'' \beta'' > \; = |^2T_2 U' \varkappa >$$
$$|^2T_2 E' \alpha' > \quad = - |^2T_2 U' \lambda >$$
$$|^2T_2 E' \beta' > \quad = |^2T_2 U '\mu > \tag{74}$$

The ket on the left is labeled according to D_4^* and the ket on the right, according to O*. There are two sets of functions which transform according to E″ in D_4^*; hence it is necessary to introduce an additional label in order to distinguish between them. As was done in connection with Eqs. (66). a J-index is employed which in the present case has values of 1/2 and 3/2. Eqs. (74) may be written in other forms which are obtainable directly from Eqs. (67).

Precisely the same procedure may be employed for Eqs. (64, 65 and 66) to determine the manner in which the states comprising $^6A_1$ and $^4T_1$ span the representations of D_4^*. We shall then get relationships of the type shown in Eqs. (74). However, for later calculations it is useful to list the basis functions in O which ultimately become associated with a particular representation in D_4^*. For example, in Eqs. (64) $|^6A_1 \tfrac{1}{2} a_1 >$ which is a basis function in O, is associated with $|^6A_1 U' \lambda >$, a basis function in O*. The latter, according to Table 8, goes over to $- |^6A_1 E' \alpha' >$ in D_4^*. Hence $|^6A_1 \tfrac{1}{2} a_1 >$ may be associated with $|^6A_1 E' \alpha' >$. Alternatively, it is seen that when $|^6A_1 E' \alpha' >$ (in D_4^*) is expanded in terms of basis functions in O, the expansion contains $|^6A_1 \tfrac{1}{2} a_1 >$ (multiplied by a numerical coefficient). The list is as follows:

E' α'	E' β'	E" α"	E" β"
$^6A_1 \frac{1}{2} a_1$	$^6A_1 - \frac{1}{2} a_1$	$^6A_1 \frac{5}{2} a_1$	$^6A_1 \frac{3}{2} a_1$
		$^6A_1 - \frac{3}{2} a_1$	$^6A_1 - \frac{5}{2} a_1$
$^4T_1 \frac{3}{2} - 1$	$^4T_1 \frac{1}{2} - 1$	$^4T_1 \frac{3}{2} 1$	$^4T_1 \frac{3}{2} 0$
$^4T_1 - \frac{1}{2} 1$	$^4T_1 - \frac{3}{2} 1$	$^4T_1 - \frac{1}{2} - 1$	$^4T_1 \frac{1}{2} 1$
$^4T_1 \frac{1}{2} 0$	$^4T_1 - \frac{1}{2} 0$	$^4T_1 - \frac{3}{2} 0$	$^4T_1 - \frac{3}{2} - 1$

We shall find (Ch. VII) that the above classification leads to considerable simplification of the spin-orbit coupling matrix.

VI. Electron Spin Resonance in Low Spin Hemoglobin

Gibson and *Ingram* *(14)* measured the principal g-values in (ferric) hemoglobin azide. They found (Table 2).

$$g_x = 1.72, \quad g_y = 2.22, \quad g_z = 2.80,$$

referred to x- and y-axes in the plane of the heme and the z-axis perpendicular to it. The first theoretical analysis of the electron spin resonance data was provided by *Griffith* *(19)*; a substantially similar analysis was given by *Kotani* *(35)*.

As will be discussed in Ch. VIII, measurements of magnetic susceptibility indicated that hemoglobin azide behaved as if it had a spin of 1/2; it is therefore described as a "low spin" compound. The fact that S = 1/2 indicates that the separation between the two-fold degenerate e-orbitals and the three-fold degenerate t_2-orbitals is large compared with pairing energies and all five electrons reside in the t_2-orbitals; the electronic configuration is therefore t_2^5. This is a fortunate circumstance because the total capacity of the t_2-orbitals is six electrons, so that occupation by five electrons leaves a single hole which, in all important respects, behaves precisely as a single electron. Thus, the very complicated $(3d)^5$ system is replaced by the relatively simple $(3d)^1$ system with the understanding that we are dealing with hole states rather than electron states. More specifically an electron configuration such as $|\xi^- \eta^2 \zeta^2 >$ is replaced by an equivalent hole described by $|\xi^+ >$. Several conclusions may be drawn immediately:

(1) The g-values of hemoglobin azide are all substantially different from the free spin value of 2.0023. Hence the hole does not behave as a free spin; a contribution from orbital angular momentum must be present.

(2) The principal g-values are all different, that is, $g_x \neq g_y \neq g_z$. The symmetry at the position of the iron atom must therefore be lower than cubic and in fact lower than tetragonal. We could assume rhombic symmetry (D_{2h}); with six-fold coordination this implies (in the ligand field approach) a distribution of charges situated at $x = \pm a$, $y = \pm b$ and $z = \pm c$ with $a \neq b \neq c$. The ligands in the heme plane are nitrogens; hence the assumption of rhombic symmetry requires the principal axis system of the g-tensor to coincide with a set of axes, two of which point at the porphyrin nitrogens. Experimentally this would mean that the directions of the magnetic fields which yield the observed values of g_x and g_y pass through the nitrogens. *Griffith* (23) notes that the validity of this assumption has been called into question by recent x-ray data; however he demonstrates that the major conclusions based on the assumption of rhombic symmetry remain intact. We shall therefore assume rhombic symmetry.

(3) The $(3d)^5$ system (or the equivalent d^1 hole) is subject to Kramers' theorem which states that when a system is composed of an odd number of electrons (or holes) it is not possible for electric fields to remove degeneracies completely — at least two-fold degeneracies must remain.

We are led, then, to the following picture: In low spin $(S = 1/2)$ hemoglobin the environment of the iron is predominantly cubic but contains tetragonal and rhombic components. That is, there are three types of ligand field potentials in the Hamiltonian which are, respectively, invariant under the operations of the groups O_h, D_{4h} and D_{2h}. Under O_h, the t_2-orbitals ξ, η, ζ, are degenerate; reduction of the symmetry to D_{4h} separates the ζ-orbital from ξ and η which still remain degenerate. Further reduction of the symmetry to D_{2h} separates the ξ- and η-orbitals thereby lifting the orbital degeneracy completely. Thus, under the influence of the low symmetry environment ξ, η, ζ are each orbital singlets with energies ε_ξ, ε_η, and ε_ζ respectively. Each orbital still has a 2-fold spin degeneracy which, according to Kramers' theorem, cannot be removed by any combination of electric fields. Finally we must include spin-orbit coupling to allow for mixing of the orbitals; the mixed orbitals will combine into a set of Kramers doublets of which the one lying lowest in energy will be the ground state and will account for the observed electron spin resonance. Kramers' theorem does not apply to systems with an even number of electrons; hence it is not applicable to ferrous hemoglobin and a sufficiently asymmetric ligand field may remove all degeneracies. Application of a magnetic field cannot remove degeneracies any further; it can only shift the relative energies. In such a system electron spin resonance is not likely to be observed except when two levels fortuitously happen to get sufficiently close to one another for a microwave photon to induce a transition. This is probably the basis for the lack of observation of electron spin resonance in high spin (S = 2) ferrous hemo-globin.

A. Kramers Doublets

As a preliminary step in the construction of the Kramers doublets we shall compute the spin-orbit coupling matrix within the t_2-orbitals ξ, η, ζ. The e-orbitals do not enter into the discussion since they are presu-med to be far removed in energy, as evidenced by the fact that hemoglo-bin azide is a low spin compound. The pertinent part of the spin-orbit coupling operator is given by

$$\vec{\ell} \cdot \vec{s} = \ell_z s_z + \tfrac{1}{2} (\ell_+ s_- + \ell_- s_+) \tag{75}$$

in which all the operators are one-electron operators and

$$\ell_\pm = \ell_x \pm i \ell_y \ , \ \ s_\pm = s_x \pm i s_y \tag{76}$$

are shift operators which satisfy relations of the form of Eq. (12). It is therefore possible to construct the matrix of $\vec{\ell} \cdot \vec{s}$ within the set of d-orbi-tals; this is given in Table 9. We note that the diagonal elements are due

Table 9. Matrix $\vec{\ell} \cdot \vec{s}$ within the set of d-orbitals, for a single electron.

$$\vec{\ell} \cdot \vec{s} = \ell_z s_z + \tfrac{1}{2} (\ell_+ s_- + \ell_- s_+)$$

$\vec{\ell} \cdot \vec{s}$	2+	1+	0+	−1+	−2+	2−	1−	0−	−1−	−2−
2+	1									
1+		$\frac{1}{2}$				1				
0+			0				$\left(\frac{3}{2}\right)^{\frac{1}{2}}$			
−1+				$-\frac{1}{2}$				$\left(\frac{3}{2}\right)^{\frac{1}{2}}$		
−2+					-1				1	
2−	1					-1				
1−		$\left(\frac{3}{2}\right)^{\frac{1}{2}}$					$-\frac{1}{2}$			
0−			$\left(\frac{3}{2}\right)^{\frac{1}{2}}$					0		
−1−					1				$\frac{1}{2}$	
−2−										1

M. Weissbluth

to $\ell_z\, s_z$, the elements above the diagonal are due to $1/2\ \ell_-\, s_+$ and the elements below the diagonal are due to $1/2\ \ell_+\, s_-$. By means of Eqs. (48) we obtain the matrix of $\vec{\ell}\cdot\vec{s}$ within the set of t_2-orbitals (Table 10). Tables given by *Ballhausen* (2) are also helpful.

Table 10. *Matrix of $\vec{\ell}\cdot\vec{s}$ within the set ξ, η, ζ, for a single electron.*

$\vec{\ell}\cdot\vec{s}$	ξ^+	η^+	ζ^-	ξ^-	η^-	ζ^+
ξ^+		$\frac{i}{2}$	$-\frac{1}{2}$			
η^+	$-\frac{i}{2}$		$\frac{i}{2}$			
ζ^-	$-\frac{1}{2}$	$-\frac{i}{2}$				
ξ^-					$-\frac{i}{2}$	$\frac{1}{2}$
η^-				$\frac{i}{2}$		$\frac{i}{2}$
ζ^+				$\frac{1}{2}$	$-\frac{i}{2}$	

It is instructive to have the matrix of $\vec{\ell}\cdot\vec{s}$ in the basis set $|t_2\,1>$, $|t_2\,0>$ and $|t_2-1>$ as defined by Eqs. (56). The matrix elements are readily computed from Table 9 and are shown in Table 11.

Table 11. *Matrix of $\vec{\ell}\cdot\vec{s}$ within the set $|t_2 1>$, $|t_2 0>$, $|t_2-1>$, for a single electron.*

$\vec{\ell}\cdot\vec{s}$	$|t_2 1^+>$	$|t_2 0^->$	$|t_2-1^+>$	$|t_2-1^->$	$|t_2 0^+>$	$|t_2 1^->$
$|t_2 1^+>$	$-\frac{1}{2}$					
$|t_2 0^->$			$-\frac{1}{\sqrt{2}}$			
$|t_2-1^+>$		$-\frac{1}{\sqrt{2}}$	$\frac{1}{2}$			
$|t_2-1^->$				$-\frac{1}{2}$		
$|t_2 0^+>$						$-\frac{1}{\sqrt{2}}$
$|t_2 1^->$					$-\frac{1}{\sqrt{2}}$	$\frac{1}{2}$

Finally we may compute the matrix of $\vec{\ell}\cdot\vec{s}$ within the basis functions in O* (Eqs. 67). As expected, by analogy with the atomic case, the matrix of $\vec{\ell}\cdot\vec{s}$ (Table 12) is diagonal in this basis set and, moreover, has only two distinct values. These are 1 and $-\frac{1}{2}$ associated with the two-fold

Table 12. *Matrix of $\vec{\ell} \cdot \vec{s}$ within basis functions in O^* for a single electron in a t_2-orbital.*

$\vec{\ell} \cdot \vec{s}$	$^2T_2\, E''\, \alpha''$	$^2T_2\, E''\, \beta''$	$^2T_2\, U'\, \varkappa$	$^2T_2\, U'\, \lambda$	$^2T_2\, U'\, \mu$	$^2T_2\, U'\, \nu$
$^2T_2\, E''\, \alpha''$	1					
$^2T_2\, E''\, \beta''$		1				
$^2T_2\, U'\, \varkappa$			$-\frac{1}{2}$			
$^2T_2\, U'\, \lambda$				$-\frac{1}{2}$		
$^2T_2\, U'\, \mu$					$-\frac{1}{2}$	
$^2T_2\, U'\, \nu$						$-\frac{1}{2}$

degenerate representation E'' and the four-fold degenerate representation U', respectively. Alternatively, we may say that under the influence of spin-orbit coupling, the originally six-fold degenerate t_2-manifold of stakes breaks up into a doublet and a quartet associated with the representations E'' and U' respectively.

Examination of the matrix in Table 10 reveals that ξ^+, η^+ and ζ^- have non-vanishing matrix elements among themselves and similarly for the set ξ^-, η^- and ζ^+. Matrix elements of orbitals from one set with orbitals from the other vanish. The general structure of the Kramers doublet will therefore be

$$\psi_1 = a_1\xi^+ + b_1\eta^+ + c_1\zeta^-$$
$$\psi_2 = a_2\xi^- + b_2\eta^- + c_2\zeta^+ \qquad (77)$$

Although there appear to be six constants, they are not all independent. According to Kramers' theorem *(21, 26)*, ψ_2 must satisfy the relation

$$\psi_2 = i\, \psi_1^* \qquad (78)$$

which is based on the definition

$$|n\, \ell\, m_\ell\, m_s >^* = (-1)^{\, m_\ell + m_s} |n\, \ell - m_\ell - m_s > \qquad (79)$$

From Eqs. (48) we may verify that

$$
\begin{aligned}
|\xi^+>^* &= i|\xi^-> & |\xi^->^* &= -i|\xi^+> \\
|\eta^+>^* &= i|\eta^-> & |\eta^->^* &= -i|\eta^+> \\
|\zeta^+>^* &= i|\zeta^-> & |\zeta^->^* &= -i|\zeta^+>
\end{aligned}
\qquad (80)
$$

The lowest Kramers doublet may therefore be written (*35*)

$$\psi_1^+ = A_1\,\xi^+ + i\,B_1\,\eta^+ + C_1\,\zeta^-$$

$$\psi_1^- = -\,A_1\,\xi^- + i\,B_1\,\eta^- + C_1\,\zeta^+ \qquad (81)$$

in which the coefficients A_1, B_1 and C_1 are taken to be real. Eqs. (44) satisfy

$$\psi_1^- = i(\psi_1^+)^* \qquad (82)$$

but the inverse relation is

$$\psi_1^+ = -\,i(\psi_1^-)^* \qquad (83)$$

as may be readily seen from Eqs. (80) and (81).

B. Interaction with a Magnetic Field

The interaction Hamiltonian may be written

$$\mathscr{H}_m = \beta\,\vec{H}\cdot(\vec{\ell} + \vec{2s}) \qquad (84)$$

In order to interpret the g-values associated with spin resonance experiments, it is therefore necessary to construct the matrix of $\vec{\ell} + \vec{2s}$ within the basis set ψ_1^+ and ψ_1^-. This construction is facilitated by the prior calculation of matrix elements of $\vec{\ell} + \vec{2s}$ within the set of d-orbitals (Eqs. 4). The matrix elements of $\ell_z + 2s_z$ are easily calculated, those of $\ell_x + 2s_x$ and $\ell_y + 2s_y$ are best handled by means of the shift operators (Eqs. 75) and the defining relations given by Eqs. (12). The resulting matrices are given in Table 13. The conversion of the matrices to the basis set ξ, η, ζ is then accomplished by means of Eqs. (48), and are given in Table 14. Finally the matrices of $\vec{\ell} + \vec{2s}$ in the basis set ψ_1^+, ψ_1^- are computed (Table 15).

For a field in the z-direction, the eigenvalues of $\mathscr{H}_m$ as read directly from Table 15 are

$$E_1^{(z)} = \beta\,H_z[(A_1 - B_1)^2 - C_1^2]$$

$$E_2^{(z)} = -\,\beta\,H_z[(A_1 - B_1)^2 - C_1^2] \qquad (85)$$

indicating that the two-fold degeneracy of the doublet is lifted by the magnetic field. The separation in energy of the two components of the doublet is

$$\Delta E = E_1^{(z)} - E_2^{(z)} = 2\,\beta\,H_z[(A_1 - B_1)^2 - C_1^2] \qquad (86)$$

Table 13. *Matrices of* $\vec{\ell} + 2\vec{s}$. *Upper and lower entries refer to x-and y-components respectively, single entries refer to z-components.*

$\vec{\ell}+2\vec{s}$	2+	1+	0+	−1+	−2+	2−	1−	0−	−1−	−2−
2+	3	1 / −i				1 / −i				
1+	1 / i	2	$(\frac{3}{2})^{\frac12}$ / $-i(\frac{3}{2})^{\frac12}$				1 / −i			
0+		$(\frac{3}{2})^{\frac12}$ / $i(\frac{3}{2})^{\frac12}$	1	$(\frac{3}{2})^{\frac12}$ / $-i(\frac{3}{2})^{\frac12}$				1 / −i		
−1+			$(\frac{3}{2})^{\frac12}$ / $i(\frac{3}{2})^{\frac12}$	0	1 / −i				1 / −i	
−2+				1 / i	−1	0 / 0				1 / −i
2−	1 / i				0 / 0	1	1 / −i			
1−		1 / i				1 / i	0	$(\frac{3}{2})^{\frac12}$ / $-i(\frac{3}{2})^{\frac12}$		
0−			1 / i				$(\frac{3}{2})^{\frac12}$ / $i(\frac{3}{2})^{\frac12}$	−1	$(\frac{3}{2})^{\frac12}$ / $-i(\frac{3}{2})^{\frac12}$	
−1−				1 / i				$(\frac{3}{2})^{\frac12}$ / $i(\frac{3}{2})^{\frac12}$	−2	1 / −i
−2−					1 / i				1 / i	−3

Table 14. *Matrices of $\vec{\ell}+2\vec{s}$ within the set ξ, η, ζ, for a single electron. Upper, middle and lower values are matrix elements of x, y, and z components respectively.*

$\vec{\ell}+2\vec{s}$	ξ^+	η^+	ζ^+	ξ^-	η^-	ζ^-
ξ^+	 1	 i	 −i 	1 −i 		
η^+	 −i	 1	i 		1 −i 	
ζ^+	 i 	−i 	 1			1 −i
ξ^-	1 i 			 −1	 i	 −i
η^-		1 i 		 −i	 −1	i
ζ^-			1 i 	 i 	−i 	 −1

Table 15. *Matrices of $\vec{\ell}+2\vec{s}$ for the lowest Kramers doublet.*

ℓ_z+2s_z	ψ_1^+	ψ_1^-
ψ_1^+	$(A_1-B_1)^2-C_1^2$	0
ψ_1^-	0	$-[(A_1-B_1)^2-C_1^2]$

(a)

ℓ_x+2s_x	ψ_1^+	ψ_1^-
ψ_1^+	0	$(B_1+C_1)^2-A_1^2$
ψ_1^-	$(B_1+C_1)^2-A_1^2$	0

(b)

ℓ_y+2s_y	ψ_1^+	ψ_1^-
ψ_1^+	0	$i[(A_1-C_1)^2-B_1^2]$
ψ_1^-	$-i[(A_1-C_1)^2-B_1^2]$	0

(c)

and electron spin resonance will be observed when the electromagnetic energy satisfies the condition

$$\hbar\omega = g_z\,\beta\,H_z = \Delta\,E \tag{87}$$

or

$$g_z = 2\,|\,(A_1 - B_1)^2 - C_1^2\,| \tag{88}$$

The eigenvalues of $\ell_x + 2s_x$ and $\ell_y + 2s_y$ are obtained from the solution of the secular determinant associated with the corresponding matrices in Table 15. They are

$$E^{(x)} = \pm\,\beta\,H_x[(B_1 + C_1)^2 - A_1^2]$$

$$E^{(y)} = \pm\,\beta\,H_y[(A_1 - C_1)^2 - B_1^2] \tag{89}$$

Therefore the principal components of the g-tensor are

$$g_x = 2|(B_1 + C_1)^2 - A_1^2|$$

$$g_y = 2|(A_1 - C_1)^2 - B_1^2|$$

$$g_z = 2|(A_1 - B_1)^2 - C_1^2| \tag{90}$$

where absolute values are used to ensure that g_x, g_y and g_z are positive quantities.

Inserting the observed g-values into Eqs. (89) and requiring that

$$A_1^2 + B_1^2 + C_1^2 = 1 \tag{91}$$

yields the following values (35)

$$A_1 = 0.973$$

$$B_1 = -0.209$$

$$C_1 = -0.097 \tag{92}$$

It is assumed, and borne out by subsequent calculation, that the observed g-values are associated with the splitting of the lowest Kramers doublet. Therefore the numerical values of the coefficients in Eqs. (92) serve to identify the lowest Kramers doublet which now becomes

$$\psi_1^+ = 0.973\ \xi^+ - 0.209\ i\ \eta^+ - 0.097\ \zeta^-$$

$$\psi_1^- = -0.973\ \xi^- - 0.209\ i\ \eta^- - 0.097\ \zeta^+ \tag{93}$$

We note that the coefficient of ξ^+ in ψ_1^+ (or the coefficient of ξ^- in ψ_1^-) has an absolute value close to unity so that the lowest Kramers doublet closely resembles the ξ-orbital. The departure from exact correspondence or the reason for the appearance of the terms in η and ζ is due to spin-orbit coupling. Thus the main effect of spin-orbit coupling is to mix the orbitals. Although spin-orbit coupling also contributes to the energy separation among the orbitals, the major cause of the energy separation is the low symmetry crystal field.

C. Energetics of the Hole

The separation in energy among the orbitals ξ, η, ζ may be extracted from Eqs. (93). It is instructive to do so, at first, by use of perturbation methods. For this purpose, we consider ψ_1^+ (or ψ_1^-) as a wave function correct to first order with the terms in η and ζ as the first order corrections. The interaction Hamiltonian is given by the spin-orbit coupling operator for a one particle system

$$\mathscr{H}_s = -\lambda \vec{\ell} \cdot \vec{s}, \quad \lambda > 0 \tag{94}$$

where we have used λ as the coefficient in place of ζ (Eq. 26) to avoid confusion with the ζ-orbital. The negative sign in the Hamiltonian arises from the fact that the system under discussion is described by a hole which carries an equivalent positive charge. It is therefore possible to express ψ_1^+ by

$$\psi_1^+ = 0.973\,\xi^+ + \frac{< \eta^+ |-\lambda \vec{\ell} \cdot \vec{s}|\xi^+ >}{\varepsilon_\xi - \varepsilon_\eta}\,\eta^+ + \frac{< \zeta^- |-\lambda \vec{\ell} \cdot \vec{s}|\xi^+ >}{\varepsilon_\xi - \varepsilon_\zeta}\,\zeta^- \tag{95}$$

where ε_ξ, ε_η, ε_ζ are the hole orbital energies for either spin orientation. An analogous expression may be written for ψ_1^-. Comparison with Eqs. (93) then gives

$$\frac{< \eta^+ |-\lambda \vec{\ell} \cdot \vec{s}|\xi^+ >}{\varepsilon_\xi - \varepsilon_\eta} = -0.209\,i$$

$$\frac{< \zeta^- |-\lambda \vec{\ell} \cdot \vec{s}|\xi^+ >}{\varepsilon_\xi - \varepsilon_\zeta} = -0.097 \tag{96}$$

The matrix elements may be read directly from Table 10. We obtain

$$< \eta^+ |-\lambda \vec{\ell} \cdot \vec{s}|\xi^+ > = \frac{i\lambda}{2}$$

$$< \zeta^- |-\lambda \vec{\ell} \cdot \vec{s}|\xi^+ > = \frac{\lambda}{2} \tag{97}$$

Substitution in Eqs. (96) then gives

$$\varepsilon_\xi - \varepsilon_\eta = -2.392\,\lambda$$

$$\varepsilon_\xi - \varepsilon_\zeta = -5.155\,\lambda \tag{98}$$

An exact calculation of the orbital separations may be performed without resorting to perturbation theory. Let the combined Hamiltonian of the spin-orbit coupling and the low symmetry ligand field, V, be

$$\mathscr{H} = -\lambda \vec{l} \cdot \vec{s} + V \qquad (99)$$

We presume that ξ, η, ζ are eigenfunctions of V with eigenvalues ε_ξ, ε_η, ε_ζ. Then the condition that

$$\mathscr{H}\,\psi_1^+ = E\,\psi_1^+ \qquad (100)$$

where ψ_1^+ is given by Eqs. (81) leads to the secular equations

$$A_1 \left[<\xi^+ |\mathscr{H}|\xi^+> - E \right] + i\,B_1 <\xi^+ |\mathscr{H}|\eta^+> + C_1 <\xi^+ |\mathscr{H}|\zeta^-> = 0$$

$$A_1 <\eta^+ |\mathscr{H}|\xi^+> + i\,B_1 \left[<\eta^+ |\mathscr{H}|\eta^+> - E \right] + C_1 <\eta^+ |\mathscr{H}|\zeta^-> = 0$$

$$A_1 <\zeta^- |\mathscr{H}|\xi^+> + i\,B_1 <\zeta^- |\mathscr{H}|\eta^+> + C_1 \left[<\zeta^- |\mathscr{H}|\zeta^-> - E \right] = 0 \qquad (101)$$

From Table 10 we obtain the matrix elements of the spin-orbit coupling; hence Eqs. 101 become

$$A_1\,(\varepsilon_\xi - E) - (i\,B_1)\,\frac{i}{2}\,\lambda + C_1\,\frac{\lambda}{2} = 0$$

$$A_1\,\frac{i}{2}\,\lambda + (i\,B_1)\,(\varepsilon_\eta - E) - C_1\,\frac{i}{2}\,\lambda = 0$$

$$A_1\,\frac{\lambda}{2} + (i\,B_1)\,\frac{i}{2}\,\lambda + C_1\,(\varepsilon_\zeta - E) = 0 \qquad (102)$$

A_1, B_1, C_1 are the coefficients in the lowest Kramers doublet and are given by Eqs. (92), E is the energy of the lowest Kramers doublet; we may set E = 0 (or any other arbitrarily chosen energy) since differences in energy are all that matter. Thus

$$\varepsilon_\xi = -\left[\frac{B_1 + C_1}{A_1} \right]\,\frac{\lambda}{2} = 0.157\,\lambda$$

$$\varepsilon_\eta = \left[\frac{C_1 - A_1}{B_1} \right]\,\frac{\lambda}{2} = 2.56\,\lambda$$

$$\varepsilon_\zeta = \left[\frac{B_1 - A_1}{C_1} \right]\,\frac{\lambda}{2} = 6.09\,\lambda \qquad (103)$$

and

$$\varepsilon_\xi - \varepsilon_\eta = -2.403\,\lambda$$

(hole)

$$\varepsilon_\xi - \varepsilon_\zeta = -5.936\,\lambda \qquad (104)$$

which may be compared with Eqs. (98) obtained from the perturbation calculation.

At this stage of the calculation we know the coefficients which describe the lowest Kramers doublet and we know the orbital separations associated with the single hole. As a final step we compute the coefficients describing the remaining two Kramers doublets. Let the general doublet be described as in Eqs. (81)

$$\psi^+ = A\,\xi^+ + i\,B\,\eta^+ + C\,\zeta^+$$

$$\psi^- = -A\,\xi^- + i\,B\,\eta^- + C\,\zeta^+ \tag{105}$$

As before, the Hamiltonian Eq. (99) and the condition

$$\mathscr{H}\,\psi^+ = E\,\psi^+ \tag{106}$$

lead to secular equations of the form of Eqs. (102) with arbitrary coefficients A, B, C. The condition for the existence of solutions is the vanishing of the secular determinant:

$$\begin{vmatrix} \varepsilon_\xi - E & -\frac{i}{2}\,\lambda & \frac{\lambda}{2} \\[2mm] \frac{i}{2}\,\lambda & \varepsilon_\eta - E & -\frac{i}{2}\,\lambda \\[2mm] \frac{\lambda}{2} & \frac{i}{2}\,\lambda & \varepsilon_\zeta - E \end{vmatrix} = 0 \tag{107}$$

The three roots of the secular determinant are the energies of the three Kramers doublets. Since ε_ξ, ε_η, and ε_ζ are known (Eqs. 103) the secular determinant may be solved in terms of λ. Each root, when substituted into the secular equations will yield a set of coefficients for that particular Kramers doublet. The final results as given by *Kotani* (35) are

k	Excitation E	A_k	B_k	C_k
1	0	0.973	-0.209	-0.097
2	$2.613\,\lambda$	0.219	0.970	0.108
3	$6.200\,\lambda$	0.071	-0.126	0.990

We note that each of the Kramers doublets corresponds very closely to one of the orbitals ξ, η, ζ.

From Eqs. (103, 104) we see that the ξ-orbital lies lowest in energy, with the η- and ζ-orbital lying above, in that order. It is necessary to emphasize once again that these orbital energies refer to a single hole which is complementary to the five electron system. If the hole were to be replaced by an electron, the orbital energies would be reversed in sign and we would have, in place of Eqs. (104)

$$\varepsilon_\xi - \varepsilon_\eta = 2.403\,\lambda$$

(electron)

$$\varepsilon_\xi - \varepsilon_\zeta = 5.936\,\lambda \tag{108}$$

with the ζ-orbital lying lowest. The results are summarized in Fig. 23.

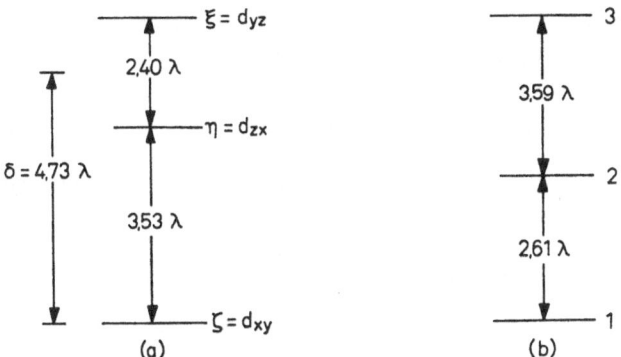

Fig. 23a and b. (a) Orbital energies for one-electron functions. For hole functions the orbital energies are inverted. (b) Kramers doublets. λ is the spin-orbit coupling parameter (∼ 435 cm⁻¹), δ is the tetragonal splitting.

With λ approximately equal to 435 cm⁻¹, the orbital separations are $\varepsilon_\xi - \varepsilon_\eta = 1040$ cm⁻¹ and $\varepsilon_\xi - \varepsilon_\zeta = 2580$ cm⁻¹. The tetragonal splitting (δ) and the rhombic splitting (μ) are

$$\delta = (\varepsilon_\xi - \varepsilon_\zeta) - \tfrac{1}{2}(\varepsilon_\xi - \varepsilon_\eta)$$

$$= 2060 \text{ cm}^{-1}$$

$$\mu = \varepsilon_\xi - \varepsilon_\eta = 1040 \text{ cm}^{-1} \tag{109}$$

Griffith (*19*) worked with the set of orbitals |t₂ 1 >, |t₂ 0 > and |t₂ − 1 > which are related to the ξ, η, ζ orbitals by Eqs. (56). The procedure is entirely analogous to that described above. For reference we give the $\vec{l} \cdot \vec{s}$ matrix in *Griffith's* orbitals (Table 11).

D. Discussion

The results of the previous analysis indicate that the observed electron spin resonance data on low spin, ferric, hemoglobin azide are explainable on the basis of a low symmetry ligand field which completely removes the orbital degeneracy of the t₂-orbitals, and spin-orbit coupling which contaminates the lowest lying orbital with contributions from the higher ones. A more complete understanding of the situation would require an answer to the following question: What are the structural features of the molecule which are responsible for a low symmetry ligand field having characteristics capable of producing orbital separations as in Fig. (23)? Attempts to answer this question have invoked (a) the histidine attachment, (b) the displacement of the iron atom from the porphyrin plane, and (c) the orientation of the azide ion. Each of these features tends to lower the symmetry from tetragonal.

Histidine (Fig. 5) contains imidazole, a five-membered, planar ring structure. The attachment to iron is shown in Fig. 9, where it is seen that the plane of the porphyrin and the plane of the imidazole are perpendicular to one another. The normals to the two planes define two perpendicular axes; a third axis may be taken perpendicular to the first two. We see then that in the plane of the porphyrin, the two axes (x and y) are no longer equivalent. Thus, a rotation through 90° about the z-axis is no longer a symmetry element and the system is not describable by the group D_{4h} but rather by a lower symmetry group such as D_{2h}. The immediate consequence is that the ξ-and η-orbitals, which are degenerate in D_{4h} (Fig. 22), are separated in D_{2h}. Also the non-equivalence of x and y implies that $g_x \neq g_y$ as actually observed. It is possible to carry the argument a bit further as was done by *Griffith* (*19*). If the imidazole plane is oriented with its normal parallel to the y-axis (Fig. 24), the imidazole nitrogen $2p\pi$-orbital will also have its maxima along the y-axis.

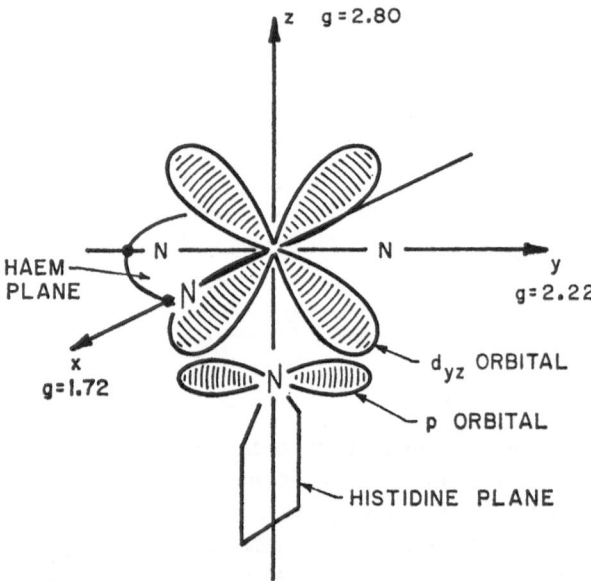

Fig. 24. Orientation of the histidine plane. The d_{yz}-orbital of Fe interacts with the p-orbital of imidazole N (*29*).

By analogy with the discussion presented in Ch. IV-C, the electrostatic repulsion of an electron in the aforementioned $2p\pi$-orbital with an electron in a ξ-orbital will be greater than with an electron in an η-orbital. The ξ-orbital is therefore driven to a higher energy. As will be discussed in Ch. IX, ξ- and η-orbitals may form π-bonds with porphyrin nitrogens, whereas a ζ-orbital may not do so; none of the three orbitals may form σ-bonds. The possibility of π-bonding would tend to lower the ξ- and η-orbitals below the ζ-orbital. The fact that ESR data show that the reverse is true would then suggest that the imidazole interaction overcomes that due to π-bonding.

Actually, the above argument was used in the reverse order to determine the orientation of the histidine on the basis of ESR data before x-ray determinations were available. Nevertheless, even though subsequent x-ray studies did not invalidate the conclusions based on ESR data, there is some doubt in attibuting the low symmetry field to the position of the histidine (37). The magnitude of the effect does not appear to be sufficient.

Griffith (23) sought another explanation based on the fact that the iron atom does not lie precisely in the plane of the porphyrin but is displaced along the four-fold axis by approximately $\frac{1}{4}$ A°. The effect of this displacement may be pictured as follows: The ζ-orbital moves out of the porphyrin plane, the electrostatic interaction with in-plane ligands is reduced thus lowering the energy of the ζ-orbital. By the same token the ξ- and η-orbitals are displaced into positions where their electrostatic interactions with the in-plane ligands is increased, thus raising their energy. The net effect, again, is to put the ζ-orbital at a lower energy than the ξ, η-orbitals despite the reverse tendencies due to bonding. *Griffith* estimated that this effect could give rise to a separation which is of the right sign, and of approximately the right magnitude as deduced from the ESR data. This effect may therefore be a significant contributor to the splitting. However, the displacement of the iron atom along the four-fold axis preserves the four-fold symmetry, leaving the ξ, η-orbitals degenerate. It would therefore be necessary to invoke another asymmetric interaction with a sufficiently low symmetry to split the ξ- and η-orbitals.

A further possibility is an effect due to the azide ion N_3^- analogous to that of the histidine. *Kotani* (37) discussed the possibility that the N_3^- ion is not oriented along the z-axis and thereby becomes the major cause of anisotropy. *Stryer* et al. (55) determined the position of the azide ion from an x-ray crystallographic study of metmyoglobin azide. Their results are shown in Fig. 25. As conjectured by *Kotani*, the azide ion is inclined at 21° to the prophyrin plane. Based on *Kotani's* estimate, it now appears quite likely that the azide ion is the major contributor to the anisotropy.

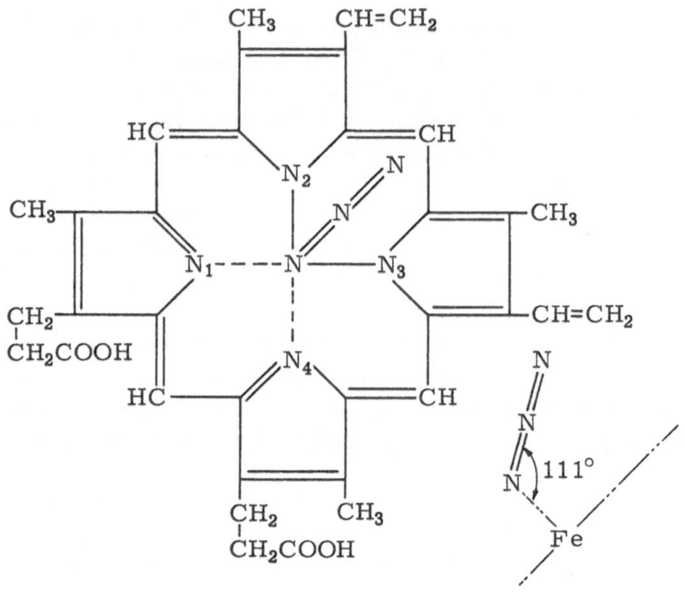

Fig. 25. Orientation of azide in ferrimyoglobin azide (55).

VII. Electron Spin Resonance in High Spin Hemoglobin

The experiments of *Gibson, Ingram* and *Schonland* (15) yielded $g_x = g_y = 6, g_z = 2$ for high spin ferric derivatives such as methemoglobin and ferrihemoglobin fluoride. Since $g_x = g_y \neq g_z$, the symmetry of the environment surrounding the iron atom must have a strong tetragonal component (D_{4h}).

We have seen that a ferric ion with an electronic configuration described by $(3d)^5$ gives rise to a ^6S ground state. The application of an octahedral ligand field or, indeed, a field of lower symmetry, produces no splitting; the ground state remains 6-fold degenerate. The application of a magnetic field, with or without a ligand field, lifts the 6-fold degeneracy and produces a set of equally spaced levels with a separation of $2\beta H$. The selection rule $\Delta M_s = \pm 1$ permits transitions only between adjacent levels resulting in $g = 2$, independently of the orientation of the magnetic field. Such an isotropic value of g is clearly contradictory to the experimental facts.

As in the previous discussion of low spin hemoglobin, it will be necessary to invoke spin-orbit coupling. However, we note immediately that there are no first order effects, that is, there can be no splitting due to spin-orbit coupling within the ^6S manifold. Lowering the symmetry does not alter the situation; it is therefore concluded that the calculation must be carried to a higher order to permit some of the excited states to interact through spin-orbit coupling with the ground state.

A. Spin-Orbit Coupling

We shall first discuss the calculation from the standpoint of a ligand field with cubic symmetry (O_h). Since in the free ion ^4P is the only term which interacts via-spin orbit coupling with ^6S we might expect that in a cubic field, the terms which evolve from ^4P would be the important ones. This is indeed the case. In a cubic field the free ion term $3d^5$, ^6S goes over into $t_2^3 e^2$, $^6A_1$ and $3d^5$, ^4P goes into $t_2^4 e$, $^4T_1$. According to the selection rules for spin-orbit coupling, $^6A_1$ has non-vanishing matrix elements with $^4T_1$ only; matrix elements of the spin-orbit coupling with excited terms other than $^4T_1$ must vanish.

The computation of matrix elements of the spin-orbit coupling is accomplished by methods analogous to those employed for free ions (Eq. 24). Racah's formalism, originally developed for atomic systems, has been extended to low symmetry groups appropriate for the description of molecules. *Griffith* (22) gives a detailed treatment of these methods. His expression for the spin-orbit coupling matrix element is the following:

$$< S \ h \ M \ \theta \ |\mathscr{H}_s \ |S' \ h' \ M' \ \theta' > \ = \ < S \ h\|\mathscr{H}_s\|S' \ h' > \Sigma \ (- \ 1)^{1+1+S-M} \ [-1]^{h+\theta}$$

$$\times \ \bar{v} \begin{pmatrix} S & S' & 1 \\ -M & M' & i \end{pmatrix} \times v \begin{pmatrix} h & h' & T_1 \\ -\theta & \theta' & -i \end{pmatrix} \tag{110}$$

S and S' are the spins of the intial and final states respectively, h and h' are the representations of the initial and final states respectively. In the cubic group O, h and h' may each stand for A_1, A_2, E_1, T_1 or T_2.

M and M' are the projection quantum numbers of S and S' respectively, i. e., M takes on the values S, S—1 ... — S.

θ and θ' are components of the representations h and h' respectively, $\mathscr{H}_s$ is the spin-orbit coupling operator.

i is a number which takes on the values 1, 0 or — 1; it labels the components of the T_1 representation.

$[- 1]^{h+\theta}$ is a special symbol with the following meaning:

$$[- 1]^{h+\theta} = [- 1]^h [- 1]^\theta$$

when h is A_1, A_2 or E, $[-1]^h = 1$,

when θ is a component of A_1, A_2 or E, $[-1]^\theta = 1$,

when h is T_1 or T_2, $[-1]^h = -1$,

when θ is a component of T_1 or T_2

$$[-1]^\theta = -1 \text{ for } \theta \text{ a component of } T_1$$

$$[-1]^\theta = 1 \text{ for } \theta \text{ a component of } T_2$$

$\bar{V} \begin{pmatrix} S & S' & 1 \\ -M & M' & i \end{pmatrix}$ is a 3-j symbol. It differs from the 3-j symbol defined by *Rotenberg* et al. (*51*) by a phase factor. The relation between the two is

$$\bar{V} \begin{pmatrix} S & S' & 1 \\ -M & M' & i \end{pmatrix} = (-1)^{S+S'+1} \begin{pmatrix} S & S' & 1 \\ -M & M' & i \end{pmatrix} \tag{111}$$

$V \begin{pmatrix} h & h' & T_1 \\ -\theta & \theta' & -i \end{pmatrix}$ is defined in terms of the coupling coefficients. These quantities are tabulated by *Griffith* (*22*) for the cubic group.

$< S \, h \| \mathscr{H}_s \| S' \, h' >$ is a reduced matrix element; it is independent of M and θ. The specific expressions for the reduced matrix element depend on the electronic configurations. For d-electrons, the reduced matrix element when taken between the same electron configurations $t_2^m \, e^n$ is

$$< t_2^m(S_1 \, h_1) \, e^n \, (S_2 \, h_2) \, Sh \| \mathscr{H}_s \| t_2^m(S_1' \, h_1') \, e^n(S_2' \, h_2') \, S' \, h' >$$

$$= (-1)^{S_1+S_2+h_1+h_2'} \, (2S+1)^{\frac{1}{2}} \, (2S'+1)^{\frac{1}{2}} \, \lambda \, (h)^{\frac{1}{2}} \, \lambda(h')^{\frac{1}{2}}$$

$$\times (-1)^{S'+h'} \, \delta_{S_2 S_2'} \, \delta_{h_2 h_2'} \, \bar{W} \begin{pmatrix} S_1' & S_1 & 1 \\ S & S' & S_2 \end{pmatrix} \times W \begin{pmatrix} h_1' & h_1 & T_1 \\ h & h' & h_2 \end{pmatrix}$$

$$\times G_n(S_1 \, h_1, \, S_1' \, h_1') < \tfrac{1}{2} \, t_2 \| su \| \tfrac{1}{2} \, t_2 > \tag{112}$$

In Eq. (112), S and h are the spin and representation respectively of a term such as $^6A_1$ or $^4T_1$. S_1, h_1 and S_2, h_2 are spins and representations arising from the electron configurations t_2^m and e^n respectively. It is understood that S_1 and S_2 are coupled together to form S while h_1 and h_2 are coupled to form h. In the same manner, S_1' and S_2' give rise to S', h_1' and h_2' to h'. We shall shortly have several examples of this process.

The symbol $(-1)^h$ is defined by

$$(-1)^{A_1} = (-1)^E = (-1)^{T_2} = +1$$

$$(-1)^{A_2} = (-1)^{T_1} = -1 \tag{113}$$

$\lambda(h)$ and $\lambda(h')$ are the degrees of representations h and h' respectively, e. g., $\lambda(A_1) = 1$, $\lambda(A_2) = 1$, $\lambda(E) = 2$, $\lambda(T_1) = 3$, $\lambda(T_2) = 3$.

$\bar{W} \begin{pmatrix} S_1' & S_1 & 1 \\ S & S' & S_2 \end{pmatrix}$ is a 6-j symbol tabulated by *Rotenberg* et al. (*51*). The remaining quantities on the right of Eq. (112) are all found in tabulations given by *Griffith* (*22*).

We shall now apply Eq. (112) to compute the reduced matrix element within the $^4T_1$ term, the latter arsising from the electronic configuration t_2^4 e. A single electron in an e-orbital can produce only term, namely, ^2E. The four electrons comprising t_2^4 may be shown to give rise to $^1A_1$, ^1E, $^1T_2$ and $^3T_1$. The ^2E cannot combine with any of the singlet terms to form the quartet term $^4T_1$; this may only be accomplished in combination with the $^3T_1$ term. Thus the reduced matrix element has the form

$$M_1 = <t_2^4 \, (^3T_1) \, e \, (^2E) \, ^4T_1 \| \mathscr{H}_s \| t_2^4 \, (^3T_1) \, e \, (^2E) \, ^4T_1 > \qquad (114)$$

$$S_1 = 1 \qquad S_1' = 1 \qquad h_1 = T_1 \qquad h_1' = T_1$$

$$S_2 = \tfrac{1}{2} \qquad S_2' = \tfrac{1}{2} \qquad h_2 = E \qquad h_2' = E$$

$$S = \tfrac{3}{2} \qquad S' = \tfrac{3}{2} \qquad h = T_1 \qquad h' = T_1$$

$$(-1)^{S_1 + S_2 + h_1 + h_2'} \, (2S + 1)^{\frac{1}{2}} \, (2S' + 1)^{\frac{1}{2}} \, \lambda(h)^{\frac{1}{2}} \, \lambda(h')^{\frac{1}{2}} \, (-1)^{S' + h'}$$

$$= (-1)^{1 + \frac{1}{2} + T_1 + E} \, (4)^{\frac{1}{2}} \, (4)^{\frac{1}{2}} \, (3)^{\frac{1}{2}} \, (3)^{\frac{1}{2}} \, (-1)^{\frac{3}{2} + T_1}$$

$$= - \, i(-1)^{T_1 + E} \cdot 4 \cdot 3 \cdot - i(-1)^{T_1}$$

$$= - \, 12(-1) \, (+1) \, (-1) = - \, 12$$

From *Rotenberg's* tables

$$\bar{W} \begin{pmatrix} S_1' & S_1 & 1 \\ S & S' & S_2 \end{pmatrix} = \bar{W} \begin{pmatrix} 1 & 1 & 1 \\ \tfrac{3}{2} & \tfrac{3}{2} & \tfrac{1}{2} \end{pmatrix} = \tfrac{1}{6} \, (\tfrac{5}{2})^{\frac{1}{2}}$$

and from *Griffith's* tables

$$W \begin{pmatrix} h_1' & h_1 & T_1 \\ h & h' & h_2 \end{pmatrix} = W \begin{pmatrix} T_1 & T_1 & T_1 \\ T_1 & T_1 & E \end{pmatrix} = \tfrac{1}{6}$$

$$G_n(S_1 \, h_1, \, S_1' \, h)' = G_4(^3T_1, \, ^3T_1) = - \, G_2(^3T_1, \, ^3T_1)$$

$$= - \, 1$$

$$< \tfrac{1}{2} \, t_2 \, \| su \| \, \tfrac{1}{2} \, t_2 > \, = 3$$

$$M_1 = - \, 12 \cdot \tfrac{1}{6} \, (\tfrac{5}{2})^{\frac{1}{2}} \cdot \tfrac{1}{6} \cdot - 1 \cdot 3 = (\tfrac{5}{2})^{\frac{1}{2}} \qquad (115)$$

The second case of importance in the present calculation is the reduced matrix element between a term belonging to a configuration $t_2^{m-1}e^n$ and a term belonging to $t_2^m e^{n-1}$. *Griffith (22)* gives

$$< t_2^{m-1}(S_1 h_1)\ e^n\ (S_2 h_2)\ Sh \| \mathscr{H}_s \| t_2^m\ (S_1' h_1')\ e^{n-1}\ (S_2' h_2')\ S'h' >$$

$$= (-1)^{S_1 - S_1' - S_2 + S_2' + h_1 + h_1' + h_2 + h_2'}$$

$$\times\ [mn(2S+1)\ (2S'+1)\ (2S_1'+1)\ (2S_2+1)\lambda(h)\lambda(h')\lambda(h_1')\lambda(h_2)]^{1/2}$$

$$\times\ < t_2^{m-1}\ (S_1 h_1),\ t_2 \}\ t_2^m\ S_1'\ h_1' >\ < e^n\ S_2 h_2\ \{|e,\ e^{n-1}(S_2' h_2') >$$

$$\times\ \bar{X} \begin{bmatrix} S_1 & S_2 & S \\ S_1' & S_2' & S' \\ \frac{1}{2} & \frac{1}{2} & 1 \end{bmatrix} \times \begin{bmatrix} h_1 & h_2 & h \\ h_1' & h_2' & h' \\ t_2 & e & T_1 \end{bmatrix}\ < \tfrac{1}{2}\ e\|su\|\ \tfrac{1}{2}\ t_2 > \qquad (116)$$

$\bar{X}$ is a 9-j symbol; the remaining terms on the right, including the coefficients of fractional parentage are given in *Griffith's* tables. We illustrate the application of Eq. (116) to the calculation of the reduced matrix element between $^6A_1$ and $^4T_1$. It has already been shown that the $^4T_1$ term originates with the configuration $t_2^4(^3T_1)\ e(^2E)$. Similarly, $^6A_1$ originates with $t_2^3(^4A_2)\ e^2(^3A_2)$. To see how this comes about we note that to form a sextet state the three electrons in t_2^3 and the two electrons in e^2 must have a maximum spin alignment which corresponds to $S = \frac{3}{2}$ and $S = 1$ respectively. The only term from t_2^3 with $S = \frac{3}{2}$ is $^4A_2$ while the only term from e^2 with $S = 1$ is $^3A_2$. The reduced matrix element may now be written

$$M_2 = < t_2^3(^4A_2)\ e^2\ (^3A_2)\ ^6A_1\|\mathscr{H}_s\|t_2^4(^3T_1)\ e\ (^2E)\ ^4T_1 > \qquad (117)$$

in which

$$
\begin{array}{lll}
S_1 = \frac{3}{2} & S_2 = 1 & S = \frac{5}{2} \\
S_1' = 1 & S_2' = \frac{1}{2} & S' = \frac{3}{2} \\
h_1 = A_2 & h_2 = A_2 & h = A_1 \\
h_1' = T_1 & h_2' = E & h' = T_1
\end{array}
$$

$$(-1)^{S_1 - S_1' - S_2 + S_2' + h_1 + h_1' + h_2 + h_2'} = (-1)^{\frac{3}{2} -1 -1 + \frac{1}{2} + A_2 + T_1 + A_2 + E} = -1$$

$$[mn(2S+1)\ (2S'+1)\ (2S_1'+1)\ (2S_2+1)\lambda(h)\lambda(h')\lambda(h_1')\lambda(h_2)]^{\frac{1}{2}}$$

$$= [4 \cdot 2 \cdot 6 \cdot 4 \cdot 3 \cdot 3 \cdot 1 \cdot 3 \cdot 3 \cdot 1]^{\frac{1}{2}} = 72\sqrt{3}$$

$$< t_2^{m-1}(S_1 h_1),\ t_2\ |\}t_2^m S_1' h_1' >\ =\ < t_2^3(^4A_2),\ t_2 |\}\ t_2^4(^3T_1) >$$

$$= -\frac{1}{\sqrt{3}}$$

$$< e^n\ S_2 h_2\ \{\ |e,\ e^{n-1}(S_2' h_2') >\ =\ < e^2(^3A_2)\ \{\ |e,\ e(^2E) >\ =\ 1$$

$$\bar{X} \begin{bmatrix} S_1 & S_2 & S \\ S_1' & S_2' & S' \\ \frac{1}{2} & \frac{1}{2} & 1 \end{bmatrix} = \bar{X} \begin{bmatrix} \frac{3}{2} & 1 & \frac{5}{2} \\ 1 & \frac{1}{2} & \frac{3}{2} \\ \frac{1}{2} & \frac{1}{2} & 1 \end{bmatrix}$$

The $\bar{X}$ coefficients are 9-j symbols which are expressible in terms of 6-j symbols (*51*).

$$\bar{X} \begin{bmatrix} \frac{3}{2} & 1 & \frac{5}{2} \\ 1 & \frac{1}{2} & \frac{3}{2} \\ \frac{1}{2} & \frac{1}{2} & 1 \end{bmatrix} = \begin{Bmatrix} \frac{3}{2} & 1 & \frac{5}{2} \\ 1 & \frac{1}{2} & \frac{3}{2} \\ \frac{1}{2} & \frac{1}{2} & 1 \end{Bmatrix}$$

$$= \sum_j (-1)^{2j} (2j+1) \begin{Bmatrix} \frac{3}{2} & 1 & \frac{1}{2} \\ \frac{1}{2} & 1 & j \end{Bmatrix} \begin{Bmatrix} 1 & \frac{1}{2} & \frac{1}{2} \\ 1 & j & \frac{3}{2} \end{Bmatrix} \begin{Bmatrix} \frac{5}{2} & \frac{3}{2} & 1 \\ j & \frac{3}{2} & 1 \end{Bmatrix} \qquad (118)$$

In the general 6-j symbol $\begin{Bmatrix} j_1 & j_2 & j_3 \\ \ell_1 & \ell_2 & \ell_3 \end{Bmatrix}$ it is necessary for $(j_1 j_2 j_3)$, $(\ell_1\ \ell_2\ j_3)$, $(j_1\ \ell_2\ \ell_3)$ and $(\ell_1\ j_2\ \ell_3)$ to form triangles. This means that e. g., $(j_1\ j_2\ j_3)$ forms a triangle when

$$j_1 + j_2 - j_3 \geq 0;\ j_1 - j_2 + j_3 \geq 0;\ -j_1 + j_2 + j_3 \geq 0 \qquad (119)$$

and

$$j_1 + j_2 + j_3$$

is an integer.

The j-values over which we need to sum are those that result in nonvanishing 6-j symbols. For

$$\begin{Bmatrix} \frac{3}{2} & 1 & \frac{1}{2} \\ \frac{1}{2} & 1 & j \end{Bmatrix} \neq 0$$

it is necessary for $(\frac{3}{2}\ 1\ j)$ and $(\frac{1}{2}\ 1\ j)$ to form triangles. The first case is satisfied by $j = \frac{1}{2}, \frac{3}{2}, \frac{5}{2}$ and the second case by $j = \frac{1}{2}, \frac{3}{2}$. Hence we will get non-zero values of the 6-j symbol only when $j = \frac{1}{2}$ or $\frac{3}{2}$. When $j = \frac{5}{2}$ the second triad, $(\frac{1}{2}\ 1\ j)$ cannot form a triangle and the 6-j symbol vanishes.

The second 6-j symbol may be rearranged so as to become identical to the first. The third 6-j symbol has non-zero values for $j = \frac{1}{2}, \frac{3}{2}, \frac{5}{2}$. We conclude that the summation extends over two values of j, namely $\frac{1}{2}$ and $\frac{3}{2}$.

From a table of 6-j coefficients such as *Rotenberg* et al. (*51*) we find

$$j = \frac{1}{2}$$

$$\begin{Bmatrix} \frac{3}{2} & 1 & \frac{1}{2} \\ \frac{1}{2} & 1 & \frac{1}{2} \end{Bmatrix} = \begin{Bmatrix} 1 & \frac{1}{2} & \frac{1}{2} \\ 1 & \frac{1}{2} & \frac{3}{2} \end{Bmatrix} = -\frac{1}{3}$$

$$\begin{Bmatrix} \frac{5}{2} & \frac{3}{2} & 1 \\ \frac{1}{2} & \frac{3}{2} & 1 \end{Bmatrix} = -\frac{1}{4}$$

$$j = \tfrac{3}{2}$$

$$\left\{\begin{matrix} \tfrac{3}{2} & 1 & \tfrac{1}{2} \\ \tfrac{1}{2} & 1 & \tfrac{3}{2} \end{matrix}\right\} = \left\{\begin{matrix} 1 & \tfrac{1}{2} & \tfrac{1}{2} \\ 1 & \tfrac{3}{2} & \tfrac{3}{2} \end{matrix}\right\} = \tfrac{1}{6}\left(\tfrac{5}{2}\right)^{\tfrac{1}{2}}$$

Therefore

$$\bar{X} = (-1)^1 \cdot 2 \cdot -\tfrac{1}{3} \cdot -\tfrac{1}{3} \cdot -\tfrac{1}{4}$$

$$+ (-1)^3 \cdot 4 \cdot \tfrac{1}{6}\left(\tfrac{5}{2}\right)^{\tfrac{1}{2}} \cdot \tfrac{1}{6}\left(\tfrac{5}{2}\right)^{\tfrac{1}{2}} \cdot -\tfrac{1}{10} \tag{120}$$

$$= \tfrac{1}{18} + \tfrac{1}{36} = \tfrac{1}{12}$$

$$X \begin{bmatrix} h_1 & h_2 & h \\ h_1' & h_2' & h' \\ t_2 & e & T_1 \end{bmatrix} = X \begin{bmatrix} A_2 & A_2 & A_1 \\ T_1 & E & T_1 \\ T_2 & E & T_1 \end{bmatrix}$$

$$= X \begin{bmatrix} T_1 & E & T_1 \\ T_2 & E & T_1 \\ A_2 & A_2 & A_1 \end{bmatrix}$$

The last form is obtained by interchanging rows according to the symmetry properties of the X-coefficients. When an X-coefficient contains an A_1 it reduces to a W-coefficient according to

$$X \begin{bmatrix} a & b & c \\ d & e & f \\ g & h & A_1 \end{bmatrix} = (-1)^{b+d+f+h}\, \lambda(c)^{-\tfrac{1}{2}}\, \lambda(g)^{-\tfrac{1}{2}}\, \delta_{cf}\, \delta_{gh}\, W\begin{pmatrix} a & b & c \\ e & d & g \end{pmatrix} \tag{121}$$

Therefore

$$X \begin{bmatrix} T_1 & E & T_1 \\ T_2 & E & T_1 \\ A_2 & A_2 & A_1 \end{bmatrix} = (-1)^{E+T_2+T_1+A_2}\, \lambda(T_1)^{-\tfrac{1}{2}}\, \lambda(A_2)^{-\tfrac{1}{2}}\, W\begin{pmatrix} T_1 & E & T_1 \\ E & T_2 & A_2 \end{pmatrix} \tag{122}$$

$$= \tfrac{1}{\sqrt{3}} \cdot -\tfrac{1}{\sqrt{6}} = -\tfrac{1}{3}\sqrt{\tfrac{1}{2}}$$

$$<\tfrac{1}{2}\,e\|su\|\,\tfrac{1}{2}\,t_2> = -3\sqrt{2} \tag{123}$$

Therefore

$$M_2 = -1 \cdot 72\sqrt{3} \cdot -\tfrac{1}{\sqrt{3}} \cdot \tfrac{1}{12} \cdot -\tfrac{1}{3}\sqrt{\tfrac{1}{2}} \cdot -3\sqrt{2} = 6 \tag{124}$$

The matrix elements of spin-orbit coupling within the manifold of states spanned by $^6A_1$ and $^4T_1$ are now readily calculated with the aid of *Griffith's* table of V-coefficients and a table of 3-j symbols. The six states of $^6A_1$ and the twelve states of $^4T_1$ give rise to an 18×18 matrix.

However, it has been shown that the substates of $^6A_1$ and $^4T_1$ group themselves according to representations in D_4^*. The effect is that the 18×18 matrix may be rearranged into 4 blocks along the main diagonal. The blocks are labeled $E'\alpha'$, $E'\beta'$, $E''\alpha''$, $E''\beta''$ and consist of matrix elements between states belonging to that particular representation. Matrix elements between a state belonging to one D_4^* representation and a state belonging to another necessarily vanish. The four blocks are shown in Tables 16 and 17. We note that the two blocks $E'\alpha'$ and $E'\beta'$ are identical 4×4 matrices while the $E''\alpha''$ and $E''\beta''$ are identical 5×5 matrices.

Table 16. *Matrix of spin-orbit coupling among components of* $^6A_1$ *and* $^4T_1$ *belonging to* $E'\alpha'$ *and* $E'\beta'$ *in* D_4^*.

(in units of ζ)

$E'\alpha'$ $\longrightarrow$		$^6A_1\,\frac{1}{2}\,a_1$	$^4T_1\,\frac{1}{2}\,0$	$^4T_1-\frac{1}{2}\,1$	$^4T_1\,\frac{3}{2}-1$
$E'\beta'$ $\longrightarrow$		$^6A_1-\frac{1}{2}\,a_1$	$^4T_1-\frac{1}{2}\,0$	$^4T_1\,\frac{1}{2}-1$	$^4T_1-\frac{3}{2}\,1$
$^6A_1\,\frac{1}{2}\,a_1$	$^6A_1-\frac{1}{2}\,a_1$	0	$-\left(\frac{6}{5}\right)^{\frac{1}{2}}$	$-\left(\frac{3}{5}\right)^{\frac{1}{2}}$	$-\left(\frac{1}{5}\right)^{\frac{1}{2}}$
$^4T_1\,\frac{1}{2}\,0$	$^4T_1-\frac{1}{2}\,0$	$-\left(\frac{6}{5}\right)^{\frac{1}{2}}$	0	$-\frac{1}{3}\left(\frac{1}{2}\right)^{\frac{1}{2}}$	$-\frac{1}{2}\left(\frac{1}{6}\right)^{\frac{1}{2}}$
$^4T_1-\frac{1}{2}\,1$	$^4T_1\,\frac{1}{2}-1$	$-\left(\frac{3}{5}\right)^{\frac{1}{2}}$	$-\frac{1}{3}\left(\frac{1}{2}\right)^{\frac{1}{2}}$	$\frac{1}{12}$	0
$^4T_1\,\frac{3}{2}-1$	$^4T_1-\frac{3}{2}\,1$	$-\left(\frac{1}{5}\right)^{\frac{1}{2}}$	$-\frac{1}{2}\left(\frac{1}{6}\right)^{\frac{1}{2}}$	0	$\frac{1}{4}$

Tables 16 and 17 contain all the matrix elements between states of the general form $|^{2S+1}\,h\,M\,\theta>$ where h is a representation in O. We have used the properties of the group D_4^* to help us organize the matrix in the most efficient way. It is also instructive, though not essential, to construct the spin-orbit matrices between states having the form $|^{2S+1}\,h\,t\,\tau>$ where t is a representation in O*. This is readily accomplished with the aid of Eqs. (64, 65 and 66); the matrices are given in Tables 18 and 19. As to be expected, these matrices are much simpler; indeed, the original 18×18 matrix has been reduced so that is no secular equation of degree higher than two.

71

Table 17. *Matrix of spin-orbit coupling among components of $^6A_1$ and $^4T_1$ belonging to $E''\alpha''$ and $E''\beta''$ in D_4^*.*

(in units of ζ)

E"α"		$^6A_1 \frac{5}{2} a_1$	$^6A_1 -\frac{3}{2} a_1$	$^4T_1 -\frac{3}{2} 0$	$^4T_1 \frac{3}{2} 1$	$^4T_1 -\frac{1}{2} -1$
E"β"		$^6A_1 -\frac{5}{2} a_1$	$^6A_1 \frac{3}{2} a_1$	$^4T_1 \frac{3}{2} 0$	$^4T_1 -\frac{3}{2} -1$	$^4T_1 \frac{1}{2} 1$
$^6A_1 \frac{5}{2} a_1$	$^6A_1 -\frac{5}{2} a_1$	0	0	0	$-\sqrt{2}$	0
$^6A_1 -\frac{3}{2} a_1$	$^6A_1 \frac{3}{2} a_1$	0	0	$-\frac{2}{\sqrt{5}}$	0	$-\left(\frac{6}{5}\right)^{\frac{1}{2}}$
$^4T_1 -\frac{3}{2} 0$	$^4T_1 \frac{3}{2} 0$	0	$-\frac{2}{\sqrt{5}}$	0	0	$-\frac{1}{2}\left(\frac{1}{6}\right)^{\frac{1}{2}}$
$^4T_1 \frac{3}{2} 1$	$^4T_1 -\frac{3}{2} -1$	$-\sqrt{2}$	0	0	$-\frac{1}{4}$	0
$^4T_1 -\frac{1}{2} -1$	$^4T_1 \frac{1}{2} 1$	0	$-\left(\frac{6}{5}\right)^{\frac{1}{2}}$	$-\frac{1}{2}\left(\frac{1}{6}\right)^{\frac{1}{2}}$	0	$-\frac{1}{12}$

It should be mentioned that there exists a direct method for computing spin-orbit matrix elements relative to states in O* without first doing the computation relative to states in O. The method is described by *Griffith* (22).

B. Fine Structure and Spin Hamiltonian

The matrices in Table 16 and 17 (or Table 18 and 19 enable) us to ascertain the effect upon $^6A_1$ when there is an admixture from $^4T_1$ by spin-orbit coupling.

Assume first that the $^4T_1$ state has a three-fold orbital degeneracy as it would if the symmetry were truly cubic. Let this zero order energy be $E(T_1)$ and let the corresponding zero order energy of $^6A_1$ be $E(A_1)$. Further let

$$E(T_1) - E(A_1) = \Delta E. \tag{125}$$

Table 18. *Spin-orbit coupling matrix among basis functions of O* belonging to E'α' and E'β' in D_4^*.*

(in units of ζ)

E'α'	E'β'	$^6A_1$ U'λ	$^4T_{1\frac{5}{2}}$ U'λ	$^4T_1$ E'α'	$^4T_{1\frac{3}{2}}$ U'λ
		$^6A_1$ U'μ	$^4T_{1\frac{5}{2}}$ U'μ	$^4T_1$ E'β'	$^4T_{1\frac{3}{2}}$ U'μ
$^6A_1$ U'λ	$^6A_1$ U'μ	0	$-\sqrt{2}$	0	0
$^4T_{1\frac{5}{2}}$ U'λ	$^4T_{1\frac{5}{2}}$ U'μ	$-\sqrt{2}$	$-\frac{1}{4}$	0	0
$^4T_1$ E'α'	$^4T_1$ E'β'	0	0	$\frac{5}{12}$	0
$^4T_{1\frac{3}{2}}$ U'λ	$^4T_{1\frac{3}{2}}$ U'μ	0	0	0	$\frac{1}{6}$

Table 19. *Spin-orbit coupling matrix among basis functions of O* belonging to E'' α'' and E'' β'' in D_4^*.*

(in units of ζ)

E'' α''	E'' β''	$^6A_1$ E''α''	$^4T_1$ E'' α''	$^6A_1$ U'ν	$^4T_{1\frac{5}{2}}$ U'ν	$^4T_{1\frac{3}{2}}$ U'ν
		$^6A_1$ E''β''	$^4T_1$ E'' β''	$^6A_1$ U'ϰ	$^4T_{1\frac{5}{2}}$ U'ϰ	$^4T_{1\frac{3}{2}}$ U'ϰ
$^6A_1$ E'' α''	$^6A_1$ E'' β''	0	$-\sqrt{2}$	0	0	0
$^4T_1$ E'' α''	$^4T_1$ E'' β''	$-\sqrt{2}$	$-\frac{1}{4}$	0	0	0
$^6A_1$ U'ν	$^6A_1$ U'ϰ	0	0	0	$-\sqrt{2}$	0
$^4T_{1\frac{5}{2}}$ U'ν	$^4T_{1\frac{5}{2}}$ U'ϰ	0	0	$-\sqrt{2}$	$-\frac{1}{4}$	0
$^4T_{1\frac{3}{2}}$ U'ν	$^4T_{1\frac{3}{2}}$ U'ϰ	0	0	0	0	$\frac{1}{6}$

Since there is no spin-orbit coupling within the $^6A_1$ term, there is no first order correction to the energy. The second order corrections to the energy ($E^{(2)}$) may be read directly from Tables 16 and 17. In terms of ζ, the spin-orbit coupling constant, they are

$$E^{(2)}\left(^6A_1\ \tfrac{1}{2}\ a_1\right) = -\left[\frac{6\zeta^2}{5\Delta E} + \frac{3\zeta^2}{5\Delta E} + \frac{\zeta^2}{5\Delta E}\right] = -\frac{2\zeta^2}{\Delta E}$$

$$= E^{(2)}\left(^6A_1 - \tfrac{1}{2}\ a_1\right)$$

$$E^{(2)}\left(^6A_1\ \tfrac{3}{2}\ a_1\right) = -\left[\frac{4\zeta^2}{5\Delta E} + \frac{6\zeta^2}{5\Delta E}\right] = -\frac{2\zeta^2}{\Delta E}$$

$$= E^{(2)}\left(^6A_1 - \tfrac{3}{2}\ a_1\right)$$

$$E^{(2)}\left(^6A_1\ \tfrac{5}{2}\ a_1\right) = -\frac{2\zeta^2}{\Delta E}$$

$$= E^{(2)}\left(^6A_1 - \tfrac{5}{2}\ a_1\right) \tag{126}$$

The six components of $^6A_1$ are each shifted in energy by precisely the same amount and the $^6A_1$ term remains six-fold degenerate. Despite the fact that this conclusion has been reached on the basis of the second order energy corrections, it is true to any order. The simplest way to see this is to diagonalize the matrices in Tables 18 and 19 and to note that the lowest eigenvalue, which is identifiable with the $^6A_1$ manifold is six-fold degenerate. Thus, provided the $^4T_1$ term remains 3-fold spatially degenerate, it will have no effect whatsoever insofar as removal of degeneracies in $^6A_1$ are concerned. An electron spin resonance experiment would, under these circumstances, disclose $g_x = g_y = g_z = 2$ which contradicts the actual observations. We are thus led to suppose that the symmetry at the site of the iron atom is lower than cubic and that the $^4T_1$ term does not have a three-fold spatial (orbital) degeneracy.

We have seen (Eqs. 73) that when the cubic symmetry is reduced to tetragonal the $^4T_1$ term is decomposed into $^4A_2$ and ^4E. The $^6A_1$ term, whose degeneracy is entirely due to spin, remains unaffected by the reduction in symmetry. *Griffith* (21) assumed that the energy separation between $^4A_2$ (the lower component) and ^4E (the higher component) was sufficiently large that the dominant interaction, via spin-orbit coupling, was between $^4A_2$ and $^6A_1$. The second order energy corrections to $^6A_1$ due to $^4A_2$ are readily obtained from Tables 16 and 17. It is only necessary to recognize that, according to Table 8, $^4A_2$ in D_4 corresponds to ^4T_1z in O and ^4T_1z, from Eqs. (61), is the same as $-i\ ^4T_1\ 0$. If we define

$$\Delta E_0 = E(^4T_1\ 0) - E(^6A_1) \qquad (127)$$

then

$$E^{(2)}\left(^6A_1\ \tfrac{1}{2}\ a_1\right) = -\frac{6\zeta^2}{5\Delta E_0} = E^{(2)}\left(^6A_1 - \tfrac{1}{2}\ a_1\right)$$

$$E^{(2)}\left(^6A_1\ \tfrac{5}{2}\ a_1\right) = 0 = E^{(2)}\left(^6A_1 - \tfrac{5}{2}\ a_1\right)$$

$$E^{(2)}\left(^6A_1\ \tfrac{3}{2}\ a_1\right) = -\frac{4\zeta^2}{5\Delta E_0} = E^{(2)}\left(^6A_1 - \tfrac{3}{2}\ a_1\right) \qquad (128)$$

It is seen that $^6A_1$ is no longer six-fold degenerate but is split into three doublets which have different energies. The states with $+\ M_s$ and $-\ M_s$ are still degenerate and, according to Kramers' theorem, there can be no further removal of degeneracies without the use of magnetic fields. It is convenient to define

$$D = \tfrac{1}{5}\frac{\zeta^2}{\Delta E_0} \qquad (129)$$

whence the energy separations among the components of $^6A_1$ become

$$E^{(2)}\left(^6A_1 \pm \tfrac{3}{2}\ a_1\right) - E^{(2)}\left(^6A_1 \pm \tfrac{1}{2}\ a_1\right) = 2D$$

$$E^{(2)}\left(^6A_1 \pm \tfrac{5}{2}\ a_1\right) - E^{(2)}\left(^6A_1 \pm \tfrac{1}{2}\ a_1\right) = 6D \qquad (130)$$

as shown on the energy level diagram (Fig. 26a).

The level system may also be described by a spin Hamiltonian

$$\mathscr{H}(S) = D(S_z^2 - C) \qquad (131)$$

in which the constant C is determined by our choice of the zero of energy. For

$$C = \tfrac{1}{3}\ S(S + 1)$$

$$= \tfrac{35}{12} \text{ when } S = \tfrac{5}{2}$$

$$\mathscr{H}(S) = D\left(S_z^2 - \tfrac{35}{12}\right) \qquad (132)$$

and

$$E = D\left(M_s^2 - \tfrac{35}{12}\right)$$

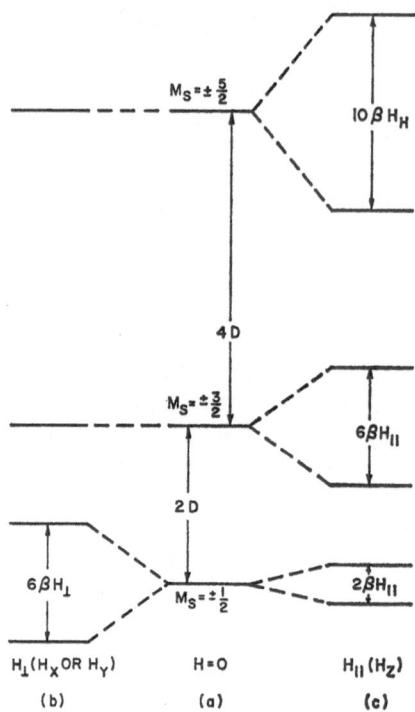

Fig. 26a-c. Splitting of $^6A_1$ in (a) zero magnetic field, (b) magnetic field perpendicular to the 4-fold axis and (c) parallel to the 4-fold axis.

or

$$E = \begin{cases} \frac{10}{3}\,D, & M_s = \pm\frac{5}{2} \\ -\frac{2}{3}\,D & \pm\frac{3}{2} \\ -\frac{8}{3}\,D & \pm\frac{1}{2} \end{cases} \tag{133}$$

This puts $E = 0$ at the center of gravity of the three levels. Alternatively we may put $C = \frac{1}{4}$; then

$$E = \begin{cases} 6D, & M_s = \pm\frac{5}{2} \\ 2D & \pm\frac{3}{2} \\ 0 & \pm\frac{1}{2} \end{cases} \tag{134}$$

which is convenient for some purposes.

Kotani (*36*) employed a somewhat modified approach. If one assumes that the ligand field is of sufficiently low symmetry, say D_{2h}, the $^4T_1$ will be split into three components $^4T_1\,x$, $^4T_1\,y$ and $^4T_1\,z$ defined by Eqs.

(61 und 62). The energies of the three components, relative to $^6A_1$, may be designated by E_x, E_y and E_z respectively. Since the spin components remain unchanged we have, for example,

$$|^4T_1 \tfrac{1}{2} x > \; = \tfrac{i}{\sqrt{2}} \left[\, | \, ^4T_1 \tfrac{1}{2} 1 > - |^4T_1 \tfrac{1}{2} \bar{1} > \, \right] \tag{135}$$

with similar relations for the other spin components.

Using the new basis set, the spin-orbit coupling matrix may be constructed from Tables 16 and 17 and Eqs. (61 and 62).

The matrix elements of components of $^4T_1$ with components of $^6A_1$ are given in Table 20, whence the second order energy corrections to the components of $^6A_1$ become

$$E^{(2)} \left(^6A_1 \, \tfrac{1}{2} \, a_1 \right) = - \frac{\zeta^2}{5} \left[\frac{6}{E_z} + \frac{2}{E_x} + \frac{2}{E_y} \right]$$

$$= E^{(2)} \left(^6A_1 - \tfrac{1}{2} \, a_1 \right)$$

$$E^{(2)} \left(^6A_1 \, \tfrac{3}{2} \, a_1 \right) = - \frac{\zeta^2}{5} \left[\frac{4}{E_z} + \frac{3}{E_x} + \frac{3}{E_y} \right]$$

$$= E^{(2)} \left(^6A_1 - \tfrac{3}{2} \, a_1 \right)$$

$$E^{(2)} \left(^6A_1 \, \tfrac{5}{2} \, a_1 \right) = - \zeta^2 \left[\frac{1}{E_x} + \frac{1}{E_y} \right]$$

$$= E^{(2)} \left(^6A_1 - \tfrac{5}{2} \, a_1 \right) \tag{136}$$

Again the $^6A_1$ term is split into three doublets each of which preserves $\pm M_s$ degeneracy. Formally the results are the same as those obtained previously. Here too we may define a zero field splitting parameter, D, given by

$$D = \frac{\zeta^2}{10} \left[\frac{2}{E_z} - \frac{1}{E_x} - \frac{1}{E_y} \right] \tag{137}$$

which leads to the same energy separations as those given in Eqs. (133, 134) and a spin Hamiltonian of the same form as Eq. (131).

A more general spin Hamiltonian which describes the effect of spin-orbit coupling to second order is

$$\mathcal{H}(S) = D[S_z^2 - \tfrac{1}{3} S(S + 1)] + E(S_x^2 - S_y^2) \tag{138}$$

with D as in Eq. (137) and E given by

$$E = \frac{\zeta^2}{10} \left[\frac{1}{E_x} - \frac{1}{E_y} \right] \tag{139}$$

Table 20. *Matrix elements of spin-orbit coupling among components of $^6A_1$ and real components of $^4T_1$.*

(in units of ζ)

$^6A_1$ \ $^4T_1$	$\frac{3}{2}x$	$\frac{3}{2}y$	$\frac{3}{2}z$	$\frac{1}{2}x$	$\frac{1}{2}y$	$\frac{1}{2}z$	$-\frac{1}{2}x$	$-\frac{1}{2}y$	$-\frac{1}{2}z$	$-\frac{3}{2}x$	$-\frac{3}{2}y$	$-\frac{3}{2}z$
$\frac{5}{2}$	$-i$	-1										
$\frac{3}{2}$			$i\frac{2}{\sqrt5}$	$-i\left(\frac{3}{5}\right)^{\frac12}$	$-\left(\frac{3}{5}\right)^{\frac12}$							
$\frac{1}{2}$	$i\left(\frac{1}{10}\right)^{\frac12}$			$i\left(\frac{3}{10}\right)^{\frac12}$	$-\left(\frac{3}{10}\right)^{\frac12}$	$i\left(\frac{6}{5}\right)^{\frac12}$	$-i\left(\frac{3}{10}\right)^{\frac12}$	$-\left(\frac{3}{10}\right)^{\frac12}$				
$-\frac{1}{2}$									$i\left(\frac{6}{5}\right)^{\frac12}$	$-i\left(\frac{1}{10}\right)^{\frac12}$	$-\left(\frac{1}{10}\right)^{\frac12}$	
$-\frac{3}{2}$							$i\left(\frac{3}{5}\right)^{\frac12}$	$-\left(\frac{3}{5}\right)^{\frac12}$				$i\frac{2}{\sqrt5}$
$-\frac{5}{2}$										i	-1	

If the crystal field is rigorously cubic both D and E are zero and there is no zero field splitting; however, there may be small contributions from a spin Hamiltonian containing quartic terms. For a tetragonal field $E = 0$. In the present case $\left| \frac{E}{D} \right| < < 1$ (37).

Therefore, according to Eq. (138), to second order in spin-orbit coupling the three levels of $^6A_1$ are almost pure eigenstates of S_z with $M_s = \pm \frac{1}{2}, \pm \frac{3}{2}, \pm \frac{5}{2}$ and $S = \frac{5}{2}$. The eigenstates may be designated by $\left| \pm \frac{1}{2} >, \right| \pm \frac{3}{2} >$ and $\left| \pm \frac{5}{2} > \right.$ respectively. Should the ligand field have a rhombic component, E will be non-vanishing and the second term in the spin Hamiltonian will contribute. $^6A_1$ will still consist of three doublets but they will no longer be pure eigenstates of S_z. It will now be shown that the level system for $E = 0$ (Fig. 26) is capable of yielding the observed g-values.

C. Interaction with a Magnetic Field

The interaction with an external magnetic field, when $L = 0$, is $2\beta \, \vec{H} \cdot \vec{S}$. Therefore the spin Hamiltonian which includes the magnetic field interaction may be written as an extension of Eq. (138)

$$
\begin{aligned}
\mathscr{H}(S) &= D[S_z^2 - \tfrac{1}{3} \, S(S + 1)] + E(S_x^2 - S_y^2) \\
&\quad + 2\beta \, \vec{H} \cdot \vec{S} \\
&= D[S_z^2 - \tfrac{1}{3} \, S(S + 1)] + \tfrac{1}{2} \, E(S_+^2 + S_-^2) \\
&\quad + 2\beta [H_z \, S_z + \tfrac{1}{2} \, (H_+ \, S_- + H_- \, S_+)]
\end{aligned}
\tag{140}
$$

To calculate g-values it is necessary to compute matrix elements of $2\beta \, \vec{H} \cdot \vec{S}$ within the substates of $^6A_1$. Consider first a magnetic field in the z-direction, taken along the fourfold axis perpendicular to the porphyrin plane. The non-vanishing matrix elements (Table 21) are all on the diagonal.

Therefore the energies of the components of $^6A_1$ may be written

$$
E = \begin{cases} 0 \pm \beta \, H_{\parallel} \\ 2D \pm 3\beta \, H_{\parallel} \\ 6D \pm 5\beta \, H_{\parallel} \end{cases}
\tag{141}
$$

where $H_z = H_{\parallel}$ and where the energy of the lowest doublet, $\left| \pm \frac{1}{2} > \right.$, in zero field, has been taken to be zero as in Eqs. (134). It is seen that each doublet of $^6A_1$ is split into two levels and the degeneracy of $^6A_1$

Table 21. *Matrices of* $2\beta\,\vec{H}\cdot\vec{S}$. *Upper and lower entries refer to* x- *and* y-*components respectively, single entries refer to* z-*components of the magnetic field.*

$M_s \rightarrow$ $\downarrow$	$\frac{1}{2}$	$-\frac{1}{2}$	$\frac{3}{2}$	$-\frac{3}{2}$	$\frac{5}{2}$	$-\frac{5}{2}$
$\frac{1}{2}$	βH_z	$3\beta\,H_x$ $-i3\beta H_y$	$2\sqrt{2}\,\beta H_x$ $i2\sqrt{2}\,\beta H_y$			
$-\frac{1}{2}$	$3\beta H_x$ $3i\beta H_y$	$-\beta H_z$		$2\sqrt{2}\,\beta H_x$ $-\varphi i2\varphi\sqrt{2}\,\beta H_y$		
$\frac{3}{2}$	$2\sqrt{2}\,\beta H_x$ $-\varphi i2\varphi\sqrt{2}\,\beta H_y$		$3\beta H_z$		$\sqrt{5}\,\beta H_x$ $i\sqrt{5}\,\beta H_y$	
$-\frac{3}{2}$		$2\sqrt{2}\,\beta H_x$ $i2\sqrt{2}\,\beta H_y$		$-3\beta H_z$		$\sqrt{5}\,\beta H_x$ $-i\sqrt{5}\,\beta H_y$
$\frac{5}{2}$			$\sqrt{5}\,\beta H_x$ $-i\sqrt{5}\,\beta H_y$		$5\beta H_z$	
$-\frac{5}{2}$				$\sqrt{5}\,\beta H_x$ $i\sqrt{5}\,\beta H_y$		$-5\beta H_z$

has been completely removed. The resulting level structure is shown in Fig. 26c. Since the matrix of $2\beta\,H_z\,S_z$ is diagonal, each level is a pure eigenstate of S_z, there is no mixing of eigenstates by the magnetic field and the appropriate selection rules are

$$\Delta M_s = \pm 1 \tag{142}$$

These selection rules permit transitions between the magnetic substates of the lowest doublet which are separated in energy by $2\beta\,H_z$. Hence

$$g_z = g_{\parallel} = 2 \tag{143}$$

Transition between the magnetic substates of the doublets with $M_s = \pm\frac{3}{2}$ and $M_s = \pm\frac{5}{2}$ are forbidden. Conceivably, if the magnetic field were high enough transitions between $|-\frac{3}{2}>$ and $|-\frac{1}{2}>$ would appear, but these have not been observed.

For a magnetic field in the x-or y-direction ($H_\perp$) we find only off-diagonal matrix elements. These are calculated by means of the shift operators S_+ and S_- which satisfy relations of the form of Eq. (12). To first order, there is no splitting of the doublets $|\pm \frac{3}{2} >$ and $|\pm \frac{5}{2} >$ while the splitting in $|\pm \frac{1}{2} >$ is obtained by diagonalizing

$$\begin{pmatrix} 0 & 3\beta\, H_x \\ 3\beta H_x & 0 \end{pmatrix}$$

or

$$\begin{pmatrix} 0 & -i\, 3\beta\, H_y \\ i\, 3\beta\, H_y & 0 \end{pmatrix}$$

In either case we obtain an energy separation of $6\beta\, H_\perp$ (Fig. 26b), and

$$g_\perp = 6 \tag{144}$$

To second order, table 21 shows that H_x or H_y can mix $|\pm \frac{1}{2} >$ with $|\pm \frac{3}{2} >$ and $|\pm \frac{3}{2} >$ with $|\pm \frac{5}{2} >$. The energies are

$$E = \begin{cases} 0 \pm 3\beta\, H_\perp - \dfrac{4\beta^2\, H_\perp^2}{D} \\[2ex] 2D + \dfrac{11}{4}\, \dfrac{\beta^2\, H_\perp^2}{D} \\[2ex] 6D + \dfrac{5}{4}\, \dfrac{\beta^2\, H_\perp^2}{D} \end{cases} \tag{145}$$

D. Discussion

We have seen that the observed g-values are explained on the basis of a spin Hamiltonian having the form of Eq. (131) and an associated level structure shown in Fig. 26. Moreover, the electron spin resonance appears to be entirely associated with the lowest Kramers doublet $|\pm \frac{1}{2} >$, which gives $g_\parallel = 2$ and $g_\perp = 6$ in agreement with experiment. These observations imply that D must be large compared to magnetic field energies. For if this were not the case, transitions between levels arising from different doublets would be expected. Indeed, as D becomes smaller than magnetic field energies and approaches zero, g approaches an isotropic value of 2. Experiments have been carried to 50,000 Mc/s (29) with no departures in linearity between frequency and magnetic field. Since 50,000 Mc/s corresponds to 1.67 cm^{-1} we conclude that $D >> 1.67$ cm^{-1}.

It is difficult to do more than to estimate a lower limit for D on the basis of ESR experiments. However, the temperature dependence of the static susceptibility is inherently capable of measuring D itself. The basis for this method will be discussed in Ch. VIII. *Beetlestone* performed such measurements over a limited region of temperature and obtained a value of 28 cm⁻¹ for ferrimyoglobin fluoride (*12*). Various authors (*23, 37*) have placed the value of D at about 5 cm⁻¹.[3]

Whatever the precise value of D in ferric high spin hemoglobin may be, it is large compared with inorganic, ferric high spin compounds containing Fe^{3+} or Mn^{2+}, all of which have a $3d^5$ electron configuration. In the inorganic compounds D lies in the range of 0.01 to 0.1 cm⁻¹. From a theoretical standpoint the low values of D in the inorganic compounds seem more natural since, as we have seen, ^6S is insensitive to low symmetry fields and any interaction which contributes to the doublet splitting is of second order or higher. Why, then, is D so much larger in hemoglobin?

Kotani (*37*) obtained an estimate of D based on the defining equation, Eq. (137) in which E_x, E_y and E_z are, respectively, the energies of the components $^4T_1$ x, $^4T_1$ y and $^4T_1$ z relative to $^6A_1$. Each energy is made up of two terms; thus

$$E_x = E(^4T_1) + \Delta E_x \tag{146}$$

in which $E(^4T_1)$ is the energy of the center of gravity of $^4T_1$ relative to $^6A_1$ and ΔE_x is the energy of $^4T_1$ x relative to the center of gravity of $^4T_1$. Analogous equations are written for E_y and E_z.

An expression for $E(^4T_1)$ has been given in Eqs. (60) in terms of the Racah parameters B and C and the cubic splitting parameter, Δ,

$$E(^4T_1) = 10\,B + 6\,C - \Delta \tag{147}$$

From Eq. (8) B = 1133 cm⁻¹ and C = 3883 cm⁻¹; the value of Δ is less certain. *Kotani* assumed a value of Δ at which $^2T_2$ crosses $^6A_1$, that is, at the point beyond which the low spin state ($^2T_2$) is of lower energy than the high spin state ($^6A_1$). The assumption is largely motivated by the desire to use the information deduced from the electron spin resonance of the low spin ferrihemoglobin azide. The crossover between $^2T_2$ and $^6A_1$ is seen to occur, according to Eqs. (60), when

$$E(^2T_2) = 15B + 10C - 2\Delta = 0 \tag{148}$$

or when

$$\Delta = \tfrac{15}{2}\,B + 5C$$

---

[3] Recent measurements by far-infrared spectroscopy in heme chloride gave 2D = 13.9 cm⁻¹ (*G. Feher*, and *P. R. Richards:* International Conference on Magnetic Resonance in Biological Systems. Stockholm 1966).

Inserting this value of Δ into the expression for $E(^4T_1)$ we obtain

$$E(^4T_1) = \tfrac{5}{2} B + C$$

$$= 6720 \text{ cm}^{-1} \tag{149}$$

Next, it is necessary to estimate ΔE_x, ΔE_y and ΔE_z. It is recalled that $^4T_1$ arises from the electronic configuration t_2^4 e. If the t_2-orbitals are not degenerate, as would be the case in a low symmetry field, the three spatial components of $^4T_1$ will have different energies corresponding to the different ways in which four electrons can distribute themselves among the three t_2-orbitals. These arrangements are $\xi^2 \eta \zeta$, $\xi \eta^2 \zeta$ and $\xi \eta \zeta^2$ with the fifth electron remaining in an e-orbital. The orbital energies relative to the center of gravity of the t_2-orbitals are (Fig. 22)

$$\varepsilon_\xi = \tfrac{1}{3} \delta + \tfrac{1}{2} \mu$$

$$\varepsilon_\eta = \tfrac{1}{3} \delta - \tfrac{1}{2} \mu$$

$$\varepsilon_\zeta = -\tfrac{2}{3} \delta \tag{150}$$

where δ is the tetragonal splitting and μ the rhombic splitting. If we identify ΔE_x, the displacement of $^4T_1$ x relative to the unshifted $^4T_1$ (also its center of gravity), with $E(\xi^2 \eta \zeta)$, the sum of the orbital energies of the four t_2-electrons, then

$$\Delta E_x = E(\xi^2 \eta \zeta) = 2\varepsilon_\xi + \varepsilon_\eta + \varepsilon_\zeta$$

$$= \varepsilon_\xi = \tfrac{1}{3} \delta + \tfrac{1}{2} \mu \tag{151}$$

Similarly

$$\Delta E_y = E(\xi \eta^2 \zeta) = \varepsilon_\xi + 2\varepsilon_\eta + \varepsilon_\zeta$$

$$= \varepsilon_\eta = \tfrac{1}{3} \delta - \tfrac{1}{2} \mu$$

$$\Delta E_z = E(\xi \eta \zeta^2) = \varepsilon_\xi + \varepsilon_\eta + 2\varepsilon_\zeta$$

$$= \varepsilon_\zeta = -\tfrac{2}{3} \delta \tag{152}$$

From the data on ferrihemoglobin azide (Eqs. 109) we had

$$\delta = 2060 \text{ cm}^{-1}, \mu = 1040 \text{ cm}^{-1}$$

Combining Eqs. (151, 152) with Eqs. (149) we obtain

$$E_x = 7920 \text{ cm}^{-1}, E_y = 6880 \text{ cm}^{-1}, E_z = 5340 \text{ cm}^{-1}$$

Substitution in Eq. (137) with $\zeta = 435 \text{ cm}^{-1}$ finally gives $D = 1.9 \text{ cm}^{-1}$.

The anisotropy in low spin ferrihemoglobin azide has been attributed mainly to the orientation of the azide ion relative to the heme plane. There is no comparable source of anisotropy in high spin ferrihemoglobin fluoride. We must therefore consider it to be a somewhat questionable procedure to use data derived from ESR on the azide compound in the analysis of ESR on fluoride. Nevertheless, it is interest to note that the value of D derived above is quite insensitive to the rhombic splitting. In fact, if we set $\mu = 0$, the value of D is increased by no more than about 1%. This seems to indicate that the significant quantity that is being transferred from the azide to the fluoride is the tetragonal splitting, which may, conceivably, have a common origin in both compounds.

VIII. Magnetic Susceptibility

A. Definitions and General Expressions

To develop an expression for the paramagnetic susceptibility we assume that the energy of a state E_n may be expressed as a power series in H, the magnetic field in a particular direction,

$$E_n = E_n^{(0)} + HE_n^{(1)} + H^2E_n^{(2)} + \cdots \tag{153}$$

The magnetic moment in the direction of the applied field is defined by

$$\mu = -\frac{\partial E_n}{\partial H} \tag{154}$$

$$= -E_n^{(1)} - 2H\,E_n^{(2)} \tag{155}$$

and the total magnetic moment per mole, M, obtained by performing a statistical average over the thermal distribution of magnetic dipoles, becomes

$$M = N\,\frac{\displaystyle\sum_n \mu\,e^{-E_n/kT}}{\displaystyle\sum_n e^{-E_n/kT}} \tag{156}$$

where N is Avogadro's number. The molar susceptibility, χ, defined by

$$\chi = \frac{M}{H} \tag{157}$$

may now be obtained by combining Eqs. (153, 155 and 156). If in Eq. (156) the expansions are carried only to linear terms in H, the molar susceptibility will be independent of H and will be given by

$$\chi = N \frac{\sum\limits_{n} \left[\frac{(E_n^{(1)})^2}{kT} - 2E_n^{(2)} \right] e^{-\frac{E_n^{(0)}}{kT}}}{\sum\limits_{n} e^{-\frac{E_n^{(0)}}{kT}}} \tag{158}$$

under the condition that $M \rightarrow 0$ as $H \rightarrow 0$.

The interaction with a magnetic field is described by the Hamiltonian

$$\mathscr{H}_m = \beta \, \vec{H} \cdot (\vec{L} + 2\vec{S}) \tag{159}$$

in which β is the Bohr magneton:

$$\beta = \frac{e\,\hbar}{2mc} = 0.9271 \times 10^{-20} \text{ erg/gauss} \tag{160}$$

Eq. (159) also defines a magnetic moment operator, $\vec{\mu}$, given by

$$\vec{\mu} = \vec{\mu}_L + \vec{\mu}_S = -\beta(\vec{L} + 2\vec{S}) \tag{161}$$

For a system in which

$$< 0|\vec{L}|0 > = 0 \tag{162}$$

indicating a complete quenching, or non-existence, of orbital angular momentum, the Hamiltonian operator of Eq. (159) becomes

$$\mathscr{H} = 2\beta \, H_z \, S_z \tag{163}$$

in which the spin and the magnetic field are both assumed to be oriented in the z-direction. For this case

$$\vec{\mu} = \vec{\mu}_S \tag{164}$$

and the energy is

$$E_n = 2\beta \, H_z < \psi_n|S_z|\psi_n > \; = 2\beta \, H_z \, E_n^{(1)} \tag{165}$$

in which the index n labels the $2S+1$ degenerate wave functions, ψ_n. The matrix elements are equal to M_s which has the values $S, S-1 \ldots -S$. Also, for this case, $E_n^{(2)} = E_n^{(0)} = 0$ and the susceptibility as given by Eq. (158) becomes

$$\chi = N \frac{(2\beta)^2[S^2 + (S-1)^2 + \cdots (-S)^2]}{kT(2S+1)} \tag{166}$$

$$= \frac{4N\beta^2}{3kT} \, S \, (S+1) \tag{167}$$

Eq. (167) is Curie's Law when the magnetic moments are associated entirely with spin alignments. It is convenient to define an effective magnetic moment μ_e, such that

$$\chi = \frac{N \, \mu_e^2}{3kT} \tag{168}$$

Comparison with Eq. (167) gives

$$\mu_e = 2\beta[S(S+1)]^{\frac{1}{2}} \tag{169}$$

An effective Bohr magneton number, n_e, may be defined by

$$\mu_e = n_e\beta \tag{170}$$

or

$$n_e = 2[S(S+1)]^{\frac{1}{2}} \tag{171}$$

The effective Bohr magneton numbersfor several values of S are given in Table 22.

Table 22. *Effective Bohr magneton numbers.*

S	n_e
$\frac{1}{2}$	1.73
1	2.83
$\frac{3}{2}$	3.87
2	4.90
$\frac{5}{2}$	5.92

In the free ion the presence of an orbital angular momentum changes Eq. (167) into

$$\chi = N \frac{g^2\beta^2}{3kT} J(J + 1) \tag{172}$$

where J is the quantum number of total angular momentum,

$$\vec{J} = \vec{L} + \vec{S} \tag{173}$$

and g is the Lande g-factor or the spectroscopie splitting g-factor given by

$$g = \tfrac{3}{2} + \frac{S(S + 1) - L(L + 1)}{2J(J + 1)} \tag{174}$$

Eq. (172) is analogous to Eq. (167) and is a statement of Curie's Law for the more general case when both orbital and spin angular momenta are present. Also by analogy with Eqs. (169, 171) an effective magnetic moment is given by

$$\mu_e = g\beta \, [J(J + 1)]^{\tfrac{1}{2}} \tag{175}$$

and an effective Bohr magneton number by

$$n_e = 2 \, [J(J + 1)]^{\tfrac{1}{2}} \tag{176}$$

The general effect of a ligand field is to remove spatial degeneracies; this tends to quench orbital angular momentum because for an orbital singlet the expectation values of L_x, L_y and L_z are zero. On the other hand spin-orbit coupling restores some contribution from orbital angular momentum. When both effects are present we may expect the magnetic susceptibility to depart from the value that would be obtained when spin angular momenta alone are present. This is often described as incomplete quenching.

B. Ferrihemoglobin

1. High Spin

For a magnetic field parallel to the fourfold axis (z-direction), the magnetic substates of $^6A_1$ are given by Eqs. (141). Comparing with Eq. (153).

$$E_n^{(0)} = 0, \; 2D, \; 6D$$

$$E_n^{(1)} = \pm \, \beta, \; \pm \, 3\beta, \; \pm \, 5\beta$$

$$E_n^{(2)} = 0 \tag{177}$$

We may therefore substitute directly into Eq. (158) to obtain

$$\chi = N\beta^2 \frac{1 + 9e^{-2x} + 25e^{-6x}}{kT\,(1 + e^{-2x} + e^{-6x})} \tag{178}$$

with

$$x = \frac{D}{kT}$$

By writing, as in Eqs. (168, 170)

$$\chi = \frac{N\beta^2\, n_e^2}{3kT} \tag{179}$$

and comparing with Eq. (178) we have

$$n_e^2(\parallel) = \frac{3\,(1 + 9e^{-2x} + 25e^{-6x})}{1 + e^{-2x} + e^{-6x}} \tag{180}$$

It is instructive to investigate this expression in the limits of high and low temperatures. At high temperature $x \rightarrow 0$ and

$$n_e(\parallel) = \sqrt{35} = 5.92 \qquad (T \rightarrow \infty) \tag{181}$$

From Table 22 we see that this corresponds to the spin-only value for $S = 5/2$. In the opposite limit, at low temperature $x \rightarrow \infty$ and

$$n_e(\parallel) = \sqrt{3} = 1.73 \qquad (T \rightarrow 0) \tag{182}$$

which corresponds to the spin-only value for $S = 1/2$. These results are readily understood from the level diagram Fig. 26. In the high temperature limit, the populations in the three doublets tend to equalize so that, in spite of their energy separations, the three doublets contribute to the susceptibility precisely as a six-fold degenerate state with $S = 5/2$. In the low temperature limit, only the level with $M_S = \pm \frac{1}{2}$ is occupied; the others are vacant. Therefore the system behaves as if $S = \frac{1}{2}$.

For a magnetic field lying in the porphyrin plane the energies are given by Eqs. (145). Therefore

$$E_n^{(0)} = 0,\ 2D,\ 6D$$

$$E_n^{(1)} = \pm\ 3\beta,\ 0,\ 0$$

$$E_n^{(2)} = -\frac{4\beta^2}{D},\ \frac{11\beta^2}{4D},\ \frac{5\beta^2}{4D} \tag{183}$$

From Eqs. (158 and 179) we obtain

$$n_e^2(\perp) = \frac{3\left[9 + \frac{8}{x} - \frac{11}{2x}\, e^{-2x} - \frac{5}{2x}\, e^{-6x}\right]}{1 + e^{-2x} + e^{-6x}} \tag{184}$$

In the limits of high and low temperature

$$n_e(\perp) = n_e(\|) = \sqrt{35} = 5.92 \qquad (T \to \infty)$$
$$n_e(\perp) = 3\sqrt{3} = 5.19 \qquad (T \to 0) \qquad (185)$$

For polycrystalline samples $\dot{n}_e^2$ is a weighted average of $n_e^2(\|)$ and $n_e^2(\perp)$ with weights of $\frac{1}{3}$ and $\frac{2}{3}$ respectively. This gives

$$n_e^2(av) = \frac{19 + \frac{16}{x} + \left(9 - \frac{11}{x}\right)e^{-2x} + \left(25 - \frac{5}{x}\right)e^{-6x}}{1 + e^{-2x} + e^{-6x}} \qquad (186)$$

and

$$n_e(av) = \sqrt{35} = 5.92 \qquad (T \to \infty)$$
$$n_e(av) = \sqrt{19} = 4.36 \qquad (T \to 0) \qquad (187)$$

Plots of Eqs. (180, 184, 186) are to be found in *Kotani* (35). Eq. (186) was used by *Beetlestone* to obtain a value of D from measurements of the static susceptibility at several temperatures. The mean value of D was 28 cm^{-1} (12).[4]

2. Low Spin

For the low spin case we refer to the set of Kramers doublets whose spacing is shown in Fig. 23. The second doublet is situated at an energy of more than 1000 cm^{-1} above the first (lowest) doublet. This is far in excess of thermal energy at room temperature — therefore the population in the second doublet will be small, and even smaller in the third doublet. The contribution of the latter to the susceptibility will be ignored.

Confining our attention to the first and second doublet, the energies in a magnetic field are obtained from Eqs. (85 and 89) to first order in H. They are

$$H = H_x \qquad E_1^x = \pm \beta H_x \left[(B_1 + C_1)^2 - A_1^2\right]$$
$$E_2^x = 2.61 \lambda \pm \beta H_x \left[(B_2 + C_2)^2 - A_2^2\right]$$

$$H = H_y \qquad E_1^y = \pm \beta H_y \left[(A_1 - C_1)^2 - B_1^2\right]$$
$$E_2^y = 2.61 \lambda \pm \beta H_y \left[(A_2 - C_2)^2 - B_2^2\right]$$

$$H = H_z \qquad E_1^z = \pm \beta H_z \left[(A_1 - B_1)^2 - C_1^2\right]$$
$$E_2^z = 2.61 \lambda \pm \beta H_z \left[(A_2 - B_2)^2 - C_2^2\right] \qquad (188)$$

---

[4] See footnote p. 82.

Numerical values for the coefficients are given in the table following Eq. (107). As before, substitution in Eq. 158 leads to

$$H = H_x \qquad n_e^2 = \frac{3}{1+q} [(0.853)^2 + (1.112)^2 \, q]$$

$$H = H_y \qquad n_e^2 = \frac{3}{1+q} [(1.101)^2 + (0.928)^2 \, q]$$

$$H = H_z \qquad n_e^2 = \frac{3}{1+q} [(1.388)^2 + (0.551)^2 \, q] \qquad (189)$$

where

$$q = e^{-2\,61\lambda/kT}$$

As $T \to 0$, $q \to 0$ and the average value of n_e becomes

$$n_e(av) = \sqrt{3.87} = 1.97 \qquad (T \to 0) \qquad (190)$$

Since most of the population resides in the lowest doublet even at room temperature, the value of n_e obtained above is not far from the room temperature value $n_e(T = 300° \text{ K}) = 2.13$ which may be compared with the spin-only value of 1.73 for $S = \frac{1}{2}$.

The calculation may be carried to second order in H by computing the 6×6 matrix of $\vec{l} + \vec{2s}$ within the set $\psi_1^{\pm}$, $\psi_2^{\pm}$ and $\psi_3^{\pm}$. The basic information is contained in Table 14. *Kotani (35)* gives the expressions for n_e^2 when the energies are expressed to second order in H.

C. Ferrohemoglobin

Both oxyhemoglobin (Hb O_2) and reduced (deoxygenated) hemoglobin (Hb) are in the ferrous state. In the case of the former we regard the six d-electrons as comprising a t_2^6 configuration thereby completely filling the t_2-orbitals. The only term possible (table 4) is $^1A_1$; we should therefore not expect to find any paramagnetic properties and none have been found. Nevertheless, an explanation based entirely on the electronic configuration of the Fe ion cannot be regarded as complete. The ground state of O_2 is known to be a triplet (S = 1). Why, then, does not the complex exhibit paramagnetism attributable to the presence of O_2?

This question was discussed by *Griffith (18)*. The general answer is associated with the likelihood that there is some covalent bonding between O_2 and Fe (or more generally, between O_2 and iron porphyrin). The axis of the oxygen molecule is probably inclined to the heme plane, perhaps parallel to it, as *Griffith's* calculations suggest, to facilitate the bonding. Such an orientation is energetically favored and has the effect of reducing the symmetry about the $O-O$ axis. To see the implication of the reduction in symmetry, it is recalled that the two molecular orbitals π_x^* and π_y^* of O_2 are ordinarily degenerate. Each orbital contains a single electron; the exchange energy favors a parallel alignment of the spins leading to $S = 1$. Upon reduction of the symmetry about the $O-O$ axis, the orbitals π_x^* and π_y^* are no longer degenerate and the state with $S = 1$ is not necessarily the one of lowest energy.

Measurements of magnetic susceptibility in reduced hemoglobin result in effective Bohr magneton numbers (n_e) of $5.2 - 5.5$ (Table 2). The spin-only value of n_e for $S = 2$ is 4.90. It is therefore natural to conclude that reduced hemoglobin is a high spin complex with $S = 2$ in which there is incomplete quenching of orbital angular momentum to account for the increase in the measured value of n_e over the spin-only value. For the ferrihemoglobin derivatives we were able to derive expressions for the susceptibility or the magneton number based on an energy level structure deduced from electron spin resonance experiments. For ferrohemoglobin there are no such data and one must be guided entirely by theoretical considerations (*21, 38*).

From Fig. 20 it is seen that the ground state in a cubic field for a high spin ferrous complex is $t_2^4 e^2\, {}^5T_2$. It will, of course, be necessary to apply spin-orbit coupling to obtain departures from the spin-only value. It appears that a cubic field with spin-orbit coupling is sufficient without the necessity of invoking lower symmetry fields (*38*).

The effect of spin-orbit coupling on ${}^5T_2$ is to split it into three levels as if we were dealing with an atomic ${}^5P$ term. In the latter case there are three values of J, namely, 3, 2 and 1 and the levels have degeneracies of 7, 5 and 3 respectively. The energies follow the Landé interval rule. Precisely the same thing occurs with ${}^5T_2$. *Griffith* (*21*) computes n_e^2 for this level structure:

$$n_e^2 = \frac{3(49x + 108) + 5(27x - 20)\, e^{-\frac{1}{2}x} + 56(3x - 4)\, e^{-\frac{5}{4}x}}{2x\left(3 + 5e^{-\frac{1}{2}x} + 7e^{-\frac{5}{4}x}\right)} \tag{191}$$

$$x = \frac{\zeta}{kT}$$

In the limit of high temperature, $n_e \to 5.3$; as the temperature is lowered, n_e rises slowly to a peak value of about 5.8 and then eventually drops to the spin-only value. At room temperature, the prediction based on Eq. (191) and the observed value are in satisfactory agreement.

IX. Molecular Orbitals

A. General Discussion

Thus far the discussion of hemoglobin has been presented from the standpoint of ligand field theory which deals entirely with the properties of the orbitals associated with the central ion. The effect of the ligands is reflected in an electrostatic potential energy term added to the Hamiltonian of the free ion. In the molecular orbital approach both the ligands and the central ion are treated on an equal footing. One constructs linear combinations of ligand orbitals with central ion orbitals to produce a set of molecular orbitals which may be bonding, non-bonding, or antibonding. The characterization of a complex as "ionic" or "covalent" now becomes a matter of the relative magnitude of the coefficients of cation vs. ligand orbitals. Both the ligands and the central ion contribute electrons to populate the molecular orbitals. Such electrons generally can no longer be considered as localized either on the ligands or on the central ion. Rather, the coefficients in the molecular orbitals determine the probalitiy of finding an electron in a particular location. One then attempts to understand the physical properties of the complex in terms of the molecular orbitals and the distribution of electrons within them.

The molecular orbital method applied to complexes may be illustrated by reference to a complex of the form MX_6 with symmetry O_h. The six ligands $X_1 \ldots X_6$ may form both σ and π coordinations with the central ion M which we take to be Fe. The previous discussions dealt entirely with the 3d-orbitals; for molecular orbital calculations we must augment these with 4s- and 4p-orbitals which may also form σ and π bonds with ligand orbitals. The classification according to O_h is given by

$$a_{1g}: 4s$$
$$t_{1u}: 4p_x, 4p_y, 4p_z$$
$$e_g: d_{z^2}, d_{x^2-y^2}$$
$$t_{2g}: d_{xy}, d_{yz}, d_{zx}$$

The ligand orbitals are similarly organized into linear combinations which transform according to representations of O_h; such linear combinations are often called symmetry orbitals. Metal ion orbitals and ligand symmetry orbitals are finally combined into molecular orbitals by the usual method of solving the appropriate secular determinants. A schematic energy level diagram for the formation of σ-bonds in an octahedral complex is shown in Fig. 27, in which we distinguish three types of orbitals: bonding (a_{1g}, e_g and t_{1u}), antibonding (e_g^* and t_{1u}^*) and nonbonding (t_{2g}).

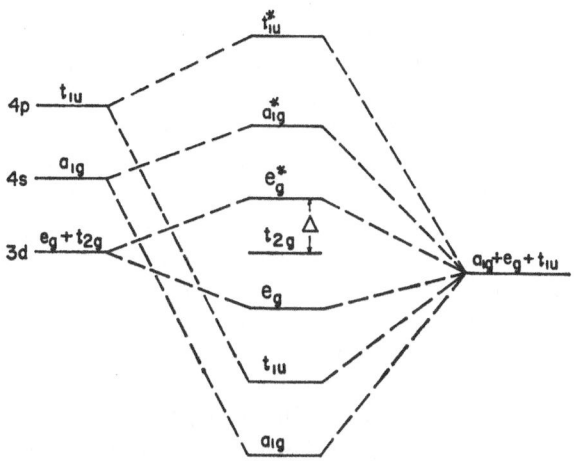

Fig. 27. Energy levels in an octahedral complex. On the left are the symmetry orbitals of the metal and on the right those for the ligands. Molecular orbitals are shown in the center.

In a previous discussion (Ch. IV) it was pointed out that the electro-static repulsion between electrons in an orbital like $d_{x^2-y^2}$ (or d_{z^2}) and the negatively charged ligands raises the energy of such orbitals relative to the set d_{xy}, d_{yz} and d_{zx}. On the basis of the molecular orbital approach we would say that the increase in energy is associated with the formation of an antibonding e_g^*-orbital. The quantity Δ which previously chara-terized the splitting between the set of e-orbitals and the set of t_2-orbitals in a cubic field is reinterpreted in the molecular orbital approach to represent the separation in energy between the antibonding e_g^*-orbitals and the nonbonding t_{2g}-orbitals. When π-bonding is included, the value of Δ may be increased or decreased depending on the position of the ligand orbitals and their occupation. In most cases, Δ is still treated as an empirical parameter to be deduced from experimental data.

Electrons originally belonging to the metal ion as well as those be-longing to the ligands are placed into the lowest orbitals until all the electrons have been accommodated. Typically, in such complexes, each ligand usually supplies two σ electrons. The twelve ligand electrons will completely occupy the bonding orbitals. Electrons from the metal ion are accommodated in the nonbonding t_{2g}- and the antibonding e_g^*-orbitals. It is worth noting that whereas the t_{2g}-orbitals are almost pure metal ion orbitals, the e_g^*-orbitals contain contributions from the ligands. Although on the molecular orbital picture we have a somewhat different interpretation as compared with the ligand field approach the general structure of the energy levels as far as the metal ion electrons are concerned remains essentially unaltered.

B. Iron-Porphyrin Complexes

Molecular orbital calculations on iron-porphyrin complexes have been performed by *Pullman, Spanjaard* and *Berthier (50)*. *Veillard* and *Pullman (59)* subsequently extended the calculations to an iron-porphyrin with two additional nitrogen ligands, one above and one below the porphyrin plane. Such a model resembles the prosthetic group in cytochrome in which the fifth and sixth ligands are histidines. *Ohno, Tanabe* and *Sasaki (45)* worked with a model which more nearly resembles heme in hemoglobin.[5] This model, too, consists of an iron porphyrin with two out-of-plane ligands; the fifth ligand is nitrogen but the sixth is either a point charge or a dipole. The description that follows is based on the calculations of *Ohno, Tanabe* and *Sasaki*.

As in all theoretical work of this type, the real molecular structure must be simplified and idealized in certain respects in order to make the calculations at all tractable. Confining our discussion for the moment to iron porphyrin, without the out-of-plane ligands, the model consists of an iron atom located in the center of a porphyrin skeleton. The peripheral substituents, which distinguish one porphyrin from another are assumed to have been removed. The system thus consists of one atom of iron, four of nitrogen and twenty of carbon with bond lengths and angles as shown in Fig. 28 on page 95. The symmetry group is D_{4h}.

Both σ- and π-bonds need to be considered. It is assumed that each of the four (in-plane) nitrogen atoms forms a σ-bond with iron; the nitrogen orbitals are taken as hybridized sp^2 σ-orbitals. For π-bonds, it is assumed that each of the four nitrogen atoms contributes a $2p\pi$-orbital and each of the 20 carbon atoms also contributes a $2p\pi$-orbital.

The set of σ-orbitals and the set of π-orbitals must now be organized into linear combinations which are basis functions for irreducible representations of the group D_{4h} whose characters are listed in table 6. The four nitrogen σ-orbitals constitute a basis set for a reducible representation of D_{4h} whose characters are

D_{4h}	E	$2C_4$	C_2	$2C_2'$	$2C_2''$	i	$2S_4$	σ_h	$2\sigma_v$	$2\sigma_d$
$\chi(\sigma)$	4	0	0	2	0	0	0	4	2	0

The reduction is now easily performed to give

$$\Gamma(\sigma) = A_{1g} + B_{1g} + E_u \tag{192}$$

The 24 π-orbitals must first be subdivided into sets such that, within a given set, the atoms are equivalent, that is, they are connected by a symmetry transformation. Each set will then be a bsis set for a reducible representation of D_{4h}. There are four such sets of equivalent π-orbitals:

---

[5] See also *M. Zerner, M. Gouterman*, and *H. Kobayashi*: Theoret. Chim. Acta *6*, 363 (1966).

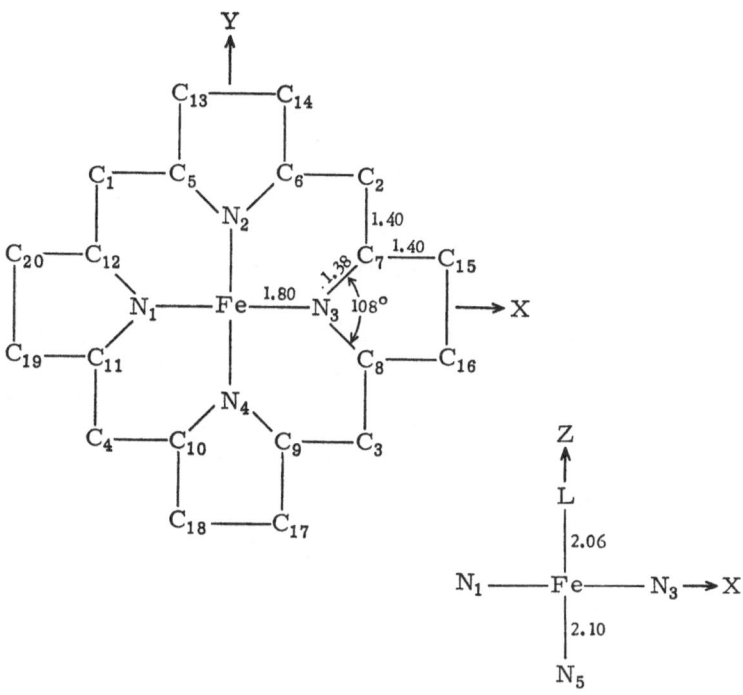

Fig. 28. The numbering system and dimensions (in Angstroms) of the prophyrin model used in the molecular orbital calculations.

$N_1 \dots N_4$, $C_1 \dots C_4$, $C_5 \dots C_{12}$, and $C_{13} \dots C_{20}$ where the numbering system is that of Fig. (28). The characters of these reducible representations are the following:

D_{4h}	E	$2C_4$	C_2	$2C_2'$	$2C_2''$	i	$2S_4$	σ_h	$2\sigma_v$	$2\sigma_d$
$\chi(\pi_{N_1 \dots N_4})$	4	0	0	-2	0	0	0	-4	2	0
$\chi(\pi_{C_1 \dots C_4})$	4	0	0	0	-2	0	0	-4	0	2
$\chi(\pi_{C_5 \dots C_{12}})$	8	0	0	0	0	0	0	-8	0	0
$\chi(\pi_{C_{13} \dots C_{20}})$	8	0	0	0	0	0	0	-8	0	0

The reduction into irreducible representations of D_{4h} gives

$$
\begin{aligned}
\Gamma(\pi_{N_1 \dots N_4}) &= A_{2u} + B_{2u} + E_g \\
\Gamma(\pi_{C_1 \dots C_4}) &= A_{2u} + B_{1u} + E_g \\
\Gamma(\pi_{C_5 \dots C_{12}}) &= A_{1u} + A_{2u} + B_{1u} + B_{2u} + 2E_g \\
\Gamma(\pi_{C_{13} \dots C_{20}}) &= A_{1u} + A_{2u} + B_{1u} + B_{2u} + 2E_g
\end{aligned}
\tag{193}
$$

95

and the linear combinations which belong to each of the irreducible representations are shown in Table 23.

The fifth ligand, N_5, is a nitrogen atom with an sp^2 orbital pointing in the direction of the iron atom (along the z-axis); the sixth ligand is either a dipole or a point charge. When the out-of-plane ligands are added the symmetry is no longer D_{4h} since operations like C_2' and C_2'' (Table 6) are no longer symmetry elements. In place of D_{4h} the appropriate symmetry group is C_{4v} which contains the operations E, $2C_4$, C_2, $2\sigma_v$ and $2\sigma_d$ (defined in Table 6). We note that these operations are included in D_{4h} so the characters of C_{4v} may be read directly from Table 6. For example, the representations A_{1g} and A_{2u} have the same characters for the set of elements in C_{4v}. Therefore A_{1g} and A_{2u} in D_{4h} both correspond to A_1, the totally symmetric representation is C_{4v}. Continuing in this fashion we find that every representation in D_{4h} can be given a label in C_{4v} according to the following scheme:

D_{4h}	C_{4v}
A_{1g}, A_{2u}	A_1
A_{2g}, A_{1u}	A_2
B_{1g}, B_{2u}	B_1
B_{2g}, B_{1u}	B_2
E_g, E_u	E

As a consequence, the symmetry orbitals listed in table 23 according to their D_{4h} classification, may also be classified according to C_{4v}. A summary of this classification scheme, including the sp^2 orbital of N_5 which belongs to the A_1-representation of C_{4v} is the following:

C_{4v}	D_{4h}	Number of orbitals
A_1	A_{1g}	3
	A_{2u}	5
	$sp^2(N_5)$	1
A_2	A_{2g}	0
	A_{1u}	2
B_1	B_{1g}	2
	B_{2u}	3
B_2	B_{2g}	1
	B_{1u}	3
E	E_g	14
	E_u	4
		Total 38

Ohno, *Tanabe* and *Sasaki* (*45*) computed the molecular orbitals using the 38 orbitals listed above. The sixth ligand was either a dipole or a point charge. Two values of the dipole moment were used: 1 a. u. ($= 2.54 \times 10^{-18}$ esu-cm) and 0.5 a. u.; the point charge was one unit of electronic charge ($= -1.0$ a. u.). Molecular orbital calculations on a structure as complicated as the one under consideration are rather involved, notwithstanding the fact that nothing more elaborate than the Hückel method is employed. The reader is referred to the original paper for details regarding various approximations, evaluation of integrals, choice of parameters etc. We shall be mainly concerned with the results of these calculations and their bearing on the interpretation of experimental observations on hemoglobin.

Fig. 29 shows the energies of a number of molecular orbitals for two choices of the sixth ligand. The orbitals are labelled according to the representations of D_{4h} or C_{4v} and their energies have been arbitrarily normalized to $E(1a_{1u}) = 0$ since differences in energy, rather than the absolute energies, are the significant quantities. The figure also shows that certain ones of the molecular orbitals are predominantly orbitals of the metal ion. This information is obtained from the coefficients in the linear combination of symmetry orbitals which go to make up the molecular orbitals. Thus the $1b_{2g}$ orbital is identical with $3d_{xy}$ since this orbital forms neither σ- nor π-bonds as is evident from table 23. The $2b_{1g}$ and $6a_1$ molecular orbitals are predominantly $3d_{x^2-y^2}$ and $3d_{z^2}$ respectively with little admixture from ligand orbitals. The degenerate pair $3d_{yz}$ and $3d_{zx}$ are more intimately mixed with ligand orbitals. It is interesting to compare Fig. 22 with Fig. 29. We find that in both cases, the orbitals are ordered, in increasing energy according to $3d_{xy}$, $(3d_{yz}, 3d_{zx})$, $3d_{z^2}$, $3d_{x^2-y^2}$, so the molecular orbital description and the ligand field description are consistent to this extent. However, if we take the hemoglobin azide data as our guide, the energy separations predicted by the molecular orbital calculations are far too large. For example, in Fig. 23 we find that $E(d_{zx}) - E(d_{xy}) \sim 1500$ cm^{-1}. The molecular orbital calculation gives a value which is an order of magnitude greater.

The molecular orbitals will be occupied by electrons originally belonging to either the ligands or the iron atom. To decide which electrons are involved we consider the atomic orbitals out of which the molecular orbitals are constructed and count the number of electrons that originally occupied those orbitals. Referring to Fig. 28, each of the carbon atoms $C_1 \ldots C_{20}$ contributes a single electron originally in a $2p\pi$-orbital. Protoporhyrin (Fig. 3) has two nitrogens in a pyridine-type structure

Table 23. *Symmetry orbitals.*

The numbering system is based on Fig. 28

	Fe	$\sigma_{N_1 \,..\, N_4}$	$\pi_{N_1 \,..\, N_4}$	$\pi_{C_1 \,..\, C_4}$
A_{1g}	4s $3d_{z^2}$	$\frac{1}{2}(N_1+N_2+N_3+N_4)$		
A_{2g}				
B_{1g}	$3d_{x^2-y^2}$	$\frac{1}{2}(N_1-N_2+N_3-N_4)$		
B_{2g}	$3d_{xy}$			
E_g	$3d_{yz}$ $3d_{zx}$		$\frac{1}{\sqrt{2}}(N_2-N_4)$ $\frac{1}{\sqrt{2}}(N_1-N_3)$	$\frac{1}{2}(C_1+C_2-C_3-C_4)$ $\frac{1}{2}(C_1-C_2-C_3+C_4)$
A_{1u}				
A_{2u}	$4p_z$		$\frac{1}{2}(N_1+N_2+N_3+N_4)$	$\frac{1}{2}(C_1+C_2+C_3+C_4)$
B_{1u}				$\frac{1}{2}(C_1-C_2+C_3-C_4)$
B_{2u}			$\frac{1}{2}(N_1-N_2+N_3-N_4)$	
E_u	$4p_x$ $4p_y$	$\frac{1}{\sqrt{2}}(N_2-N_4)$ $\frac{1}{\sqrt{2}}(N_1-N_3)$		

$$\pi_{C_5 \,..\, C_{12}} \qquad\qquad\qquad \pi_{C_{13} \,..\, C_{20}}$$

$$\left.\begin{array}{l} \tfrac{1}{2}\,(C_7+C_8-C_{11}-C_{12}) \\[4pt] \tfrac{1}{2}\,(C_5+C_6-C_9-C_{10}) \end{array}\right\} \qquad\qquad \left.\begin{array}{l} \tfrac{1}{2}\,(C_{15}+C_{16}-C_{19}-C_{20}) \\[4pt] \tfrac{1}{2}\,(C_{13}+C_{14}-C_{17}-C_{18}) \end{array}\right\}$$

$$\left.\begin{array}{l} \tfrac{1}{2}\,(C_5-C_6-C_9+C_{10}) \\[4pt] \tfrac{1}{2}\,(C_7-C_8-C_{11}+C_{12}) \end{array}\right\} \qquad\qquad \left.\begin{array}{l} \tfrac{1}{2}\,(C_{13}-C_{14}-C_{17}+C_{18}) \\[4pt] \tfrac{1}{2}\,(C_{15}-C_{16}-C_{19}+C_{20}) \end{array}\right\}$$

$$\frac{1}{2\sqrt{2}}\,(C_5-C_6+C_7-C_8+C_9-C_{10}+C_{11}-C_{12}) \qquad \frac{1}{2\sqrt{2}}\,(C_{13}-C_{14}+C_{15}-C_{16}+C_{17}-C_{18}+C_{19}-C_{20})$$

$$\frac{1}{2\sqrt{2}}\,(C_5+C_6+C_7+C_8+C_9+C_{10}+C_{11}+C_{12}) \qquad \frac{1}{2\sqrt{2}}\,(C_{13}+C_{14}+C_{15}+C_{16}+C_{17}+C_{18}+C_{19}+C_{20})$$

$$\frac{1}{2\sqrt{2}}\,(C_5-C_6-C_7+C_8+C_9-C_{10}-C_{11}+C_{12}) \qquad \frac{1}{2\sqrt{2}}\,(C_{13}-C_{14}-C_{15}+C_{16}+C_{17}-C_{18}-C_{19}+C_{20})$$

$$\frac{1}{2\sqrt{2}}\,(C_5+C_6-C_7-C_8+C_9+C_{10}-C_{11}-C_{12}) \qquad \frac{1}{2\sqrt{2}}\,(C_{13}+C_{14}-C_{15}-C_{16}+C_{17}+C_{18}-C_{19}-C_{20})$$

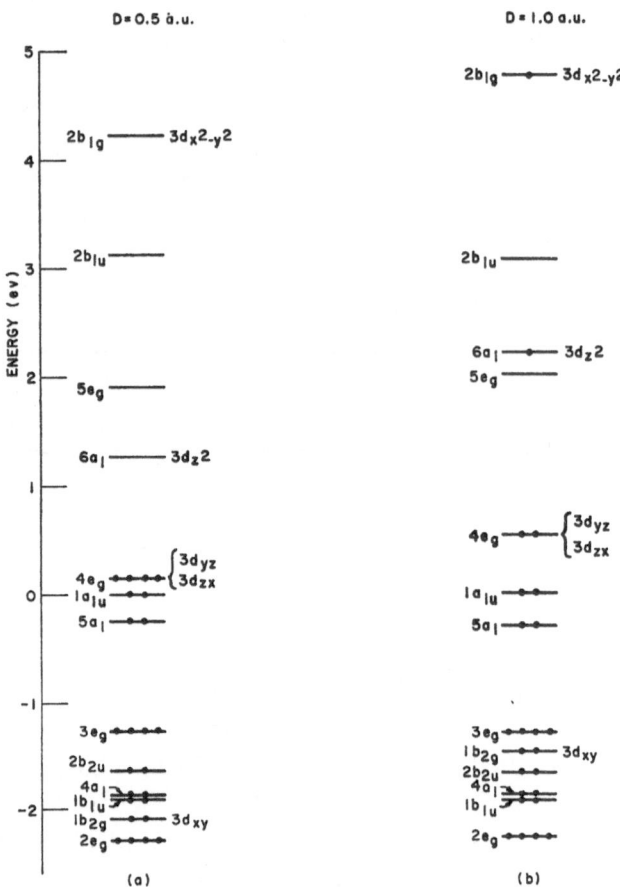

Fig. 29a and b. Molecular orbitals in ferroporphyrin with two out-of-plane ligands (*Ohno* et al. (*45*)). Occupied molecular orbitals lying below those shown in the diagram are $1b_{2u}$, $3a_1$, $1e_g$, $2a_1$, $1e_u$, $1a_1$, $1b_{1g}$. The energies are arbitrarily normalized to $E(1a_{1u}) = 0$. The ligand in the sixth position is a dipole of strength $D = 0.5$ a. u. in (a) and $D = 1.0$ a. u. in (b). The distribution of electrons corresponds to $S = 0$ in (a) and to $S = 2$ in (b).

and two nitrogens in a pyrrole-type structure

For a pyridine nitrogen the electron configuration may be written

$$(1s)^2(sp^2)^2(sp^2)^2_{nb}(2p)^1$$

which shows that there are two nonbonding electrons in an sp²-orbital, this being the orbital which points toward the Fe atom, and a single bonding electron in a 2p-orbital. These orbitals participate in σ- and π-bonds respectively. On this basis, a pyridine nitrogen therefore contributes a total of three electrons. The electron configuration for a pyrrole nitrogen is

$$(1s)^2(sp^2)^3(2p)^2_{nb}$$

In this case there is one bonding electron in an sp²-orbital pointing toward Fe and two nonbonding electrons in a 2p-orbital. Again we have a total of three electrons in orbitals which form σ- and π-bonds with Fe orbitals. It is concluded that each of the four nitrogens in prophyrin contributes three electrons although the distribution of these electrons among the atomic orbitals is different in the two types of nitrogen. For N_5, the imidazole nitrogen, we count the two nonbonding electrons in an sp²-orbital pointing toward Fe as in the case of the pyridine-type nitrogen. The 2pπ-orbital for this nitrogen does not contribute to any of the molecular orbitals, hence the electron residing in it is not included.

In counting the number of electrons contributed by Fe, it is helpful to think of the manner in which heme is formed from protoporphyrin (11). Two protons, belonging to pyrrole hydrogens, are displaced leaving behind two electrons. A ferrous ion with six d-electrons is incorporated into the structure to make heme. The original two electrons may be assigned to Fe-orbitals; thus a total of eight electrons will be contributed.

We have then the following pool of electrons for our model of ferro-hemoglobin: $C_1 \ldots C_{20}(20)$, $N_1 \ldots N_4(12)$, $N_5(2)$, Fe(8) for a total of 42 electrons to be distributed among 38 orbitals. Ferrihemoglobin will be represented by 41 electrons; in most cases the ligand in the sixth position is an ion such as F^-, CN^-, N_3^- etc. thereby preserving charge neutrality in the ferric complex.

Finally it is necessary to consider the manner in which the electrons are distributed among the molecular orbitals. For oxyhemoglobin $S = 0$; therefore the 42 electrons are simply placed in the lowest possible set of molecular orbitals. In each orbital the spins are paired and the total spin of the system is zero. The oxygen molecule has no permanent electric dipole moment but may have an induced dipole moment estimated to be 0.55 a. u. (45). We have therefore shown the electrons in oxyhemoglobin occupying the orbitals derived from the calculation in which the sixth ligand is a dipole of strength $D = 0.5$ a. u. (Fig. 29). However, this choice is for illustrative purposes only. The charge distribution within the molecule may now be obtained from the molecular orbital coefficients and the occupation numbers. In the central region of the molecule the charges are

$$Fe^{+2.07} \qquad N_{1..4}^{-0.64} \qquad N_5^{+0.08} \qquad (D = 0.5 \text{ a. u.})$$

$(S = 0)$

$$Fe^{+2.46} \qquad N_{1..4}^{-0.61} \qquad N_5^{+0.06} \qquad (D = 1.0 \text{ a. u.})$$

with smaller charges distributed among the peripheral carbons.

In high spin ferrohemoglobin the sixth position is occupied by a water molecule which has a permanent dipole moment of 0.73 a.u. The induced dipole moment is approximately the same as for oxygen; we shall therefore illustrate the occupation scheme using the level system shown in Fig. 29 for which the dipole in the sixth position has a value $D = 1.0$ a.u. Since $S = 2$, we may put 38 electrons with their spins paired, into the lowest set of orbitals while the remaining four electrons are placed in separate orbitals to permit a parallel alignment of spins. The manner in which this may be done is somewhat arbitrary. Based simply on orbital energies the configuration would be

$$\ldots (4e_g)^2 \, (5e_g)^2$$

Ohno et al. (45) suggest that the singly occupied orbitals should be localized around the Fe atom. This leads to

$$\ldots (4e_g)^2 \, (6a_1)^1 (2b_{1g})^1$$

which corresponds to

$$\ldots (3d_{yz})^1 \, (3d_{zx})^1 \, (3d_{z2})^1 \, (3d_{x2-y2})^1$$

as shown in Fig. 29. The charge distribution in the central region is

$$Fe^{+1.27} \qquad N_{1..4}^{-0.60} \qquad N_5^{+0.03} \qquad (D = 0.5 \text{ a. u.})$$

$(S = 2)$

$$Fe^{+1.60} \qquad N_{1..4}^{-0.60} \qquad N_5^{+0.02} \qquad (D = 1.0 \text{ a. u.})$$

As pointed out by Ohno et al. (45), there are difficulties in achieving internal consistency for the calculations on ferrihemoglobin. If one electron is removed from the ferrous complex, the resulting charge distribution is

$$Fe^{+2.5} \qquad N_{1..4}^{-0.6} \qquad N_5^{+0.1}$$

C. Spectra

In molecules as complex as the porphyrins a theoretical approach can, at best, account for only the gross aspects of the absorption spectra. Nontheless, molecular orbital theory when judiciously coupled with experimental information provides a number of unifying concepts. Recent reviews have been given by *Gurinovich* et al., *(25)* and by *Braterman* et al., *(7)*.

The observed spectra arise from electric dipole transitions which may be classified in three broad categories as (1) d–d, (2) charge transfer and (3) π–π*.

The d–d transitions are between two molecular orbitals both of which are predominantly 3d-orbitals localized in the region of the metal. An example from Fig. 29 is $4e_g - 6a_1$ which may also be described by $t_{2g}(3d_{yz}, 3d_{zx}) - e_g(3d_{z^2})$. We note that transitions of this type are forbidden by the parity selection rule since the dipole moment operator is odd and both the initial and final states are even. However, in a complex molecule there may well exist couplings between vibrational and electronic motions such that the center of symmetry is not preserved. In that event parity is no longer a good quantum number and the transition between d-orbitals may occur but generally with lower intensity than for an allowed transition.

When an electron is excited from a molecular orbital mainly localized on the metal to one localized on the ligands, or conversely from the ligands to the metal, the resulting absorption band is known as a charge transfer band. Finally we have the π–π* transitions which occur between molecular orbitals largely associated with the ligands.

In hemoglobin, the absorption spectra spanning the visible and near infrared region are due primarily to the allowed π–π* transitions *(45)*.

$$5a_{2u} \ (5a_1) - 5e_g$$

$$1a_{1u} - 5e_g$$

The ground state of the molecule is A_{1g}; the excited state corresponding to the electron configuration $(5a_{2u})^1 \ (5e_g)^1$ or $(1a_{1u})^1 \ (5e_g)^1$ has the symmetry E_u as may easily be verified from Table 6. Hence the transition is described by $A_{1g} - E_u$ and the matrix elements for dipole transitions, which govern the intensity of the absorption bands are

$$< A_{1g} | \overrightarrow{er} | E_u >$$

The selection rules are now obtained from the transformation properties of $\overrightarrow{r}$ in D_{4h}. Thus when $\overrightarrow{r} = z$, Table 6 may be used to show that z belongs

103

to the representation A_{2u}. Since the product representation $A_{1g} \times A_{2u} \times E_u$ does not contain the totally symmetric representation A_{1g}, the matrix element necessarily vanishes and the transition is therefore symmetry forbidden. On the other hand, when $\vec{r} = x$ or y we find that (x, y) belong to E_u and $A_{1g} \times E_u \times E_u$ contains A_{1g}. It is concluded that light polarized parallel to the porphyrin plane may be absorbed whereas light polarized perpendicular to the plane may not.

Platt (49) and *Gouterman (17)* have shown that for the interpretation of porphyrin spectra it is essential to take account of configuration interaction. This means that a particular band cannot be attributed to a transition in which a single electron is excited from one molecular orbital to another. Rather, it is necessary to construct a two-fold degenerate state having the form

$$\Psi_1 = a_1 E'_u + b_1 E''_u \tag{194}$$

where E'_u and E''_u are molecular states, both having symmetry E_u and arising from the electron configurations $(5a_{2u})^1 (5e_g)^1$ and $(1a_{1u})^1 (5e_g)^1$ respectively; a_1 and b_1 are normalized constants. The matrix element of the Hamiltonian taken between the two configurations provides a measure of the strength of the interaction.

In this way we achieve a new pair of mixed states, each two-fold degenerate and displaced in energy by an amount which depends on the interconfiguration matrix element. The state of higher energy is usually designated B and the lower one Q. A further important consequence of invoking configuration interaction is that the transition matrix elements $< A_{1g} | \vec{er} | B >$ and $< A_{1g} | \vec{er} | Q >$ are quite different; the one to the higher state (B) is much larger than to lower state (Q). On the one-electron model, without configuration mixing the transition matrix elements to each of the two excited states are the same.

The general interpretation of the hemoglobin spectra is now the following: The transition $A_{1g} - B$ which has the higher energy and the higher intensity is associated with the Soret band. Using the orbital separations for $5e_g - 1a_{1u}$ as calculated by *Ohno* et al. *(45)* of approximately 2.1 ev and the interconfiguration matrix element estimated by *Gouterman (17)* to be about 0.4 ev, the peak of the Soret band would lie at 500 mμ compared to the observed value of 400–420 mμ (Ch. II). The longer wavelength region of the spectrum is associated with the lower energy and lower intensity transition $A_{1g} - Q$ which is further split into $(0 - 0)$ and $(0 - 1)$ vibrational bands. Finer details are discussed *(7)* on the basis of charge transfer transitions.

An interesting correlation between absorption spectra and magnetic susceptibility for the ferrihemoglobin hydroxides was obtained by *George* et al. *(13)*. The extinction coefficients in the major absorption bands of the hydroxides have values which fall between those for high and low spin ferrihemoglobin derivatives. Assuming that the hydroxides are mixtures of high and low spin forms, it is possible to determine the relative concentrations of each from a measurement of the magnetic susceptibility on the mixture. In this way it is determined that ferrihemoglobin hydroxide is a 50–50 mixture. Further measurements on the extinction coefficients in the mixture make it possible to reconstruct the major features of the absorption spectra of the high and low spin hydroxides individually.

X. Mössbauer Resonance

We shall summarize the main features of the theory relevant to the interpretation of Mössbauer spectra of hemoglobins, typical examples of which are shown in Figs. 14 and 15.

A. Isomer Shift

The electrostatic interaction of s-electrons with a nucleus of finite dimensions produces shifts in the nuclear energy levels. Since nuclear charge distributions generally vary from one nuclear state to another, the magnitude of the shift will also depend on the nuclear state. It may be shown *(8)* that a gamma ray photon emitted in a transition from an excited state $|e>$ to the ground state $|g>$ will be shifted in energy by an amount

$$\Delta E = F(Z) \, |\psi_s(0)|^2 \, \frac{\delta R_\mu}{R_\mu} \qquad (195)$$

in which $F(Z)$ is a fairly complicated function of the atomic number, Z; R_μ is the radius of the equivalent uniform charge distribution; $\psi_s(0)$ is the s-electron wave function at the nucleus and

$$\delta R_\mu = (R_\mu)_e - (R_\mu)_g \qquad (196)$$

The isomer shift (in velocity units) is given by

$$\delta = \frac{c}{E_0} \, [(\Delta E)_A - (\Delta E)_S] \qquad (197)$$

105

where A and S refer to the absorber and source respectively; E_0 is the transition energy between the levels $|e>$ and $|g>$ (14.4 kev for Fe^{57}) and c is the velocity of light. Eq. (197) shows that the isomer shift is proportional to the difference in $|\psi_s(0)|^2$ between the absorber and the source.

The equivalent uniform charge distribution may be expressed in terms of the mass number A by

$$R_\mu = 1.2 \ A^{1/3} \ x \ 10^{-13} \ cm$$

For Fe^{57}

$$R_\mu = 4.6 \ x \ 10^{-13} \ cm$$

$$\frac{\delta R_\mu}{R_\mu} = - \ 1.8 \ x \ 10^{-3}$$

indicating that the radius of the equivalent charge distribution in the ground state is larger than for the excited state.

In iron compounds, a variation in the isomer shift may come about through a variation in the number of 3d-electrons or through covalency effects involving the 3d-electrons. The latter, by themselves, cannot contribute to the isomer shift since they have a vanishing amplitude at the nucleus. Nevertheless the 3d-electrons may contribute indirectly through their screening effect on inner s-electrons, mainly the 3s (60).

B. Quadrupole Interaction

1. General Expressions

The Hamiltonian which describes the interaction between the nuclear quadrupole moment and an electric field gradient is given by (1, 54, 58, 60)

$$\mathcal{H}_Q = \frac{e^2 q Q}{4I(2I-1)} \ [3I_z^2 - I(I + 1) + \eta(I_x^2 - I_y^2)] \tag{198}$$

where I_x, I_y and I_z are components of the nuclear spin operator; Q is the nuclear quadrupole moment defined by

$$Q = \ <I \ I| \ \sum_{p=1}^{z} (3z_p^2 - r_p^2)|I \ I > \tag{199}$$

The sum is taken over the Z protons in the nuleus and the matrix element is evaluated for a nuclear state described by a spin quantum number I and projection quantum number $m_I = I$. Q is positive for a cigar-shaped distribution and negative for one in the shape of a door-knob. The quantity eq is defined by

$$eq = V_{zz} = \left(\frac{\partial^2 V}{\partial z^2}\right)_0 \tag{200}$$

where the second derivative is evaluated at the origin and the coordinate system has been chosen to coincide with the principal axis system of the symmetric electric field gradient (EFG) tensor

$$\left(\frac{\partial^2 V}{\partial x_i\, \partial x_j}\right)_0$$

η is an asymmetry parameter,

$$\eta = \frac{V_{xx} - V_{yy}}{V_{zz}} \tag{201}$$

An alternative form for the Hamiltonian is

$$\mathscr{H}_Q = \frac{e^2 q Q}{4I(2I-1)} \left[3I_z^2 - I(I+1) + \frac{\eta}{2}\left(I_+^2 + I_-^2\right)\right] \tag{202}$$

with

$$I_\pm = I_x \pm i\, I_y$$

In the derivation of $\mathscr{H}_Q$ it has been assumed that *Laplace's* equation holds. The immediate consequence is that in a cubic environment

$$V_{xx} = V_{yy} = V_{zz} = 0$$

and there can be no quadrupole interaction. In the case of axial symmetry we may choose the z-axis to coincide with the axis of symmetry in which case $\eta = 0$ and a single parameter, V_{zz}, suffices to describe the interaction. When it is necessary to speak of "the electric field gradient" we shall mean $-V_{zz}$.

In Fe^{57} the nuclear ground state has a spin of $\frac{1}{2}$. The matrix elements of $\mathscr{H}_Q$ within the set $|\frac{1}{2}>$, $|-\frac{1}{2}>$ are all zero; hence there is no quadrupole interaction in the ground state. In the first excited state $I = \frac{3}{2}$; the matrix of $\mathscr{H}_Q$ within the manifold of states belonging to $I = \frac{3}{2}$ is shown in Table 24.

As seen there is only one 2×2 matrix; the eigenvalues are

$$E = \pm \frac{e^2 q Q}{4}\left(1 + \frac{\eta^2}{3}\right)^{\frac{1}{2}} \tag{203}$$

Table 24. *Matrix Elements of Quadrupole Interaction*, $\mathscr{H}_Q$, *for a Nuclear State with* $I = \frac{3}{2}$ *(in units of* $\frac{1}{4}e^2qQ$*)*

$$\mathscr{H}_Q = \frac{e^2qQ}{4I(2I-1)} \left[3I_z^2 - I(I+1) + \frac{\eta}{2}\left(I_+^2 + I_-^2\right)\right]$$

$\mathscr{H}_Q$	$\frac{3}{2}$	$\frac{1}{2}$	$-\frac{3}{2}$	$\frac{1}{2}$
$\frac{3}{2}$	1	$\frac{\eta}{\sqrt{3}}$		
$-\frac{1}{2}$	$\frac{\eta}{\sqrt{3}}$	-1		
$-\frac{3}{2}$			1	$\frac{\eta}{\sqrt{3}}$
$\frac{1}{2}$			$\frac{\eta}{\sqrt{3}}$	-1

We note that the fourfold degenerate level $I = \frac{3}{2}$ is split by the quadrupole interaction into two 2-fold degenerate states $|\pm \frac{3}{2}>$ and $|\pm \frac{1}{2}>$ separated in energy by

$$\Delta = \frac{e^2qQ}{2}\left(1 + \frac{\eta^2}{3}\right)^{\frac{1}{2}} \tag{204}$$

For Q positive and $V_{zz} > 0$ the state $|\pm \frac{3}{2}>$ lies above $|\pm \frac{1}{2}>$; for $V_{zz} < 0$ the reverse is true.

The transition between the excited state $I_e = \frac{3}{2}$ and the ground state $I_g = \frac{1}{2}$ is of the magnetic dipole type (M1), and is therefore governed by the selection rules

$$\Delta m = 0, \pm 1$$

The intensities of the six allowed transitions are proportional to

$$\begin{pmatrix} \frac{3}{2} & 1 & \frac{1}{2} \\ m_e & M & m_g \end{pmatrix}^2 F_L^M (\theta)$$

with

$$m_e + M + m_g = 0$$

The 3-j symbol represents the coupling of the angular momenta $I_e = \frac{3}{2}$ and $I_g = \frac{1}{2}$ through the magnetic dipole radiation field for which $L = 1$.

m_e, M, and m_g are projection quantum numbers for I_e, L, and I_g respectively. $F_L^M (\theta)$ is an angular factor:

$$F_1^0(\theta) = \tfrac{3}{2} \sin^2\theta$$

$$F_1^{\pm 1} (\theta) = \tfrac{3}{4} (1 + \cos^2 \theta)$$

in which θ is the angle between the axis of quantization and the direction of observation. The relative intensities may now be listed as follows:

m_e $(I_e = \tfrac{3}{2})$	m_g $(I_g = \tfrac{1}{2})$	Relative Intensity	Average Relative Intensity
$\tfrac{3}{2}$	$\tfrac{1}{2}$	$\tfrac{3}{4} (1 + \cos^2 \theta)$	1
$\tfrac{1}{2}$	$\tfrac{1}{2}$	$\sin^2 \theta$	$\tfrac{2}{3}$
$-\tfrac{1}{2}$	$\tfrac{1}{2}$	$\tfrac{1}{4} (1 + \cos^2 \theta)$	$\tfrac{1}{3}$
$\tfrac{1}{2}$	$-\tfrac{1}{2}$	$\tfrac{1}{4} (1 + \cos^2 \theta)$	$\tfrac{1}{3}$
$-\tfrac{1}{2}$	$-\tfrac{1}{2}$	$\sin^2 \theta$	$\tfrac{2}{3}$
$-\tfrac{3}{2}$	$-\tfrac{1}{2}$	$\tfrac{3}{4} (1 + \cos^2 \theta)$	1

In the absence of any perturbations the single transition between $I_e = \tfrac{3}{2}$ and $I_g = \tfrac{1}{2}$ has a relative intensity given by the sum of the intensities for the six allowed transitions. This sum is seen to be independent of θ. When a quadrupole intercation is present the excited state with $I_e = \tfrac{3}{2}$ becomes a doublet while the ground state with $I_g = \tfrac{1}{2}$ remains unaffected. The relative intensities of the two possible transitions, $\pm \tfrac{3}{2}(I_e = \tfrac{3}{2}) \rightarrow \pm \tfrac{1}{2} (I_g = \tfrac{1}{2})$ and $\pm \tfrac{1}{2} (I_e = \tfrac{3}{2}) \rightarrow \pm \tfrac{1}{2} (I_g = \tfrac{1}{2})$ become identical Finally, when a magnetic field is applied the degeneracies are completely removed and the six transitions averaged over all directions have intensities in the ratio $3:2:1:1:2:3$.

From Eq. (204) it is seen that in order to calculate the quadrupolar splitting it is necessary to calculate q and $q\eta$ defined by Eqs. (200) and (201). For an ion surrounded by ligands there will be two types of contribution; one comes from the valence electrons, which, in the case of Fe, are the 3d-electrons and the second comes from the charge distribution on the ligands. Each of these contributions is multiplied by a Sternheimer factor which accounts for the polarization of the inner core electrons by the valence and ligand charges.

M. Weissbluth

Following *Ingalls* (27) we may write

$$\frac{V_{zz}}{e} = q = (1-R)\, q_v \quad + (1-\gamma_\infty) q_1$$

$$\frac{V_{xx} - V_{yy}}{e} = q\eta = (1-R)(q\eta)_v \quad + (1-\gamma_\infty)(q\eta)_1$$

in which q_v and $(q\eta)_v$ are the contributions from the 3d-electrons, q_1 and $(q\eta)_1$ are the contributions from the ligand charges, $(1-R)$ and $(1-\gamma_\infty)$ are the Sternheimer factors.

2. Contributions from 3d-electrons

For an electron at x, y, z and a nucleus at the origin, V_{zz} at the nucleus is

$$V_{zz} = \frac{\partial^2}{\partial z^2} \left(-\frac{e}{r}\right) = -e\, \frac{3z^2 - r^2}{r^5} \tag{205}$$

$$= -e\, \frac{3\cos^2\theta - 1}{r^3}$$

When the electron is distributed over an orbital it is necessary to compute the expectation value

$$< \frac{V_{zz}}{e} > = - < 3\cos^2\theta - 1 > < \frac{1}{r^3} > \tag{206}$$

One way to perform this computation is to recognize that

$$3\cos^2\theta - 1 = 2 \left(\frac{4\pi}{5}\right)^{\frac{1}{2}} Y_2^0 \tag{207}$$

and that

$$< Y_{\ell'}^{m'} | Y_L^M | Y_\ell^m > = (-1)^{m'} \left[\frac{(2\ell' + 1)(2L + 1)(2\ell + 1)}{4\pi}\right]^{\frac{1}{2}} \tag{208}$$

$$\times \begin{pmatrix} \ell' & L & \ell \\ -m' & M & m \end{pmatrix} \begin{pmatrix} \ell' & L & \ell \\ 0 & 0 & 0 \end{pmatrix}$$

with

$$- m' + M + m = 0 \tag{209}$$

110

Confining our attention to d-orbitals,

$$< Y_2^{m'} \, | Y_2^0 \, | Y_2^m > \; = (-1)^{m'} \, 5 \left(\frac{5}{4\pi}\right)^{\frac{1}{2}}$$

$$\times \begin{pmatrix} 2 & 2 & 2 \\ -m' & 0 & m \end{pmatrix} \begin{pmatrix} 2 & 2 & 2 \\ 0 & 0 & 0 \end{pmatrix} \tag{210}$$

Condition (209) requires that

$$m = m'$$

so that all off-diagonal matrix elements vanish. Also from the tables of *Rotenberg* et al. (*51*)

$$\begin{pmatrix} 2 & 2 & 2 \\ 0 & 0 & 0 \end{pmatrix} = - 2\left(\tfrac{1}{70}\right)^{\frac{1}{2}}$$

Therefore

$$< \frac{V_{zz}}{e} > \; = 20 \left(\tfrac{1}{70}\right)^{\frac{1}{2}} (-1)^m \begin{pmatrix} 2 & 2 & 2 \\ -m & 0 & m \end{pmatrix} < \frac{1}{r^3} > \qquad \text{(d-electrons)} \tag{211}$$

and

m	$\begin{pmatrix} 2 & 2 & m \\ -m & 0 & m \end{pmatrix}$	$< \dfrac{V_{zz}}{e} > < \dfrac{1}{r^3} >^{-1}$
2	$2\left(\tfrac{1}{70}\right)^{\frac{1}{2}}$	$\tfrac{4}{7}$
1	$\left(\tfrac{1}{70}\right)^{\frac{1}{2}}$	$-\tfrac{2}{7}$
0	$-2\left(\tfrac{1}{70}\right)^{\frac{1}{2}}$	$-\tfrac{4}{7}$
-1	$\left(\tfrac{1}{70}\right)^{\frac{1}{2}}$	$-\tfrac{2}{7}$
-2	$2\left(\tfrac{1}{70}\right)^{\frac{1}{2}}$	$\tfrac{4}{7}$

From Eqs. (47, 48) we finally obtain

$$< \frac{V_{zz}}{e} > < \frac{1}{r^3} >^{-1} = \begin{cases} -\tfrac{4}{7} & \text{for } d_{z^2} \\ \tfrac{4}{7} & d_{x^2-y^2} \\ -\tfrac{2}{7} & d_{zx} \\ -\tfrac{2}{7} & d_{yz} \\ \tfrac{4}{7} & d_{xy} \end{cases} \tag{212}$$

In a manner analogous to that leading to Eq. (205)

$$V_{xx} - V_{yy} = - e \, \frac{3[\sin^2\theta(\cos^2\theta - \sin^2\varphi)]}{r^3} \tag{213}$$

M. Weissbluth

Also

$$\sin^2\theta(\cos^2\theta - \sin^2\varphi) = \left(\frac{8\pi}{15}\right)^{\frac{1}{2}}(Y_2^2 + Y_2^{-2}) \tag{214}$$

Corresponding to Eq. (210)

$$< Y_2^{m'} |Y_2^2 + Y_2^{-2}|Y_2^m > = -10\left(\frac{5}{4\pi}\right)^{\frac{1}{2}}\left(\frac{1}{70}\right)^{\frac{1}{2}}(-1)^{m'}$$

$$\times \left[\begin{pmatrix} 2 & 2 & 2 \\ -m' & 2 & m \end{pmatrix} + \begin{pmatrix} 2 & 2 & 2 \\ -m' & -2 & m \end{pmatrix}\right] \tag{215}$$

For this case Eq. (209) indicates that all diagonal elements will vanish. The non-vanishing elements are those for which

$$m' = 2 \quad 1 \quad 0 -1 -2$$
$$m = 0 - 1 \pm 2 \quad 1 \quad 0$$

From this we conclude that the expectation value of $V_{xx} - V_{yy}$ vanishes for d_{z^2}, $d_{x^2-y^2}$, d_{xy}. From

$$< Y_2^1|Y_2^2 + Y_2^{-2}|Y_2^{-1} > = < Y_2^{-1}|Y_2^2 + Y_2^{-2}|Y_2^1 >$$

$$= -\frac{2}{7}\left(\frac{15}{8\pi}\right)^{\frac{1}{2}} \tag{216}$$

we obtain

$$\left< \frac{V_{xx} - V_{yy}}{e} \right> < \frac{1}{r^3} >^{-1} = \begin{cases} 0 & \text{for } d_{z^2} \\ 0 & d_{x^2-y^2} \\ -\frac{6}{7} & d_{zx} \\ \frac{6}{7} & d_{yz} \\ 0 & d_{xy} \end{cases} \tag{217}$$

It has already been shown on general grounds that both q and qη vanish in a cubic field. This may also be verified directly from the form of the cubic potential V_c, as given in Eq. (40), by evaluating the second derivatives at the origin. We may also view this result from the standpoint of the degeneracies of the t_2- and e-orbitals in a cubic field. An electron accupying a t_2-orbital has an equal probability of residing in any one of the three substates. From Eqs. (212) and (217) both $< V_{zz} >$ and $< V_{xx} - V_{yy} >$ vanish. The same argument applies to the 2-fold degenerate e-orbitals. Thus any distribution of electrons among the t_2- and e-orbitals gives $< V_{zz} > = 0$ and $< V_{xx} - V_{yy} > = 0$ provided the t_2- and e-orbitals retain their 3-fold and 2-fold degeneracies respectively. These degeneracies are, of course, simply manifestations of the cubic symmetry of the ligand field.

To get any contribution to q_v or $(q\eta)_v$ it is necessary to reduce the symmetry, at least to tetragonal. For this case the d-orbitals have the energy level structure shown in Fig. (22). We note that the two-fold degenerate level (d_{zx}, d_{yz}) lies above d_{xy} by an energy δ. Since the t_2-levels are no longer three-fold degenerate, q_v need not vanish. However, because of the degeneracy of d_{zx} and d_{yz} the parameter $(q\eta)_v$ must still be zero. Upon inclusion of a rhombic component the degeneracy of the t_2-orbitals is completely removed and both q_v and $(q\eta)_v$ may be non-vanishing. It will be assumed that the e-levels lie some 10^4 cm^{-1} above the t_2-levels and are therefore not thermally accessible. The discussion will therefore be confined to the t_2-levels which are of the form shown in Fig. (22) for tetragonal symmetry and in Fig. (23) for rhombic symmetry. For the latter we define

$$\Delta_1 = \varepsilon_\xi - \varepsilon_\zeta = E(d_{yz}) - E(d_{xy})$$

$$\Delta_2 = \varepsilon_\eta - \varepsilon_\zeta = E(d_{zx}) - E(d_{xy}) \tag{218}$$

The magnitudes of q_v and $(q\eta)_v$ will now be determined by the distribution of electrons among the t_2-orbitals. We consider those distributions which are relevant to hemoglobin.

$S = 0$. The electron distribution is t_2^6. This is a closed shell configuration in which each t_2-orbital is doubly occupied. From Eqs. (212) and (217), the contributions to q_v and to $(q\eta)_v$ sum to zero. Hence

$$(S = 0), \; q_v = (q\eta)_v = \begin{cases} 0 \text{ (tetragonal)} \\ 0 \text{ (rhombic)} \end{cases} \tag{219}$$

$S = 2$. The electron distribution is $t_2^4 e^2$. The two electrons in the e-orbitals have their spins parallel; hence they separately occupy the two e-orbitals, d_{z^2} and $d_{x^2-y^2}$. The contributions to q_v are both zero and the contributions to $(q\eta)_v$ cancel. Of the four electrons in the t_2-orbitals three must reside in separate orbitals which again give no net contribution to either q_v or $(q\eta)_v$. The fourth electron may reside in any of the three t_2-orbitals and it is the sole contributor. Upon taking the statistical average over the t_2-orbitals we get

$$(S = 2), \; q_v = \begin{cases} \frac{4}{7} < \dfrac{1}{r^3} > \dfrac{1}{Z_t} \left[1 - e^{-\delta/kT}\right] \text{ (tetragonal)} \\[2mm] \frac{4}{7} < \dfrac{1}{r^3} > \dfrac{1}{Z_r} \left[1 - \frac{1}{2} e^{-\Delta_1/kT} - \frac{1}{2} e^{-\Delta_2/kT}\right] \text{ (rhombic)} \end{cases} \tag{220}$$

and

$$(S = 2), \ (q\eta)_v = \tfrac{6}{7} < \frac{1}{r^3} > \frac{1}{Z_r} \left[e^{-\Delta_1/kT} - e^{-\Delta_2/kT} \right] \text{ (rhombic)} \qquad (221)$$

in which

$$Z_t = 1 + 2e^{-\delta/kT}$$

$$Z_r = 1 + e^{-\Delta_1/kT} + e^{-\Delta_2/kT} \qquad (222)$$

Combining Eqs. (220) and (221)

$$q_v \left(1 + \tfrac{1}{3} \eta_v^2 \right)^{\tfrac{1}{2}} = \tfrac{4}{7} < \frac{1}{r^3} > \frac{1}{Z_r} \Big[1 + e^{-2\Delta_1/kT} + e^{-2\Delta_2/kT}$$

$$(S = 2), \qquad -e^{-\Delta_1/kT} - e^{-\Delta_2/kT} - e^{-(\Delta_1 + \Delta_2)/kT} \Big]^{\tfrac{1}{2}} \qquad (223)$$

for a ligand field with a rhombic component.

$(S = \tfrac{1}{2})$. The electron distribution is t_2^5. Again in this case the three electrons which occupy separate t_2-orbitals give no net contribution. The remaining two electrons may be distributed as follows:

$$(d_{xy})^1 \ (d_{zx})^1; \ (d_{xy})^1 \ (d_{yz})^1; \ (d_{zx})^1 \ (d_{yz})^1$$

The statistical average over these electronic configurations gives

$$(S = \tfrac{1}{2}), \ q_v = \begin{cases} \tfrac{4}{7} < \frac{1}{r^3} > \frac{1}{Z_t'} \left[1 - e^{-\delta/kT} \right] & \text{(tegragonal)} \\ & \text{(rhombic)} \\ \tfrac{2}{7} < \frac{1}{r^3} > \frac{1}{Z_r'} \left[1 + e^{-(\Delta_1 - \Delta_2)/kT} - 2e^{-\Delta_1/kT} \right] & \end{cases} \qquad (224)$$

$$(S = \tfrac{1}{2}), \qquad (q\eta)_v = \tfrac{6}{7} < \frac{1}{r^3} > \frac{1}{Z_r'} \left[-1 + e^{-(\Delta_1 - \Delta_2)/kT} \right] \qquad (225)$$

in which

$$Z_t' = 2 + e^{-\delta/kT} \qquad (226)$$

$$Z_r' = 1 + e^{-(\Delta_1 - \Delta_2)/kT} + e^{-\Delta_1/kT} \qquad (227)$$

The expression analogous to Eq. (223) now becomes

$$q_v \left(1 + \tfrac{1}{3} \eta_v^2 \right)^{\tfrac{1}{2}} = \tfrac{4}{7} < \frac{1}{r^3} > \frac{1}{Z_r'} \Big[1 + e^{-2\Delta_1/kT} - e^{-\Delta_1/kT}$$

$$+ e^{-2(\Delta_1 - \Delta_2)/kT} - e^{-(\Delta_1 - \Delta_2)/kT}$$

$$(S = \tfrac{1}{2}), \qquad - e^{-(2\Delta_1 - \Delta_2)/kT} \Big]^{\tfrac{1}{2}} \qquad (228)$$

$S = \frac{5}{2}$. The electron distribution is $t_2^3 e^2$. The five electrons are in separate orbitals and the sum of their contributions to q_v and to $(q\eta)_v$ vanishes.

In all cases, when the rhombic component collapses so that

$$\Delta_1 = \Delta_2 = \delta$$

the expressions reduce to those for tetragonal symmetry.

3. Contributions from Ligands

There are two ways in which we might calculate the contributions from the ligands. The first depends on an estimate of the charge distribution based on the coefficients of a molecular orbital calculation. Using known or assumed structural information to provide distances and angles a direct computation of the EFG parameters may be performed. The second method is based on a knowledge of the tetragonal and rhombic splittings which may, perhaps, be deduced from electron spin resonance data.

For the first method we obtain the appropriate expressions simply by summing terms of the form of Eq. (205) and (213). Thus

$$V_{zz} = \sum_{i=1}^{N} Z_i e \, \frac{3\cos^2\theta_i - 1}{r_i^3}$$

$$V_{xx} - V_{yy} = \sum_{i=1}^{N} Z_i e \, \frac{3\sin^2\theta_i \cos 2\varphi_i}{r_i^3} \tag{229}$$

where $Z_i e$ represents the charge on the i-th ligand and $(r_i, \theta_i, \varphi_i)$ are the position coordinates. For the second method we note that the tetragonal and rhombic components of the ligand field potential may be written

$$V_t + V_r = B_2^0 \, (3z^2 - r^2) + 3B_2^2 \, (x^2 - y^2)$$

in which higher order terms have been neglected. By direct differentiation we find

$$eq_1 = V_{zz} = 4B_2^0$$

$$e(q\eta)_1 = V_{xx} - V_{yy} = 12B_2^2 \tag{230}$$

It is seen that the tetragonal field does not contribute to η while the rhombic field does not contribute to q.

It has been shown that the cubic splitting Δ is related to the coefficient of the cubic term in the ligand field potential by Eq. (58). In analogous manner it is possible to relate the tetragonal splitting δ and the rhombic splitting μ to the coefficients B_2^0 and B_2^2 respectively. This is best accomplished by the operator equivalent method (40) which makes it possible to obtain matrix elements of V_t and V_r within the set of d-orbitals. It is found that

$$\delta = \frac{6}{7} < r^2 > B_2^0 \, e, \qquad \mu = \frac{12}{7} < r^2 > B_2^2 \, e \qquad (231)$$

From Eqs. (218) we then obtain

$$\Delta_1 = \delta - \tfrac{1}{2} \mu = \tfrac{6}{7} < r^2 > (B_2^0 - B_2^2) \, e$$

$$\Delta_2 = \delta + \tfrac{1}{2} \mu = \tfrac{6}{7} < r^2 > (B_2^0 + B_2^2) \, e \qquad (232)$$

and from Eq. (230)

$$q_1 = \frac{14}{3 \, e^2} \frac{\delta}{< r^2 >}, \qquad (q\eta)_1 = \frac{7}{e^2} \frac{\mu}{< r^2 >} \qquad (233)$$

C. Magnetic Interaction

The third type of interaction which enters into the interpretation of Mossbauer spectra is the interaction of the nucleus with a magnetic field. Although the field may be due to external sources we are primarily interested in magnetic fields due to electronic spin alignments. We shall suppose that the quadrupole interaction is also present so that a coordinate system is defined by the principal axis system of the EFG tensor. For a magnetic field arbitrarily oriented relative to this system of axes, the Hamiltonian for the magnetic interaction is

$$\mathscr{H}_m = -\vec{\mu}_n \cdot \vec{H} \qquad (234)$$

in which $\vec{\mu}_n$ is the nuclear magnetic moment operator,

$$\vec{\mu}_n = \gamma_n \hbar \vec{I} = \left(\frac{\mu_n}{I}\right) \vec{I} \qquad (235)$$

μ_n is defined by

$$\mu_n = < I \, I | \mu_z | I \, I >, \qquad (236)$$

γ_n is the gyromagnetic ratio which is related to the Larmor precession frequency $\vec{\omega}_L$ by

$$\vec{\omega}_L = - \gamma_n \vec{H}, \tag{237}$$

$\vec{H}$ is an effective magnetic field which represents the magnetic interaction between the nucleus and the surrounding electrons. For a single electron it is given by

$$\vec{H} = - g\beta \left[\frac{\vec{\ell}}{r^3} + \left(\frac{3\vec{r}\,(\vec{s}\cdot\vec{r})}{r^5} - \frac{\vec{s}}{r^3} \right) + \frac{8\pi}{3}\,\vec{s}\,\delta\,(\vec{r}) \right] \tag{238}$$

where the symbols have their usual meanings. When substituted into Eq. (234), the first term in $\vec{H}$ represents the interaction of the nuclear magnetic moment with the current established by the orbital motion of the electron; the second and third terms give the dipolar interaction between the nuclear and electronic spins; the third term containing the δ-function is a contact term which gives a non-vanishing contribution only for s-electrons. On the other hand when the contact term contributes, the other terms in Eq. (238) do not. When more than one electron is present $\vec{H}$ is obtained by summing over all the electrons outside of closed shells.

The Hamiltonian in Eq. (234) may be written in terms of θ and φ, the polar and azimuthal angles of $\vec{H}$ relative to the principal axes of the EFG tensor.

$$\mathscr{H}_m = - \gamma_n \hbar\, H[I_x \sin\theta\cos\varphi + I_y \sin\theta\sin\varphi + I_z \cos\theta]$$

With the help of the identity

$$\frac{\sin\theta}{2}\,(e^{-i\varphi}\,I_+ + e^{i\varphi}\,I_-) \equiv I_x \sin\theta\cos\varphi + I_y \sin\theta\sin\varphi \tag{239}$$

the Hamiltonian becomes

$$\mathscr{H}_m = - \gamma_n \hbar\, H \left[\frac{\sin\theta}{2}\,(e^{-i\varphi}\,I_+ + e^{i\varphi}\,I_-) + \cos\theta\, I_z \right] \tag{240}$$

The matrix elements of $\mathscr{H}_m$ for $I_g = \frac{1}{2}$ and $I_e = \frac{3}{2}$ are given in Table 25. When $\vec{H}$ is parallel to one of the principal axes, say the z-axis, the matrix of $\mathscr{H}_m$ simplifies considerably and it is possible to give a closed expression for the combined quadrupole and magnetic field interaction.

117

M. Weissbluth

Table 25. *Matrix elements of the magnetic interaction.*

$$\mathcal{H}_m = -\gamma \hbar H \left[\frac{\sin\theta}{2} (e^{-i\varphi} I_+ + e^{i\varphi} I_-) + \cos\theta \; I_z \right]$$

$$I_g = \tfrac{1}{2} \cdot \qquad\qquad -\gamma_g \hbar H = \alpha$$

$\mathcal{H}_m$	$\frac{1}{2}$	$-\frac{1}{2}$
$\frac{1}{2}$	$\frac{\alpha}{2}\cos\theta$	$\frac{\alpha}{2}\sin\theta\, e^{-i\varphi}$
$-\frac{1}{2}$	$-\frac{\alpha}{2}\sin\theta\, e^{i\varphi}$	$-\frac{\alpha}{2}\cos\theta$

$$I_e = \tfrac{3}{2} \cdot \qquad\qquad -\gamma_e \hbar H = \beta$$

$\mathcal{H}_m$	$\frac{3}{2}$	$-\frac{1}{2}$	$-\frac{3}{2}$	$\frac{1}{2}$
$\frac{3}{2}$	$-\frac{3\beta}{2}\cos\theta$			$\frac{\sqrt{3}\,\beta}{2}\sin\theta\, e^{-i\varphi}$
$-\frac{1}{2}$		$-\frac{\beta}{2}\cos\theta$	$\frac{\sqrt{3}\,\beta}{2}\sin\theta\, e^{-i\varphi}$	$\beta\sin\theta\, e^{i\varphi}$
$-\frac{3}{2}$		$\frac{\sqrt{3}\,\beta}{2}\sin\theta\, e^{i\varphi}$	$-\frac{3\beta}{2}\cos\theta$	
$\frac{1}{2}$	$\frac{\sqrt{3}\,\beta}{2}\sin\theta\, e^{i\varphi}$	$\beta\sin\theta\, e^{-i\varphi}$		$\frac{\beta}{2}\cos\theta$

$$(I_g = \tfrac{1}{2}) \qquad E = \pm\, \frac{\gamma_g \hbar H}{2} \tag{241}$$

$$(I_e = \tfrac{3}{2}) \qquad E = \begin{cases} \dfrac{\gamma_e \hbar H}{2} \pm \dfrac{e^2 q Q}{4}\left[\left(1 + \dfrac{4\gamma_e\hbar H}{e^2 q Q}\right)^2 + \dfrac{\eta^2}{3}\right]^{\frac{1}{2}} \\[3mm] -\dfrac{\gamma_e\hbar H}{2} \pm \dfrac{e^2 q Q}{4}\left[\left(1 - \dfrac{4\gamma_e\hbar H}{e^2 q Q}\right)^2 + \dfrac{\eta^2}{3}\right]^{\frac{1}{2}} \end{cases} \tag{242}$$

in which (*7*)

$$\frac{\gamma_g}{2\pi} = 138 \text{ c/s}, \qquad \frac{\gamma_e}{2\pi} = -77 \text{ c/s}$$

118

D. Discussion

The predominant feature of practically all hemoglobins is that in some temperature region the quadrupole splitting (Δ) lies in the region of 0.20 to 0.24 cm/sec. This is illustrated in Figs. (14) and (15). The magnitude of the splitting appears to be independent of the ionization state (whether ferrous or ferric) or the spin state (whether high or low). Further, whereas the magnitude of Δ for ferrous hemoglobin falls within the range of splittings observed in other ferrous compounds, the splitting in ferric hemoglobin is very large compared with that in other ferric compounds.

We have seen that the quadrupole splitting depends on both q and $q\eta$ and that each of these terms may, generally, be written as a sum of a valence term and a ligand term. When $S = 0$ or $\frac{5}{2}$ only the ligands contribute; when $S = \frac{1}{2}$ or 2 both ligands and valence electrons must be included. These cases are now considered individually.

$S = 0$. We use the results of the molecular orbital calculation of *Ohno* et al. (45). An extension of Eq. (229) to include a dipole as one of the ligands gives

$$q_1 = \sum_{i=1}^{5} Z_i \frac{3\cos^2\theta_i - 1}{r_i^5} - \frac{6D_6}{er_6^3} \qquad (243)$$

Referring to the charge distribution calculated by *Ohno* et al. and the structural model Fig. 28, we have

i	Ligand	Z_i	r_i	θ_i
1 ... 4	$N_1 .. N_4$	-0.60	1.80A°	$\frac{\pi}{2}$
5	N_5	$+0.05$	2.10A°	π

D_6 is a dipole on the z-axis (No. 6 position) with the negative side facing the Fe atom; $r_6 = 2.06$ A° and the strength of the dipole is taken as 0.5 a. u. to give approximate correspondence to oxyhemoglobin. Substitution in Eq. (243) gives q_1; since the model (Fig. 28) has tetragonal symmetry, $(q\eta)_1 = 0$. Using the values (27)

$$Q = 0.29 \times 10^{-24} \text{ cm}^2$$

$$1 - \gamma_\infty = 12$$

the quadrupole splitting (Eq. 204) is computed to be

$$\Delta = 0.18 \text{ cm/sec}$$

$S = \frac{5}{2}$. As pointed out by *Ohno* et al. *(45)* the molecular orbital calculation for the ferric case lacks internal consistency. Hence we turn to the information based on ESR experiments. As has been discussed in Ch. VII, the zero field splitting parameter which appears in the spin Hamiltonian (Eq. 131) may be estimated from the orbital energies derived from ESR data on the low spin hemoglobin azide (Fig. 23). Using the values

$$\delta = 2060 \text{ cm}^{-1}, \ \mu = 1040 \text{ cm}^{-1}$$

and

$$<r^2> = 1.4 \text{ a. u. } (27)$$

we obtain, from Eqs. (233)

$$q = q_1 (1 - \gamma_\infty) = 2.54 \times 10^{24} \text{ cm}^{-3}$$

$$q\eta = (q\eta)_1 (1 - \gamma_\infty) = 1.92 \times 10^{24} \text{ cm}^{-3}$$

$$\Delta = 0.12 \text{ cm/sec} \tag{244}$$

$S = \frac{1}{2}$. The basic data used in the discussion of the $S = \frac{5}{2}$ case came from ESR experiments on hemoglobin azide for which $S = \frac{1}{2}$. We therefore take the ligand contribution as given by Eqs. (244); it remains then to estimate the valence part. For this purpose we employ Eq. (224) and (225) with

$$\Delta_1 = \delta + \tfrac{1}{2}\mu = 2580 \text{ cm}^{-1}$$

$$\Delta_2 = \delta - \tfrac{1}{2}\mu = 1540 \text{ cm}^{-1}$$

These energies are much larger than thermal energies; hence the Boltzmann factors may all be neglected and we have

$$q_v = \frac{2}{7} < \frac{1}{r^3} >, \qquad (q\eta)_v = -\frac{6}{7} < \frac{1}{r^3} >$$

With

$$< \frac{1}{r^3} > = 4.8 \text{ a. u.}, \quad 1 - R = 0.68$$

as given by *Ingalls (27)*, we obtain

$$(1 - R)\, q_v = 6.35 \times 10^{24} \text{ cm}^{-3}$$

$$q = (1 - R)\, q_v + (1 - \gamma_\infty)\, q_1 = 8.89 \times 10^{24} \text{ cm}^{-3}$$

$$(1 - R)\, (q\eta)_v = -19 \times 10^{24} \text{ cm}^{-3}$$

$$q\eta = (1 - R)\, (q\eta)_v + (1 - \gamma_\infty)\, (qn)_1 = -17.1 \times 10^{24} \text{ cm}^{-3}$$

$$\Delta = 0.58 \text{ cm/sec}$$

The calculation may be improved somewhat by using the lowest Kramers doublet Eq. (93) which consists of a mixture of orbitals and reflects the effect of spin-orbit coupling. Repeating the calculation with the doublet we obtain $\Delta = 0.54$ cm/sec.

$S = 2$. The appropriate expressions are given by Eqs. (220) and (221). Since we are using the same orbital model as previously, the Boltzmann factors are negligible and we have

$$q_v = \tfrac{4}{7} < \frac{1}{r^3} >, \quad (q\eta)_v = 0$$

Hence

$$q = (1 - R)\, q_v + (1 - \gamma_\infty)\, q_1 = 15.2 \times 10^{24} \text{ cm}^{-3}$$

$$q\eta = (1 - \gamma_\infty)\, (q\eta)_1 = 1.9 \times 10^{24} \text{ cm}^{-3}$$

$$\Delta = 0.66 \text{ cm/sec}$$

In summary, the estimates of the quadrupole splitting are

S	Δ
	cm/sec
0	0.18
2	0.66
$\frac{1}{2}$	0.54
$\frac{5}{2}$	0.12

For $S = 0$ the agreement is reasonable; for $S = 2$ and $1/2$ the estimate is too high by a factor of about 3; for $S = 5/2$ the estimate is too low. As pointed out by *Ingalls* (27) and others, there are considerable uncertainties in the parameters that enter into such calculations and it may well be that the discrepancies between the theoretical and experimental values are consequences of such uncertainties and that the basic approach may still be correct.[6] Nevertheless it is well to be on the alert for contributing factors which have not been taken into account in the above calculation. In particular, a good measurement of the zero field splitting parameter D would establish whether the orbital picture of Fig. 23 may be legitimately applied to the high spin case.

In ferrihemoglobin, a marked asymmetry in the quadrupole doublet appears when the temperature is lowered below room temperature. The high velocity component (Fig. 15) broadens first and then, upon further

---

[6] Substantially better agreement with experimental values has been obtained on the basis of recent molecular orbital computations (*M. Zerner, M. Gouterman,* and *H. Kobayashi*: Theoret. Chim. Acta 6, 363 [1966]). The importance of covalency effects has been emphasized by *G. Lang* and *W. Marshall*: Proc. Phys. Soc. *87*, 3 (1966).

reduction of the temperature, the entire spectrum acquires the appearance of a broad weak resonance. In contrast, the doublet in oxyhemoglobin (Fig. 14) is symmetric at $5°$ K; a small degree of asymmetry is observed in high spin ferrohemoglobin (Fig. 14).

Two mechanisms have been proposed to explain the appearance of an asymmetric doublet in randomly oriented substances with no magnetic ordering. One mechanism is based on the combination of the directional quantities — the angular distribution function of the magnetic dipole radiation and the Debye-Waller factor which becomes anisotropic in systems of lower than cubic symmetry. This mechanism predicts an asymmetry which should decrease as the temperature is lowered, in contradiction to the experimental observations in hemoglobin. The second mechanism is based on magnetic interactions described by the general Hamiltonian Eq. (234).

Whether such interactions occur or not depends on the relative magnitude of two characteristic lifetimes. These are the relaxation time of the electronic spins (τ_R) and the period of the nuclear Larmor precession (τ_L). τ_R may be regarded as the mean time between successive changes in electronic spin orientation. It is generally composed of both spin-lattice and spin-spin contributions of which the former are temperature-dependent and the latter are not. τ_L is related to the internal magnetic field by

$$\frac{2\pi}{\tau_L} = -\gamma|H|$$

where $|H|$ is the magnitude of the internal magnetic field Eq. (238).

In high spin Fe^{3+} there are no unpaired s-electrons and there is no orbital angular momentum. The internal magnetic field arises indirectly — the unpaired 3d-electrons, through the exchange interaction, interact with s-electrons of one spin differently than with s-electrons of opposite spin. As a result, electrons with spin "up" acquire a somewhat different spatial distribution compared with that for electrons with spin "down". The contact term in Eq. (238) may then be written (after integration of the δ-function)

$$\vec{H}_c = \frac{16\pi}{3}\,\beta\,\vec{s}\left[\sum |\,\psi_u(0)\,|^2 - \sum |\,\psi_d(0)\,|^2\right]$$

where $\psi_u(0)$ and $\psi_d(0)$ are the wave functions at the nucleus of -selectrons with spin "up" and "down" respectively. More generally, the internal magnetic fields in the various hemoglobins are mainly due to the contact term in Eq. (238); the dipolar terms in the Hamiltonian contribute relatively little because the environment is predominantly cubic. Experimental values for Fe^{3+} compounds lie in the range of 400 to 500 kgauss.

In paramagnetic substances, it is necessary to compare τ_R with τ_L. If $\tau_R < < \tau_L$ so that the orientation of the electronic spins fluctuates very rapidly, the internal magnetic field "seen" by the nucleus will average to zero and there will be no magnetic interactions. This appears to be the case with the ferrous hemoglobins and one observes unbroadened, nearly symmetric, quadrupole doublets. If on the other hand $\tau_R > > \tau_L$, magnetic hyperfine interactions can occur and one may expect line broadening and perhaps even a fully resolved hyperfine spectrum.

Blume (6) has analyzed the situation for Fe^{57}. He points out that there exists a range of relaxation times which influence the $\pm \frac{3}{2} \rightarrow \pm \frac{1}{2}$ transitions differently from the $\pm \frac{1}{2} \rightarrow \pm \frac{1}{2}$ transitions which means that one transition may be broadened due to magnetic effects while the other is still relatively unaffected. As the temperature is lowered and the relaxation time is increased sufficiently, both transitions are affected and both components of the quadrupole doublet are broadened. This effect is associated with an internal magnetic field parallel to the axis of a cylindrically symmetric electric field gradient; when the magnetic field has a component perpendicular to the symmetry acis, transitions may be induced among the magnetic substates of the nucleus. The lifetimes of these substates are no longer equal; the $\pm \frac{1}{2}$ lifetime will be shorter than that of the $\pm \frac{3}{2}$. By this mechanism the $\pm \frac{1}{2}$ state will be broadened relative to the $\pm \frac{3}{2}$ state.

Acknowledgement. This work was begun during a visit to the Institut de Biologie Physico-Chimique in 1965; the hospitality of Professors *B.* and *A. Pullman* is gratefully acknowledged.

This research was supported by the National Science Foundation under Grant NSF GB 3994 and by the Office of Naval Research under Contract Nonr 225(87).

References

1. *Abragam, A.:* L Effet Mössbauer. New York: Gordon and Breach 1964.
2. *Ballhausen, C. J.:* Introduction to Ligand Field Theory. New York: McGraw-Hill Book Co. 1962.
3. *Bearden, A. J., T. H. Moss, W. S. Caughey,* and *C. A. Beaudreau:* Proc. Nat. Acad. Sci. U. S. *53*, 1246 (1965).
4. *Bennett, J. E.,* and *D. J. E. Ingram:* Nature *177*, 275 (1956).
5. —, *J. F. Gibson,* and *D. J. E. Ingram:* Proc. Roy. Soc. (London) *A420*, 67 (1957).
6. *Blume, M.:* Phys. Rev. Letters *14*, 96 (1965).
7. *Braterman, P. S., R. C. Davies,* and *R. J. P. Williams:* Advances in Chem. Phys. 7, 359 (1964).
8. *Boyle, A. J. F.,* and *H. E. Hall:* Reports on Progress in Physics *25*, 441 (1962).

M. Weissbluth

9. *Corey, R. B.*, and *L. Pauling:* Proc. Roy. Soc. *B141*, 10 (1953).
10. *Drabkin, D. L.*, in: Haematin Enzymes. Eds. *J. E. Falk, R. Lemberg*, and *R. K. Morton.* New York: Pergamon Press 1961.
11. *Fruton, J. S.*, and *S. Simonds:* General Biochemistry. New York: John Wiley and Sons 1958.
12. *George, P.:* Biopolymers Symp. *1*, 45 (1964).
13. —, *J. Beetlestone*, and *J. S. Griffith:* Rev. Mod. Phys. *36*, 441 (1964).
14. *Gibson, J. F.*, and *D. J. E. Ingram:* Nature *180*, 29 (1957).
15. —, *D. J. E. Ingram*, and *D. Schonland:* Disc. Far. Soc. *26*, 72 (1958).
16. *Gonser, U.*, and *R. W. Grant:* Biophys. J. *5*, 768 (1965).
17. *Gouterman, M.:* J. Chem. Phys. *30*, 1139 (1959).
18. *Griffith, J. S.:* Proc. Roy. Soc. (London) *A235*, 23 (1956).
19. — Nature *180*, 30 (1957).
20. — Disc. Far. Soc. *216*, 81 (1958).
21. — The Theory of Transition Metal Ions. Cambridge: Cambridge University Press 1961.
22. — The Irreducible Tensor Method for Molecular Symmetry Groups. New Jersey: Prentice-Hall, Englewood Cliffs 1962.
23. — Biopolymers Symp. *1*, 35 (1964).
24. —, in: Molecular Biophysics, p. 191. Eds. *B. Pullmann*, and *M. Weissbluth.* New York: Academic Press 1965.
25. *Gurinovich, G. P., A. N. Sevchenko*, and *K. N. Solov'ev:* Soviet Physics Uspekhi *6*, 67; Usp. Fiz. Nauk *79*, 173 (1963).
26. *Heine, V.:* Group Theory in Quantum Mechanics. New York: Pergamon Press 1960.
27. *Ingalls, R.:* Phys. Rev. *133(A)*, 787 (1964).
28. *Ingram, D. J. E., J. F. Gibson*, and *M. F. Perutz:* Nature *178*, 906 (1956).
29. —, in: Paramagnetic Resonance, Vol II, p. 809. Ed. *W. Low.* New York: Academic Press 1963.
30. *Judd, B. R.:* Operator Techniques in Atomic Spectroscopy. New York: McGraw-Hill Book Co. 1963.
31. *Kendrew, J. C., G. Bodo, H. M. Dintzis, R. G. Parrish, H. W. Wyckoff*, and *D. C. Phillips:* Nature *181*, 662 (1958).
32. — *R. E. Dickenson, P. E. Strandberg, R. G. Hart, D. R. Davies, D. C. Phillips*, and *V. C. Shore:* Nature 185, 422 (1960).
33. — Science *139*, 1259 (1963).
34. *Koster, G. F., J. O. Dimmock, R. G. Wheeler*, and *H. Statz:* Properties of the Thirty-Two Point Groups. Cambridge, Massachusetts: M. I. T. Press 1963.
35. *Kotani, M.:* Supp. of the Prog. of Theoret. Phys. *17*, 4 (1961).
36. — Rev. Mod. Phys. *35*, 717 (1963).
37. — Biopolymers Symp. *1*, 67 (1964).
38. — Advances in Chem. Phys. *7*, 159 (1964).
39. *Lemberg, R.*, and *J. W. Legge:* Hematin Compounds and Bile Pigments. New York: Interscience 1949.
40. *Low, W.:* Paramagnetic Resonance in Solids, Solid State Physics, Sup. 2. New York: Academic Press 1960.
41. *Maling, J. E.*, and *M. Weissbluth* in Electronic Aspects of Biochemistry, p. 93. Ed. *B. Pullman.* New York: Academic Press 1964.
42. — — (to be published).
43. *Muirhead, H.*, and *M. F. Perutz:* Nature *199*, 633 (1963).
44. *Nielson, C. W.*, and *G. F. Koster:* Spectroscopic Coefficients for the p^n, d^n, and f^n Configurations. Cambridge, Massachusetts: M. I. T. Press 1963.

45. *Ohno, K., Y. Tanabe*, and *F. Sasaki:* Theoret. Chim. Acta (Berl.) *1*, 378 (1963).
46. *Pauling, L.*, and *C. D. Coryell:* Proc. Nat. Acad. Sci. U. S. *22*, 210 (1936).
47. *Perutz, M. F.* in Brookhaven Symposia in Biology No. 13, Protein Structure and Function, U. S. Dept. of Commerce, Office of Technical Services Springfield, Va., 1960, p. 165.
48. — *Rossman, M. G., A. F. Cullis, H. Muirhead, G. Will*, and *A. C. T. North:* Nature *185*, 416 (1960).
49. *Platt, J. R.:* J. Chem. Phys. *18*, 1168 (1950).
50. *Pullman, B., C. Spanjaard*, and *G. Berthier:* Proc. Nat. Acad. Sci. U. S. *46*, 1011 (1960).
51. *Rotenberg, M., R. Bivins, N. Metropolis*, and *J. K. Wooten, Jr.:* The 3-j and 6-j Symbols. Cambridge, Massachusetts: The Technology Press, Massachusetts Institute of Technology 1959.
52. *Schoffa, G.:* Advances in Chem. Phys. 7, 182 (1964).
53. *Slater, J. C.:* Quantum Theory of Atomic Structure, Vol II. New York: Mc-Graw-Hill Book Co. 1960.
54. *Slichter, C. P.:* Principles of Paramagnetic Resonance. New York: Harper and Row 1963.
55. *Stryer, L., J. C. Kendrew*, and *H. C. Watson:* J. Mol. Biol. *8*, 96 (1964).
56. *Sugano, S.*, and *R. G. Shulman:* Phys. Rev. *130*, 517 (1963).
57. *Tanabe, Y.*, and *S. Sugano:* J. Phys. Soc., Japan *9*, 766 (1954).
58. *Tinkham, M.:* Group Theory and Quantum Mechanics. New York: McGraw-Hill Book Co. 1964.
59. *Veillard, A.*, and *B. Pullman:* J. Theoret. Biol. *8*, 317 (1965).
60. *Wertheim, G. K.:* Mössbauer Effect. New York: Academic Press 1964.

(Received June 14, 1966)

Chlorophyll Triplet States

Some Theoretical Considerations on Triplet Formation

Dr. G. M. Maggiora[1] and Dr. L. L. Ingraham[2]

Department of Biochemistry and Biophysics, University of California, Davis, California, USA

Table of Contents

I. Introduction

Livingstone and *Owens* have shown that the triplet state is an essential intermediate in chlorophyll-sensitized photochemical reactions occurring in homogeneous solutions (*17*). It has not yet been satisfactorily proven whether or not the triple state participates in *in vivo* chlorophyll reactions, but nevertheless there is considerable evidence that the sequence of biochemical reactions in photosynthesis is initiated by a chlorophyll molecule in its triplet state (*5*).

---

[1] National Institutes of Health Predoctoral Fellow IFI-GM-23, 276.
[2] This research was supported in part by USPHS grant GM 08285.

126

It is not our purpose here to detail the many reactions which require triplet-state chlorophyll, but rather to investigate the process of triplet formation and to ascertain the "fitness" of chlorophyll in producing triplet states. We will only deal with triplet formation in solution. In order to provide a common basis for our discussion of "fitness" we have summarized some of the important concepts in section II.

Before delving into some of the theoretical aspects of the problem, it should be pointed out that photochemical reactions which are due to *primary processes, i. e.* reactions which directly involve interaction with an electronically excited state of a molecule, are dependent, among other things, on the energy and lifetime of the excited state.

Clearly, the lifetime of a particular excited electronic state is important with respect to its ability to react. Generally, the longer the lifetime, the greater the chance of reaction. The lifetime of the triplet state is greater than that of the singlet state, sometimes many times greater. For example, the singlet state can have a lifetime of 10^{-6} sec. while the corresponding triplet state can exist for as long as 10 sec. Obviously, due to its longer lifetime, the triplet population is generally much larger than that of the excited singlets, and consequently it is the principal reacting species.

The energy of the reactive species is also of prime importance. It can be shown that for simple electronic systems where *configuration interaction* is not important the singlet and triplet states occur in pairs, which due to electron-electron repulsion are split apart, the triplet state possessing the lower energy *(18)*.

The energy of the triplet state cannot be obtained from absorption spectra unless the energy of the singlet-triplet splitting is known. It is much easier to determine this energy from emission spectra. In order that an order of magnitude of the energy of these excited triplet states be obtained consider the data of *Fernandez* and *Becker* on the emission spectra of solutions of dry chlorophyll *a* and *b* *(10)*. They observed emissions at $\lambda = 7550$ A and $\lambda = 7330$ A for chlorophyll *a* and *b*, respectively. These emissions were assigned to transitions from the triplet state to the singlet ground state. The energy difference between the triplet state and ground state is obtained from the following relation

$$E = \frac{hc}{\lambda} = \frac{2.86 \times 10^5}{\lambda} \frac{Kcal}{mole}$$

where h is *Planck's* constant and c is the velocity of light. Table 1 shows a summary of the calculations for chlorophyll *a* and *b*. The importance of these calculations lies in the fact that if, indeed, it is the triplet state

G. M. Maggiora and L. L. Ingraham

which is reacting, approximately 40 $\frac{\text{Kcal}}{\text{mole}}$ of the needed energy for reaction is supplied in the form of the excited triplet species.

Table 1. *Wavelength and energy of long-lived chlorophyll a and b emissions (10).*

Chlorophyll	λ	$E = \dfrac{2.86 \times 10^5}{\lambda} \dfrac{\text{Kcal}}{\text{mole}}$
a	7550 A	37.9
b	7330 A	39.0

II. Theoretical Considerations

1. Electronic States and Transitions

Consider the ground state electronic configuration of formaldehyde[3].

$$(1s_O)^2\ (1s_C)^2\ (2s_O)^2\ (\sigma_{CH})^2\ (\sigma^*_{CH})^2\ (\pi_{CH})^2\ (n_O)^2\ (\pi^*_{CO})^0\ (\sigma^*_{CO})^0$$

We note that the above ground state is a singlet. In order that we obtain a more compact notation we shall explicitly list only those orbitals which participate in electronic transitions in the visible and UV portions of the spectrum. Therefore, for the above configuration we would write

$$S_0 = (\pi_{CO})^2\ (n_O)^2\ (\pi^*_{CO})^0$$

Fig. 1 shows the orbital shapes of interest, and Fig. 2 shows two important transitions, whose properties are summarized in Table 2 (see p. 130). The above considerations allow us to construct a simplified energy level diagram [Fig. 3]. The vibrational substates of the various excited electronic states shown in Fig. 3 have been omitted for clarity. Due to the high frequency of collisions among molecules in liquids or solids vibrational relaxation to the zero-point vibrational level of a particular electronic state is extremely rapid. Therefore, it is generally assumed that only molecules in their lowest vibrational levels exist long enough to be important photochemically.

[3] For an excellent discussion of molecular orbitals see reference (2). Reference (23) also provides a good discussion of molecular orbitals as well as other quantum mechanical topics important to photochemistry.

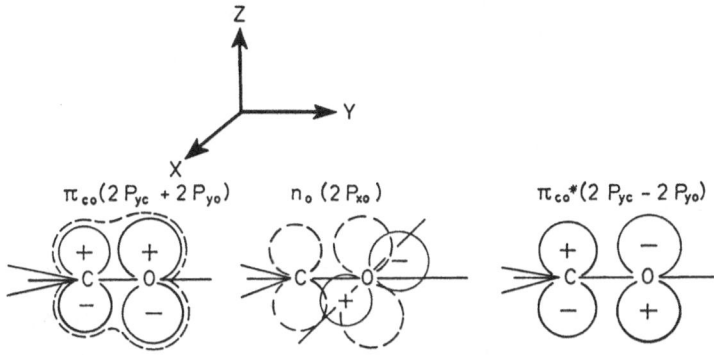

Fig. 1. Molecular orbitals of formaldehyde. Only those orbitals in the range of photo-chemical phenomena are shown.

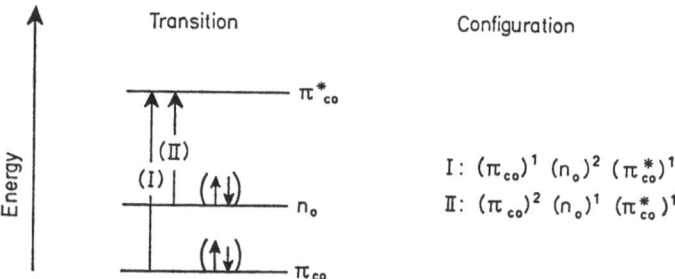

Fig. 2. Two important transitions: (I) $\pi \to \pi^*$ and (II) $n \to \pi^*$. The configurations of their excited states are shown.

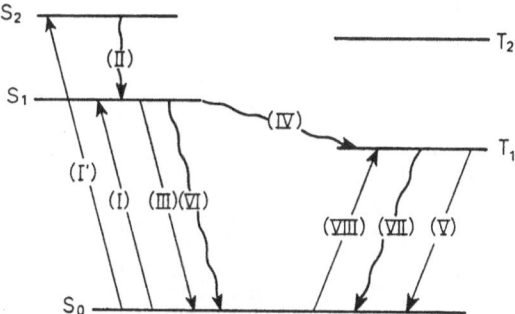

Fig. 3. The various pathways of energy absorption, emission, and radiationless transfer. See Table 3 on page 130 for detailed explanations of the various processes. Vibrational substates have been omitted for clarity.

Table 2. *Comparison of various properties of* $\pi \rightarrow \pi^*$ *and* $n \rightarrow \pi^*$ *transitions (21).*

Type of Transition (Fig. 2)	Designation of Transition	State Designation	Intensity	Conjugative Effects Lengthening Conjugation	Solvent Effects Increase Polarity	Effect of Electron Donating Substituents
I	$\pi \rightarrow \pi^*$	$^{2S+1}(\pi, \pi^*)$	Large	Red Shift	Red Shift	Red Shift
II	$n \rightarrow \pi^*$	$^{2S+1}(n, \pi^*)$	Small	Red Shift	Blue Shift	Blue Shift

* $2S + 1$ is the spin multiplicity of the state.

Table 3. *Important absorption, emission, and radiationless processes of molecular photochemistry [see also Fig. 3].*

Transition	Type of Transition*	Designation of Transition	Approximate Lifetimes (sec)**
I	Absorption	$S_0 \rightarrow S_1$	10^{-15}
I'	Absorption	$S_0 \rightarrow S_2$	10^{-15}
II	Radiationless	$S_1 \rightsquigarrow S_2$	$10^{-11} - 10^{-13}$
III	Fluorescent Emission	$S_0 \rightarrow S_1$	$10^{-6} - 10^{-9}$
IV	Radiationless	$S_1 \rightsquigarrow T_1$	$10^{-7} - 10^{-11}$
V	Phosphorescent Emission	$S_0 \leftarrow T_1$	$10 - 10^{-3}$
VI	Radiationless	$S_0 \rightsquigarrow T_1$	$10^{-2} - 10^{-6}$
VII	Radiationless	$S_0 \rightsquigarrow T_1$	$1 - 10^{-3}$
VIII	Absorption	$S_0 \rightarrow T_1$	—

* This selection is only meant to be representative and does by no means exhaust the various types of transitions occurring due to optical excitation process.

** Lifetimes are meant only to be representative of the orders of magnitude one encounters.

The various energy transferring processes shown in Fig. 3 are defined as follows:

1. *Absorption:* absorption of radiation to produce an electronically excited state [(I), (I'), (VIII)].

2. *Fluorescent emission:* emission of radiation from the first excited singlet state to the ground state [(III)].

3. *Phosphorescent emission:* emission of radiation from the first excited triplet state to the ground state [(V)].

4. *Radiationless process:* conversion of one electronic state to another without absorption or emission of radiation. The energy transferred is usually in the form of heat energy — mainly vibrational energy. Radiationless processes can be further divided into *internal conversion* processes where the multiplicity of the state does not change [(II), (VI)] and *inter system crossing* processes where the multiplicity of the state does change [(IV), (VII)].

2. Molecular Wavefunctions[4]

Consider the complete wavefunction for the Ith molecular state

$$\Psi_I = \psi_i \varphi_{i'} \sigma_{i''} \tag{1}$$

where ψ_i, $\varphi_{i'}$, and $\sigma_{i''}$ represent the wavefunctions for the Ith orbital, vibrational, and spin states, respectively. If the spin state is a "pure" spin state then we can alternatively write Eq. (1) as

$$^{2s+1}\Psi_I = {}^{2s+1}\psi_i\varphi_{i'} \tag{1'}$$

where $2S + 1$ indicates the multiplicity of the nominal spin state[5].

We are *also* assuming that ψ_i is a product function of one electron wave functions

$$\psi_i = \chi_1(1)\,\chi_2(2)\ldots\chi_n(n) \tag{2a}$$

and that $\varphi_{i'}$ is also a product of vibrational wave functions

$$\varphi_{i'} = \sigma_1(q_1)\,\sigma_2(q_2)\ldots\sigma_n(q_n) \tag{2b}$$

where the q_i are normal coordinates of the vibrational motion. Writing the complete molecular wave function, Ψ_I, as the product of the orbital,

---

[4] Several excellent books which cover in detail the topics we are about to discuss are available. In particular, see references (*8, 13, 15, 16,* and *19*).

[5] In order that a consistent notation be used we shall denote *total* molecular states with capital letters, *e. g.* I and J. The Ith state will always be considered to be lower than the corresponding J^{th} states of the same multiplicity. Generally the Ith and J^{th} states will refer to the ground and first excited molecular states. The small letters, *i. e.* i, i', and i'', represent the orbital, vibrational, and spin substates of the total molecular states.

vibrational, and spin functions is only an approximation, which implies that there is no interaction between the orbital, vibrational and spin coordinates of the total wave function.

3. Transition Moment and Intensity

The transition moment for electric dipole radiation[6] between the Ith and J^{th} molecular states is given by

$$\vec{M}_{IJ} = \int \overline{\Psi}_I \vec{\mu} \Psi_J d\tau \equiv (\Psi_I \mid \vec{\mu} \mid \Psi_J)^{[7]} \tag{3}$$

where the dipole moment operator, $\vec{\mu}$, is,

$$\vec{\mu} = e \sum_k Z_k \vec{r}_k \tag{4}$$

$\vec{r}_k$ being the position vector of the kth particle, Z_k the charge on the kth particle, and e the charge of an electron. The probability of the transition from the Ith to J^{th} state is directly proportional to $|\vec{M}_{IJ}|^2$.

The intensity of absorbed radiation is proportional to the Einstein transition probability for stimulated absorption, B_{IJ}; while that for emitted radiation is proportional to the transition probability for spontaneous emission, A_{JI} [Eqs. (5) and (6)]. The intensities are also proportional to the frequency, ν, of absorbed or emitted radiation.

$$I_{abs} = K_{abs} \nu B_{IJ} \tag{5}$$

$$I_{em} = K_{em} \nu A_{JI} \tag{6}$$

K_{abs} and K_{em} are proportionality constants[8].
B_{IJ} and A_{JI} can be given in terms of $|\vec{M}_{IJ}|^2$ as,

$$B_{IJ} \propto |\vec{M}_{IJ}|^2 \tag{7}$$

$$A_{JI} \propto \nu^3 |\vec{M}_{JI}|^2 \tag{8}$$

Combining Eqs. (5), (6), (7), and (8) we obtain,

$$I_{abs} = K'_{abs} \nu |\vec{M}_{IJ}|^2 \tag{9}$$

$$I_{em} = K'_{em} \nu^4 |\vec{M}_{JI}|^2 \tag{10}$$

---

[6] There are, of course, other types of radiation which may also be important, for example, electric quadrupole and magnetic dipole radiation. The ratio of their intensities is given by $I_{dipole} : I_{quad} : I_{mag} = 1 : 10^{-5} : 10^{-5}$ (16). Therefore in this paper we shall only consider electric dipole radiation.

[7] Throughout this paper we shall use the Bra-Ket notation of Dirac as described in reference (13).

[8] The expressions for these as well as other constants of importance can be found in references (8, 13, 16, and 19).

where K'_{abs} and K'_{em} are proportionality constants. From Eqs. (9) and (10) it is apparent that the intensities are strongly dependent on the transition moment. The intensity of emitted radiation is, however, also strongly dependent on the frequency of emitted radiation.

4. Oscillator Strength

The oscillator strength, f_{IJ}, is also a quantity of interest in that it can be related to the *experimental extinction coefficient*, ε, by the relationship,

$$f_{IJ} = 4.32 \times 10^{-9} \int \varepsilon(\tilde{v}) d\tilde{v} \qquad (11)$$

$\int \varepsilon(\tilde{v}) d\tilde{v}$ is the integrated absorption band on the wave number, $\tilde{v}$, scale. If we assume the shape of the absorption band to be gaussian we obtain the following approximate relationship,

$$f_{IJ} \simeq 4.32 \times 10^{-9} \tilde{v}_{1/2} \, \varepsilon_{max} \qquad (12)$$

ε_{max} is the extinction coefficient at the wave number of maximum absorbance, and $\tilde{v}_{1/2}$ is the half-width of the band where $\varepsilon = {}^1/_2 \, \varepsilon_{max}$ (2). Furthermore, f_{IJ} is related to the transition moment by Eq. (13), K_f being a constant.

$$f_{IJ} = K_f \, |\vec{M}_{IJ}|^2 \qquad (13)$$

Through Eqs. (11) or (12) and Eq. (13) the relationship between the experimental extinction coefficients can be related to the quantum mechanical transition moment integral,

$$|\vec{M}_{IJ}|^2 = \frac{4.32 \times 10^{-9}}{K_f} \int \varepsilon(\tilde{v}) d\tilde{v} \qquad (14)$$

$$|\vec{M}_{IJ}|^2 \simeq \frac{4.32 \times 10^{-9}}{K_f} \tilde{v}_{1/2} \, \varepsilon_{max} \qquad (14')$$

Let us now examine some of the properties of $\vec{M}_{IJ}$ which lead us to a discussion of the *spectroscopic selection rules*. First recall [Eq. (4)]

$$\vec{\mu} = e \sum_k Z_k \vec{r}_k$$

which can be rewritten as

$$\vec{\mu} = e \sum_{k'} \vec{r}_{k'} - e \sum_{k''} Z_{k''} \vec{r}_{k''} \qquad (15)$$

Electronic dipole moment operator: $\vec{\mu}_e = e \sum_{k'} \vec{r}_{k'} \qquad (16a)$

Nuclear dipole moment operator: $\quad \vec{\mu}_n = e \sum_{k''} Z_{k''} \vec{r}_{k''}$ (16b)

Substitution of the above expressions into Eq. (3) gives[9],

$$\vec{M}_{IJ} = \{(\varphi_{I'}|\varphi_{J'})(\psi_i|\vec{\mu}_e|\psi_j) - (\psi_i|\psi_j)(\varphi_{I'}|\vec{\mu}_n|\varphi_{J'})\}(\sigma_{I''}|\sigma_{J''}) \quad (17)$$

In order that a conciseness of notation be obtained, we shall for the most part use an abbreviated Bra-Ket notation[10]. In the new notation Eq. (17) can be rewritten as

$$M_{IJ} = \{(i'|j')(i|\vec{\mu}_e|j) - (i|j)(i'|\vec{\mu}_n|j')\} \, (i''|j'') \quad (17')$$

In most cases of interest to us we will be dealing with electronic transitions, $i.$ $e.$, transitions between orthogonal electronic states ψ_i and ψ_j. Therefore $(i|j) = 0$, so that the second term in the parenthesis of Eq. (17') vanishes. Eq. (17') now becomes,

$$\vec{M}_{IJ} = (i'|j')(i|\vec{\mu}_e|j)(i''|j'') \quad (18)$$

From a consideration of the three terms in Eq. (18) the various selection rules can be obtained.

5. Vibrational Selection Rule

A knowledge of the Franck-Condon principle is necessary for an understanding of the factors which influence the magnitude of the vibrational overlap, $(i'|j')$. Simply stated, it says that *an electronic transition takes place so rapidly that the nuclei do not move appreciably during such a transition*[11]. As is evident from Fig. 4, such electronic transitions will take place "vertically".

---

[9] Note that $\vec{\mu}_n$ and $\vec{\mu}_e$ operate only on the nuclear and electric coordinates, respectively. The dipole moment operator *does not* operate on the spin coordinates. This fact, coupled with the fact that Ψ_I and Ψ_J are written as product functions of *independent* orbital, vibrational, and spin coordinates allows the factoring of Eq. (3) as seen in Eqs. (17) and (17').

[10] Using the designations given in the footnote on page 131, the abbreviated notation is as follows:

$$(\Psi_I|\vec{\mu}|\Psi_J) \equiv (I|\vec{\mu}|J)$$
$$(\psi_i|\vec{\mu}_e|\psi_j) \equiv (i|\vec{\mu}_e|j)$$
$$(\varphi_{I'}|\vec{\mu}_n|\varphi_{J'}) \equiv (i'|\vec{\mu}_n|j')$$
$$(\psi_i|\psi_j) \equiv (i|j)$$
$$(\varphi_{I'}|\varphi_{J'}) \equiv (i'|j')$$
$$(\sigma_{I''}|\sigma_{J''}) \equiv (i''|j'')$$

[11] Lifetime of an electronic absorption transition: $\tau_e \simeq 10^{-15}$ sec. Lifetime of a vibrational transition: $\tau_v \simeq 10^{-12}$ sec.

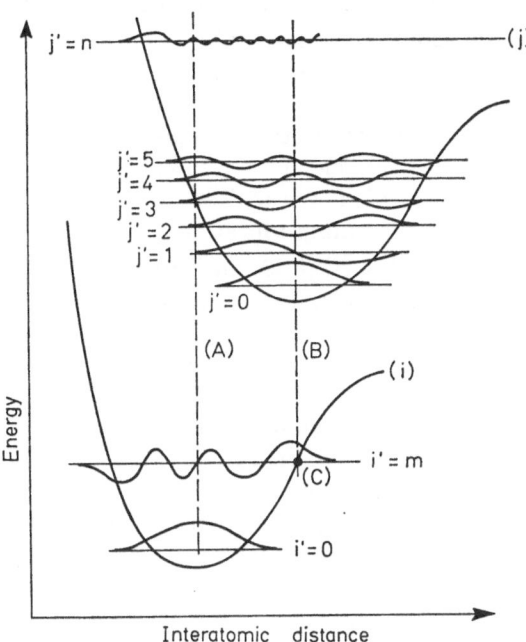

Fig. 4. Potential curves for the ith and j[th] electronic states of a diatomic molecule. Vibrational substates and their wavefunctions are also included. [Note: The shape of the potential curve for the jth state is usually flatter and its equilibrium internuclear distance is greater due to weaker bonding. A more detailed discussion of this will be found in the text.]

Fig. 4 shows the potential curves of the ground and excited states of a diatomic molecule. The excited state has a larger equilibrium internuclear distance and a "shallower" potential curve than the ground state. This might be expected if the excited state is a (π, π^*) state. In this case an electron is removed from a bonding π-orbital and placed in an antibonding π^*-orbital. This considerably weakens the bonding in the excited state and leads to the type of curve shown.

In polyatomic molecules multidimensional potential surfaces are needed, but the principles involved are the same as those for diatomic molecules. The bonding in polyatomic (π, π^*) states is, for reasons discussed above, much weaker than that in the ground state. (n, π^*) states, however, have bonding that is very similar to that in the ground state. (n, π^*) states are formed from electrons in non-bonding orbitals which are placed into antibonding π^*-orbitals, and therefore π-bonding is not weakened to too great an extent. Electronic repulsion effects are increased however, but their effects on the bonding are not as profound.

135

As is the case with electronic wave functions, vibrational wave functions which are solutions of the *same* vibrational *Hamiltonian* are orthogonal, *i. e.*, the vibrational states within each electronic energy level are mutually orthogonal. Only in the case where the vibrational *Hamiltonian* is the same for two different electronic states will the vibrational wave functions between the two electronic states satisfy the orthogonality conditions. Generally this is not true. The problem then is to determine which transitions give maximum vibrational overlap, so that the value $\vec{M}_{IJ}$ and hence I_{abs} and I_{em} [Eqs. (9) and (10)] will be maximum[12].

The vibrational integral can also be written in the following form,

$$(i'|j') \equiv \int \varphi_{I'}\varphi_{J'}dv \qquad (19)$$

the integration being over the vibrational coordinates[13].

The value of the product function $\varphi_{I'}\varphi_{J'}$, over the interval of integration is very important in determining the value of the vibrational overlap. If the value of $\varphi_{I'}\varphi_{J'}$ changes sign a number of times $(i'|j') \simeq 0$.

First let us consider vibrational transitions from $i' = 0$ to $j' = 0$, 1, 2, ..., n[14]. Fig. 4 shows that the transitions $(0 \rightarrow j')$, where $j' = 0$, 1, 2, 3, 4, 5, have sufficient vibrational overlap so that $(i'|j') > 0$. The intensities of the transitions, which are directly proportional to the vibrational overlap, are in order $(0 \rightarrow 1) > (0 \rightarrow 0) > (0 \rightarrow 2) > (0 \rightarrow 3) > (0 \rightarrow 4) > (0 \rightarrow 5)$. Fig. 5 shows the four most intense vibrational bands which contribute to the broad electronic band. In these transitions the value of $\varphi_{I'}\varphi_{J'}$ as a function of interatomic distance does not oscillate between plus and minus too often. In the case of the $(0 \rightarrow n)$ transition, however, $\varphi_{I'}\varphi_{J'}$ is seen to oscillate considerably, and consequently $(i'|j') \simeq 0$.

The same type of analysis can be applied to $(m \rightarrow j')$ transitions. Again examination of Fig. 4 shows that the $(m \rightarrow 0)$ transition possesses the greatest vibrational overlap. Transitions of type $(m \rightarrow j')$ with $j' > 0$ possess vibrational overlaps which decrease as j' increases. In cases where j' approaches n, $\varphi_{I'}\varphi_{J'}$ oscillates rapidly and $(i'|j') \simeq 0$.

In very rigid molecules (*e. g.* highly conjugated structures) the nuclear conformation in low lying electronic excited states is very similar to that of the ground state. Their potential curves as well as their equilib-

---

[12] Note that for the above statement to be true the values of $(i|\vec{\mu_e}|j)$ and $(i''|j'')$ must not be zero. The discussion of the conditions under which they are zero will be the subject of the next two selection rules.

[13] We will deal only with real vibrational and orbital wave functions in our analysis so that the complex function notation need not be used.

[14] Vibrational transitions from i' to j' will for conciseness of notation be written as $(i' \rightarrow j')$.

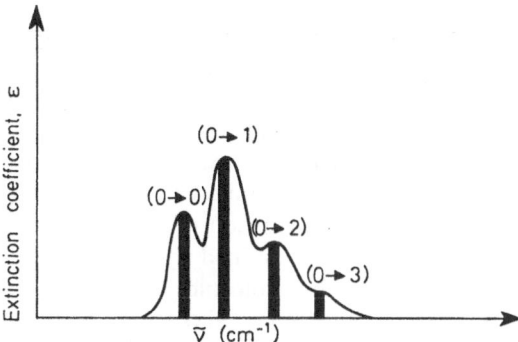

Fig. 5. Typical absorption band for an electronic transition showing some of the more prominent vibrational transitions which make up the fine structure of the broad electronic band.

rium internuclear distances would also be very similar. If they were identical, the vibration wavefunctions of the electronic states would be identical. In this case $(i'|j') = 0$ unless $i' = j'$. Under these circumstances only $(i' \rightarrow j')$ transitions where $i' = j'$ would be allowed. In the case where the potential curves are almost identical $(i' \rightarrow j')$ transitions where $j' = i'$, $i' \pm 1$ would also be allowed. As the excited state becomes more unlike the ground state $(i' \rightarrow j')$ transitions, in which i' differs greatly from j', will become more prevalent. These conclusions can also be reached by comparing the vibrational wavefunctions of the electronic ground and excited states as was done in the previous paragraph. When the potential curves are identical, it is an easy matter to show the orthogonality of the vibrational wavefunctions.

$(i' \leftarrow j')$ transitions in which vibrational energy is emitted can also be treated in the above manner. In cases where the nuclear conformation of the excited and ground electronic states are very similar $(i' \leftarrow j')$ where $i' = j'$, $j' \pm 1$ are the most strongly vibrationally allowed transitions.

From the discussion on page 135, it is clear that (n, π^*) states generally have a nuclear conformation which is closer to that of the ground state than do (π, π^*) states. Therefore, vibrational transitions to or from (n, π^*) states will usually adhere more closely to the $j' = i'$, $i' \pm 1$ selection rule for absorption and the $i' = j'$, $j \pm 1$ selection rule for emission than (π, π^*) states.

Fig. 5 shows a typical electronic absorption band with its corresponding vibrational sub-bands. The most intense vibrational transitions occur from the zero-point vibrational level of the lower electronic state to several vibrational levels of the upper electronic state. The reason that the main vibrational transitions *originate* from the zero-point vibrational level can be seen from the following argument. Vibrational levels of the lower electronic state of an ensemble of molecules have a Boltzman population distribution so that the zero-point level has the largest population; hence the intensity, which is also a function of the number of molecules undergoing transition, will be largest for $(0 \to j')$ transitions. The limitation on j' is dependent upon the factors previously discussed.

The strongest vibrational transitions which occur in emission bands are of the form $(i \leftarrow 0)$. The population of the vibrational levels of an excited electronic state is not determined by the Boltzman distribution, but rather by the fact that the molecules in higher vibrational levels undergo extremely rapid vibrational relaxation to the zero-point level [see page 128] so that essentially all molecules in excited electronic states are in their zero-point vibrational levels. One important difference between the intensity of absorbed and emitted radiation should again be noted. Both I_{abs} and I_{em} depend on $|\vec{M}_{IJ}|^2$ [Eqs. (9) and (10)], but I_{abs} depends on ν, while I_{em} depends on ν^4.

6. Space Symmetry Selection Rule

An understanding of how the dipole moment operator, $\vec{\mu}_e$, and the orbital wave function, ψ_i, affect the value of $(i|\vec{\mu}_e|j)$ is essential[15,16]. A consideration of Eq. (16) shows that we can write

$$(i|\vec{\mu}_e|j) = e \int \psi_i \sum_k \vec{r}_k \psi_j dV. \tag{20}$$

Suppose that we are dealing with an n electron i^{th} state, which for convenience we will assume to be the ground state. Let us assume that the electron excited during the transition from the ground state to the first excited state comes from the highest energy *one electron orbital* [see Eq. (2a)] of the ground state[17]. Therefore, we can write ψ_i and ψ_j as

$$\psi_i = \chi_1(1) \chi_2(2) \cdots \chi_{n-1}(n-1) \chi_n(n)$$

$$\psi_j = \chi_1(1) \chi_2(2) \cdots \chi_{n-1}(n-1) \chi_{n+1}(n) \tag{21}$$

[15] Note that the integral $(i|\vec{\mu}_e|j)$ can be written as $\int \bar{\psi}_i \vec{\mu}_e \psi_j dV$ [see Eq. (3)]. In all cases of interest to us in which we apply simple molecular orbital theory $\bar{\psi}_i$ can be written ψ_i since all wave functions used are real. If the function $\psi_i \vec{\mu}_e \psi_j$ is odd the integral is zero, while if $\psi_i \vec{\mu}_e \psi_j$ is even the integral is non-vanishing. Considerations of the "evenness" of the function do not, however, tell us the magnitude of the integral. Even though the integral is non-zero it could be very small.

[16] The powerful methods of Group Theory allow one to determine the vanishing or non-vanishing of integrals of the above form. For an excellent account of group-theoretical methods applied to chemistry see references *(14)* and *(20)*.

[17] Two electron transitions are forbidden except in cases where *configuration interaction* is present. For a proof of this statement, see *Jaffe (15,* page 123).

Substitution of Eq. (21) into Eq. (20) while noting that the one electron orbitals are orthonormal, gives upon simplification

$$(i|\vec{\mu}_e|j) = e \int \chi_n (n) \, \vec{r}_n \, \chi_{n+1} (n) \, dV_n \qquad (22)$$

The value of the electronic transition moment therefore depends on the value of the integral on the right hand side of Eq. (22) which can be rewritten as

$$\mathscr{I} \equiv \int \chi_n (n) \, \vec{r}_n \, \chi_{n+1} (n) \, dV_n \equiv \int \vec{r}_n \, \chi_n (n) \, \chi_{n+1} (n) \, dV_n \qquad (23)$$

The "order of magnitude" of $\mathscr{I}$ may sometimes be estimated by considering the value of the function

$$\Omega \equiv \chi_n (n) \, \chi_{n+1} (n) \qquad (24)$$

It should be noted that the vector $\vec{r}_n$ effects only the symmetry of $\mathscr{I}$ and not its magnitude. Therefore if Ω is small $\mathscr{I}$ generally will be small and vice-versa. Let us apply this reasoning to the case of formaldehyde discussed previously. For various states [see Figs. 1 and 2, and Table 2] we have

$$S_0 = \psi_i = (\pi_{CO})^2 \, (n_O)^2$$
$$^1(n, \pi^*) = \psi_j = (\pi_{CO})^2 \, (n_O)^1 \, (\pi_{CO}^*)^1$$
$$^1(\pi, \pi^*) = \psi_j' = (\pi_{CO})^1 \, (n_O)^2 \, (\pi_{CO}^*)^1 \qquad (25)$$

Substitutions of Eq. (25) into Eq. (20) gives integrals of the form of Eq. (22) which can be rewritten as Eq. (23).

$$\mathscr{I} \, (n \rightarrow \pi^*) = \int \vec{r} n_O \pi_{CO}^* \, dV \qquad (26a)$$

$$\mathscr{I} \, (\pi \rightarrow \pi^*) = \int \vec{r} \pi_{CO} \pi_{CO}^* \, dV \qquad (26b)$$

From an examination of Fig. 1 it is clear that $\Omega \, (n \rightarrow \pi^*)$ is approximately zero, while $\Omega \, (\pi \rightarrow \pi^*)$ is quite large. On this basis alone, providing that effects from vibrational or spin prohibition do not adversely affect the transition, the intensity of an $n \rightarrow \pi^*$ transition would be much less than the corresponding $\pi \rightarrow \pi^*$ transition. This is found to be the case [Table 3].

7. Spin Selection Rule

Probably the singly most important selection rule in triplet state chemistry is the spin selection rule. If we are dealing with "pure" spin states $\sigma_{i''}$ and $\sigma_{j''}$, then the integral $(i''|j'')$ rigorously satisfies the relationship

$$(i''|j'') = \delta_{i''j''}; \quad \delta_{i''j''} = \begin{matrix} 1 & i'' = j'' \\ 0 & i'' \neq j'' \end{matrix} \qquad (27)$$

139

In cases of interest to photochemistry the nominal spin states will be the singlet and triplet states. It is immediately obvious on inspection of Eq. (18) that if i″ ≠ j″, *i. e.* if the transition involves a change in spin state (singlet → triplet or triplet → singlet), then $\vec{M}_{IJ} = 0$. Hence, we have a *strong* spin selection rule. In many cases, however, pure spin states do not exist, but rather internal or external perturbations "mix" the nominal spin states producing states which contain both singlet and triplet components. The most important spin "mixing" perturbation-*spin-orbit-coupling* — will be discussed next[18].

8. Spin-Orbit Coupling

The spin-orbit *Hamiltonian* for a single electron can be written as (*4*, page 113)

$$\hat{H}_{so} = \frac{1}{2\,m^2\,c^2}\left(\frac{1}{r}\frac{dU\,(r)}{dr}\right)\vec{\ell}\cdot\vec{s} \tag{28}$$

For the many electron case Eq. (28) would simply become

$$\hat{H}_{so} = \frac{1}{2\,m^2\,c^2}\sum_i\left(\frac{1}{r_i}\frac{dU\,(r)}{dr_i}\right)\vec{\ell}_i\cdot\vec{s}_i \tag{29}$$

where m is the electronic mass, c is the velocity of light, r is the electronic distance, U (r) is the potential acting on the electron, and $\vec{\ell}$ and $\vec{s}$ are the orbital and spin angular momentum operators, respectively. In Eq. (29) the terms with subscripts refer to a particular electron. In most problems of interest to us we will only deal with an approximate one electron case which can be described by Eq. (28). If U (r) is approximately a central field potential we can write

$$U\,(r) = -\frac{Ze^2}{r} \tag{30}$$

where Z is the nuclear charge. Substituting Eq. (30) into Eq. (28) yields

$$\hat{H}_{so} = \frac{Ze^2}{2\,m^2\,c^2}\left(\frac{1}{r^3}\right)\vec{\ell}\cdot\vec{s} \tag{31}$$

---

[18] There are several types of spin-orbit coupling. We are interested only in that which takes place between the orbital and spin angular momentum of an electron due to its interaction with an attractive nuclear field of charge Z. Coupling between the orbital and spin angular momentum due to interaction with the repulsive field of another electron, and coupling between the spin angular momentum of one electron with the orbital angular momentum of another (spin-other-orbit interaction) will not be considered here. Their magnitudes are usually very small with respect to the first case mentioned.

The spin-orbit *Hamiltonian* of Eq. (31) is correct only for a bare nucleus. In the case of many-electron atoms where the nucleus is surrounded by a "core" of electrons, the electrostatic potential, U (r), changes more rapidly with r because of the rapid change in shielding by the core as we move closer to the nucleus, and consequently the value of $\dfrac{dU\ (r)}{dr}$ increases rapidly. Therefore, optical electrons[19] whose orbits "penetrate" into the core of atoms with large Z tend to have large spin-orbit interactions. In the case of paramagnetic atoms, the electron distribution is not spherically symmetrical as in the previously discussed cases. Paramagnetic atoms show extremely strong spin-orbit coupling. *Gouterman* has given an excellent theoretical discussion of spin-orbit coupling produced by various diamagnetic and paramagnetic metals in metalloporphyrins (*12*).

The problem of spin-orbit coupling in molecules is extremely complex, due mainly to the fact the potential function must take into account the potentials of all the nuclei. Two assumptions are made, however, that considerably simplify the problem:

1. The potential around each atom is considered to be due to essentially only that atom.

2. A large atom has a dominant effect on the magnitude of the spin-orbit interaction such that the effect on the spin-orbit interaction due to the other atoms can be neglected.

For molecules we are then able to write

$$\hat{H}_{so} = K_{so}\ \zeta\ \vec{l}\cdot\vec{s} \tag{32}$$

where K_{so} is a constant which depends on the type of molecule and ζ is the spin-orbit coupling parameter which depends on the perturbing atom. ζ is related to the $\left[\dfrac{1}{r}\ \dfrac{dU\ (r)}{dr}\right]$ term in Eq. (29). Typical values of ζ are presented in Table 4.

Table 4. *Z and ζ Values for various atoms (23).*

Atom	Z	ζ
Carbon	6	28
Nitrogen	7	78
Oxygen	8	152
Fluorine	9	272
Chlorine	17	587
Bromine	35	2,460
Iodine	53	5,060

[19] Optical electrons are electrons which can undergo optical transitions. In atoms the optical electrons are located in the outer, highest energy *atomic* orbitals. In the cases of interest to us in this paper, the optical electrons are located in the highest n and π *molecular* orbitals.

The difference between the electron distribution in (n, π^*) and (π, π^*) states is important in considerations of spin-orbit coupling. In the (π, π^*) state the optical electrons are highly *delocalized*; while in the (n, π^*) state the optical electron which remains in the n orbital is highly *localized*. If the localized n-electron is *on* or *near* a large atom spin-orbit coupling will occur. In order for spin-orbit coupling to occur in a (π, π^*) state the perturbing atom must be extremely large or paramagnetic due to the delocalized nature of the π and π^* electrons.

In cases where the perturbing atom is "semi-isolated" its importance lies mainly in its spin-orbit interaction and not in its bonding interactions. In these cases, the n, π, and π^* molecular orbitals which describe the optical electrons' motion do not directly include the semi-isolated atom. The chlorophyll molecule is a possible example of this, in that the atomic orbitals of the magnesium atom might not strongly interact with the n, π, and π^* molecular orbitals of the unsaturated ring. If, however, the molecular orbitals "penetrate" the core of the magnesium atom to a great enough degree, the optical electrons which undergo spin-orbit interaction can be treated as if they "belong" to the magnesium atom, and the ideas involved in our crude central field approximation can be used.

If the perturbing atom is attached directly to the molecule we speak of an *Internal Heavy Atom Effect*, whereas if the atom is part of a solvent molecule, we speak of an *External Heavy Atom Effect*. We shall be mainly concerned with the internal heavy-atom effect.

9. "Mixing" of Spin States

The effect of spin-orbit coupling on "mixing" of singlet and triplet states can be seen by considering first-order non-degenerate perturbation theory (8). Let us consider the *perturbed* triplet wavefunction for the Jth state where $^3\Psi_J^0$ represents the *unperturbed* triplet wavefunction for the Jth

$$^3\Psi_J = {}^3\Psi_J^0 + \Delta^1\Psi_J^0 \tag{33}$$

state and $^1\Psi_J^0$ represents the corresponding *unperturbed* singlet wavefunction for the Jth state. We have assumed for simplicity that only singlets and triplets of the same electronic configuration interact — e. g. ^{1}(n, π^*) and ^{3}(n π^*). The following analysis is identical whether or not more singlets are mixed, but the equations are slightly more cumbersome.

The mixing coefficient, Δ, is given by

$$\Delta = \frac{(^3\Psi_J^0|\hat{H}_{SO}|^1\Psi_J^0)}{|^3E_J - {}^1E_J|} \tag{34}$$

and determines the amount of singlet character that $^3\Psi'_J$ will possess. Clearly the value of the mixing coefficient is dependent on the spin-orbit interaction matrix element between the unperturbed singlet and triplet states and inversely dependent on the energy separation between the singlet and triplet states that mix. For a more convenient notation we shall denote $|\,{}^3E_J - {}^1E_J\,|$ by E_{ST}.

First, let us consider what factors influence E_{ST}. We have seen [page 127] that due to electron-electron repulsion the *paired* singlet and triplet states are split such that the triplet state is lower in energy. The energy difference is given by integrals of the form

$$E_{ST} \propto \left(\psi_{gr}\,|\,\frac{e^2}{r}\,|\,\psi_{ex}\right) \equiv e^2 \int \psi_{gr}\,\psi_{ex}\,\frac{1}{r}\,dV \tag{35}$$

where ψ_{gr} and ψ_{ex} are the spatial ground and excited state wavefunctions. $\psi_{gr}\psi_{ex}(\pi, \pi^*)$ is larger than $\psi_{gr}\psi_{ex}(n,\pi^*)$ due to the greater overlap in the (π, π^*) state over that in the (n, π^*) state. Therefore, it is clear that $E_{ST}(\pi, \pi^*) > E_{ST}(n, \pi^*)$. Of course singlet-triplet transitions may take place between different configurations so that the above argument is not always valid.

Second, let us consider the spin-orbit matrix element. If the operator form of $\hat{H}_{SO}$ [see Eq. (32)] is substituted into the integral we obtain[20].

$$(^3\Psi_J^0|\hat{H}_{SO}|^1\Psi_J^0) = K_{SO}\,\zeta\,(^3\Psi_J^0|\,\vec{\ell}\cdot\vec{s}|^1\Psi_J^0) \tag{36}$$

As was discussed previously [page 141 and Table 4] ζ is large when an optical electron is in an orbit with a high probability of being close to a nucleus of high Z ("penetration effect"). The assumption that in molecules the large perturbing atom has a spherically symmetrical field allows us to treat the $\vec{\ell}\cdot\vec{s}$ operator as if we were dealing with an atomic system. Since $\vec{\ell}\cdot\vec{s} = \ell_x s_x + \ell_y s_y + \ell_z s_z$, it can be shown that the s operators essentially "mix" the spin states while the ℓ operators act as rotation operators which operate on the spatial part of the wave function. For a

---

[20] The $\vec{\ell}\cdot\vec{s}$ operator does not operate on the vibrational coordinates of the complete molecular wave function. They can, therefore, be factored out so that a vibrational overlap integral is again obtained [see Eq. (18)]. We shall assume that this factor does *not* adversely affect the value of Eq. (36) and shall continue to use the *unfactored* total wavefunction with this consideration in mind. We have also assumed no vibronic coupling.

complete discussion of how the rotation operators effect the value of $(^3\Psi_J^0| \vec{\ell} \cdot \vec{s} |^1\Psi_J^0)$ see references *14* and *18*.

Let us now examine the effect that mixing of spin states has on the transition moment. Because the large energy difference between the ground and excited states makes the mixing coefficient very small we shall assume that the ground singlet state $^1\Psi_I$ can be represented by the unperturbed wave function.

$$^1\Psi_I \equiv {}^1\Psi_I^0 \tag{37}$$

Substitution of Eqs. (33) and (37) into Eq. (3) and simplifying gives

$$\vec{M}_{JI} = (^3\Psi_J^0|\vec{\mu}_e|^1\Psi_I^0) + \Delta \,(^1\Psi_J^0|\vec{\mu}_e|^1\Psi_I^0) \tag{38}$$

Expanding the integrals of Eq. (38) in the form of Eq. (18) shows clearly that the first integral vanishes due to the spin selection rule, while the second integral is not similarly effected because the transition is between singlet spin states. Therefore $\vec{M}_{JI}$ becomes

$$\vec{M}_{JI} = \Delta \,(^1\Psi_J^0|\vec{\mu}_e|^1\Psi_I^0) \tag{39}$$

The value of the integral can now be evaluated using previously discussed methods.

10. Radiationless Transitions

Radiationless transitions [see page 131, Fig. 3 and Table 3] as manifested in internal conversion and intersystem crossing processes play an important role in energy transfer processes. The most important radiationless transition from the standpoint of photochemistry is the intersystem crossing process, $S_1 \rightsquigarrow T_1$, which is necessary for population of the triplet state. This transition is generally represented as in Fig. 6a [see also Fig. 3]. $S_1 \rightsquigarrow T_1$ is actually an *isoenergetic* process followed by rapid relaxation of the excited vibrational substate of T_1 to its lowest vibrational substate. Since the vibrational relaxation is extremely rapid [see page 128], if δ is large enough the molecule will remain in the triplet state, because the thermal energy present is insufficient to raise it to the necessary excited vibrational state to perform the reverse process.

Consider what happens when radiation is absorbed by a diatomic molecule in its ground state [Fig. 7]. The transition from S_0 to S_1 leaves the molecule in an excited vibrational state of S_1. Then rapid vibrational relaxation to the zero-point vibrational level occurs. At point C the potential surfaces intersect[21]; the molecule at this point has the

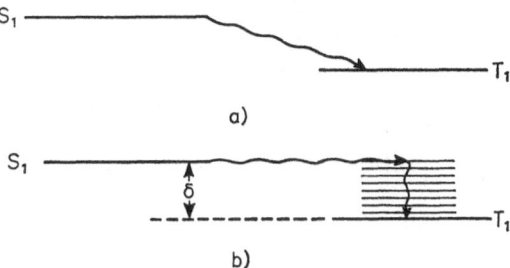

Fig. 6a and b. Radiationless transition $S_1 \rightsquigarrow T_1$. a) Usual depiction of $S_1 \rightsquigarrow T_1$. b) Actual processes occurring in $S_1 \rightsquigarrow T_1$.

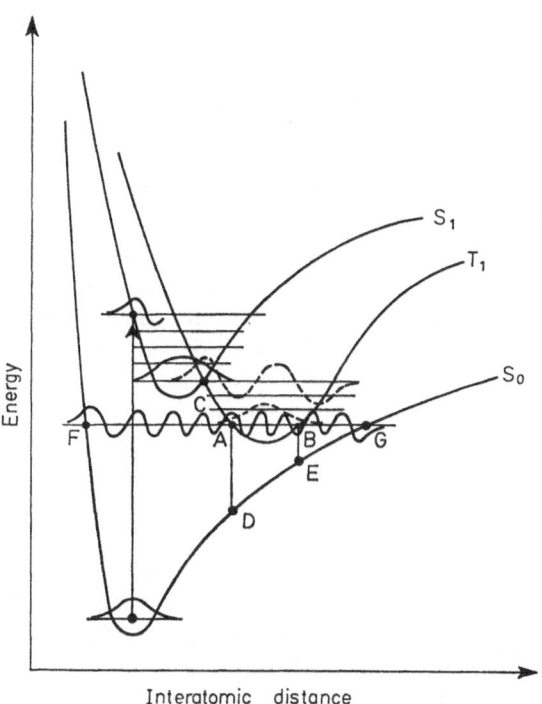

Fig. 7. Potential curves of ground and first excited singlet as well as first excited triplet states of a diatomic molecule. Some vibrational substates and their wave-functions are included. [Note: In actuality the potential curves *do not* intersect. For a comprehensive discussion of this phenomena see the section on non-adiabatic transitions of reference (16). For our purposes we will assume that the potential curves do, in fact, intersect.]

same geometry in the zero-vibrational level of S_1 as it does in an excited vibrational level of T_1. At this point a "crossing" from S_1 to T_1 is isoenergetic; however, it involves a change in spin state from singlet to triplet. Spin-orbit coupling, which mixes the singlet and triplet states, allows the molecule to "shuttle" back and forth between singlet and triplet spin states. The spin prohibition, however, slows down the process $S_1 \rightsquigarrow T_1$ by a factor of $\simeq 10^6$ with respect to other radiationless processes which do not require a spin state change.

The stronger the coupling, $i.\ e.$ the larger the mixing coefficient, the greater the probability of $S_1 \rightsquigarrow T_1$. In order to determine the quantum mechanical probability, P, of $S_1 \rightsquigarrow T_1$, we use the perturbation operator[22], $\hat{P}$, such that

$$P\ (S_1 \rightsquigarrow T_1) = (^1\Psi_1|\hat{P}|^3\Psi_1)^2 \tag{40}$$

$\hat{P}$ does not operate on the vibrational coordinates so that Eq. (40) can be rearranged to give

$$P\ (S_1 \rightsquigarrow T_1) = (^1\psi_1|\hat{P}|^3\psi_1)^2\ (j'\ (S_1)\ |\ j'\ (T_1))^2 \tag{41}$$

Therefore $P\ (S_1 \rightsquigarrow T_1)$ is directly proportional to $(j'\ (S_1)\ |\ j'\ (T_1))^2$. If we examine point C of Fig. 7, we see that $j'\ (S_1) = 0$ and $j'\ (T_1) = 3$. In this case the vibrational overlap is large as can be seen by observing the wavefunctions for the two vibrational substates at point C. From examination of the other intersystem crossing process, $S_0 \rightsquigarrow T_1$, it is clear that between points A and B the vibrational overlap is small. If we assume that the first integral of Eq. (41) does not differ markedly between $S_1 \rightsquigarrow T_1$ and $S_0 \rightsquigarrow T_1$ processes, then it is apparent that $P\ (S_1 \rightsquigarrow T_1) > P\ (S_0 \rightsquigarrow T_1)$ on the basis of vibrational overlap alone.

The same argument can be made for the $S_0 \rightsquigarrow S_1$ transition. The zero-point level of the S_1 potential curve would overlap with an extremely high vibrational state of S_0 in which the vibrational wave function would oscillate rapidly making the overlap between these vibrational states very small. Hence $P\ (S_1 \rightsquigarrow S_0)$ would also be very small.

In most cases of vibrational de-excitation in large rigid molecules we are dealing with $(i' \leftarrow 0)$ transitions where $i' >> 2$. As shown in section II.5 the vibrational overlap in these cases is very small. Hence from Eq. (41) it follows that $S_0 \rightsquigarrow S_1$ and $S_0 \rightsquigarrow T_1$ radiationless processes have a low probability of occurrence in large, rigid molecules.

---

[21] In actuality the potential curves *do not* cross. *Kauzman* (*16*, page 536) presents an excellent discussion of this phenomenon. For our purposes, we will assume that the potential curves do, in fact, cross.

[22] *Kauzman* (*16*, page 539) gives a detailed account of the probabilities of such nonadiabatic transitions.

This does not apply, however, to $S_1 \rightsquigarrow T_1$ processes. In large molecules with many nuclear degrees of freedom, there is a high probability that many crossings between the complex polydimensional potential surfaces may occur. As was shown, when potential surfaces intersect, the vibrational overlap at the points of intersection is relatively large. From Eq. (41) it follows that the probability of $S_1 \rightsquigarrow T_1$ processes is reasonably high provided that spin-orbit coupling is present. Therefore from the above, it follows that *large rigid molecules should primarily use fluorescent and phosphorescent pathways of de-excitation.* Experimentally this has been verified.

An analysis of the above processes in terms of the "classical" Franck-Condon Principle may be helpful. From the Franck-Condon Principle, we know that the nuclei do not change appreciably during an electronic transition. This implies that large amounts of electronic potential energy are not rapidly converted into kinetic energy. Consider first $S_1 \rightsquigarrow T_1$. The crossing point C is a turning point of motion for the vibrational substates of both S_1 and T_1. At this point the geometries of the two electronic states are identical. The kinetic energy at this point is zero for both S_1 and T_1. Therefore, except for the spin prohibition, this transition should occur, in that the Franck-Condon Principle is not violated. In the case of $S_0 \rightsquigarrow T_1$ the potential curves do not cross. At or near the turning points of motion for T_1, *i. e.* A and B, a change to S_0 would require either a gain in the amount of kinetic energy between AD and BE or a change in nuclear coordinates to F or G. Either of these occurrences would violate the Franck-Condon Principle so their probability of occurrence is very small. Of the two processes, $S_1 \rightsquigarrow T_1$ and $S_0 \rightsquigarrow T_1$ shown in Fig. 7, $S_1 \rightsquigarrow T_1$ is much more likely to occur.

The same type of analysis can be applied to internal conversion processes which occur between electronic states of the same multiplicity. If the potential surfaces cross, the probability of transitions of the type $S_1 \rightsquigarrow S_2$ is very large. The lifetimes of the processes are much shorter than those which are spin prohibited. If the potential surfaces do not cross then the probability of transition is greatly diminished.

One more point of importance should be emphasized. $S_1 \rightsquigarrow T_1$ transitions can take place from higher vibrational substates of S_1 if vibrational overlap with the vibrational substate of T_1 is sufficiently large. The fact, as mentioned earlier, that *intra*-state vibrational relaxation to the zero-point vibrational level occurs so rapidly, means that essentially all molecules in S_1 will be in the zero-point vibrational level. Therefore, radiationless transitions "tend" to occur from the zero-point vibrational level of the upper electronic state to a higher vibrational substate of the lower electronic state even though vibrational overlap between upper vibrational substates of S_1 and T_1 is good.

11. Lifetimes

Of primary interest in this paper is an understanding of how a large triplet state population is obtained, especially in chlorophyll. If we re-examine Fig. 3 it becomes apparent that many energy transfer processes compete for the excitation energy produced by the absorption of radiation. The problem of populating the triplet state, T_1, then becomes a kinetic problem in which the various energy transferring processes compete with one another. The faster the process, *i. e.* the shorter its lifetime, τ, the better it can compete with the other energy transferring processes present. Fig. 8 shows the de-excitation processes of primary interest. De-excitation processes from higher singlet and triplet states not pictured here are assumed to be essentially instantaneous with respect to those processes shown in Fig. 8 [see also Fig. 3 and Table 3]. In order to

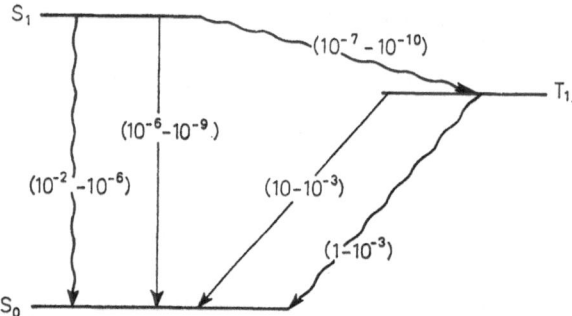

Fig. 8. Paths of radiative and radiationless de-excitation of S_1. Lifetimes (in seconds) of various processes are given in parentheses.

obtain a large triplet population $S_1 \rightsquigarrow T_1$ should be as fast as possible so that it can compete effectively with $S_0 \leftarrow S_1$. $S_0 \rightsquigarrow S_1$ is generally too slow to be a major pathway for de-excitation of S_1. Once we obtain molecules in T_1 the $S_0 \leftarrow T_1$ and $S_0 \rightsquigarrow T_1$ de-excitation processes should be slowed as much as possible in order that the population of T_1 increase. If the lifetime of T_1 is sufficiently long, the molecules in this state can be treated as distinct chemical entities which can undergo reaction. A quantitative discussion of the quantum mechanical basis of the lifetimes of the various processes is beyond the scope of this paper. However, we shall attempt to give some qualitative considerations.

The intrinsic radiative lifetime (*i. e.* the lifetime if no radiationless pathways of de-excitation exist) of a process is given by

$$\tau^0 = A_{JI}^{-1} \tag{42}$$

which when combined with Eq. (8) gives

$$\tau^0 = \frac{A_{em}}{|\vec{M}_{JI}|^2} \tag{43}$$

where A_{em} is a constant which contains ν^3.

Two cases of importance will be considered:

1. Transitions in which no change in spin multiplicity occurs.
2. Transitions in which a change in spin multiplicity occurs and spin-orbit coupling is present.

In case (1) $\vec{M}_{JI}$ is given by Eq. (18) where $(i''|j'') = 1$. Eq. (43) becomes

$$\tau^0 = \frac{A_{em}}{(j'|i')^2(j|\vec{\mu}_e|i)^2} \tag{44}$$

In case (2) $\vec{M}_{JI}$ is given by Eq. (39). In this case Eq. (43) is given by

$$\tau^0 = \frac{A_{em}}{\Delta^2(j'|i')^2(j|\vec{\mu}_e|i)^2} \tag{45}$$

Again the spin functions integrate to one because they are both of the same multiplicity.

In both cases the vibrational overlap factor is usually close to one so that its effect on τ^0 can be neglected. This can be seen if we examine Fig. 4. Recall that molecules in excited electronic states such as j are usually found in their $j' = 0$ vibrational substates. This is due to rapid vibrational relaxation from higher substates. Therefore, in electronic emission transitions from the jth to i^{th} electronic states only vibrational transitions of the form $(i' \leftarrow 0)$ need be considered. As can be seen from Fig. 4, the vibrational transition $(m \leftarrow 0)$ has good overlap such that $(0|m)$ will be close to one.

In section II.6 it was pointed out that the magnitude of $(j|\vec{\mu}_e|i)$ is dependent on the symmetry of ψ_j and ψ_i. For transitions between (n, π^*) and ground states the transition moment is close to zero, whereas it is large for transitions between (π, π^*) and ground states. Therefore from Eq. (44) of case (1) it follows that the lifetime of the $S_0 \leftarrow S_1$ (n, π^*) fluorescent transition is greater than that of the $S_0 \leftarrow S_1$ (π, π^*) transition.

The reverse is true for phosphorescent transitions $S_0 \leftarrow T_1$ (n, π^*) and $S_0 \leftarrow T_1$ (π, π^*). Unlike case (1) where a change in multiplicity does not occur T_1 (n, π^*) and T_1 (π, π^*) must undergo changes in multiplicity and hence belong to case (2).

From Eq. (45) it is clear that a large value of Δ, due to strong spin-orbit coupling, will greatly decrease τ^0. As was pointed out in section II.8 due to the highly delocalized nature of the optical electrons spin-

G. M. Maggiora and L. L. Ingraham

orbit coupling in $^3(\pi, \pi^*)$ is very small compared to that in $^3(n, \pi^*)$ states. Therefore, Δ will be larger in the $^3(n, \pi^*)$ state than in the $^3(\pi, \pi^*)$ state. Hence, this increase in Δ is large enough to compensate for the *corresponding* decrease in the transition moment [as discussed above] so that τ^0 becomes smaller for the $^3(n, \pi^*)$ state.

By combining Eqs. (8), (11), and (13) with Eq. (42) we can also write τ^0 in terms of the experimentally determinable extinction coefficient.

$$\tau^0 = K^0 \left(\frac{g_j}{g_i}\right) \frac{1}{\tilde{v}_a^2} \ [\int \varepsilon \ (\tilde{v}) \ d\tilde{v}]^{-1} \tag{46}$$

K^0 is a constant, (g_j/g_i) is the statistical multiplicity ratio between the ith and j^{th} states ($g_e = 1$ and 3 for singlet and triplet states, respectively), and $\tilde{v}_a$ is the *average* frequency of the transition (cm^{-1}). Using the same argument as that for Eq. (12) gives for Eq. (46),

$$\tau^0 \simeq K^0 \left(\frac{g_j}{g_i}\right) \frac{1}{\tilde{v}_a^2} \ [\tilde{v}_{1/2} \ \varepsilon_{max}]^{-1} \tag{47}$$

In the case of radiationless transitions there is no simple relationship such as Eq. (42). Discussion of the lifetimes of radiationless processes is very complicated and shall not be further considered. Suffice it to say that radiationless transitions occurring between states of the same multiplicity have considerably shorter lifetimes, than those occurring between states of different multiplicities. If, however, the spin-orbit coupling is strong, as with large or paramagnetic perturbing atoms, and the energy difference between the mixing states small, the mixing coefficient will be large and the spin prohibition lessened.

Intersystem crossing radiationless transitions usually take place from S_1 (n, π^*) states. There may be several reasons for this. First, the lifetime of the S_1 (n, π^*) state is much longer than that of S_1 (π, π^*), so that the intersystem crossing process can compete favorably with fluorescent de-excitation. Second, the highly delocalized nature of the optical electrons in the S_1 (π, π^*) state causes the spin-orbit coupling to be much smaller than in the more localized S_1 (n, π^*) state. Hence the spin prohibition is lessened to a greater extent in the S_1 (n, π^*) ⤳ T_1 process again allowing it to compete more favorably with fluorescent de-excitation.

Symmetry selection rules for radiationless transitions have also been derived by El-Sayed for nitrogen heterocyclics (7). He showed that to the first order no spin-orbit coupling occurs between singlet and triplet states of the same configuration, *i. e.*

$$(^1(n, \pi^*) \ |\hat{H}_{so}|^3 \ (n, \pi^*)) \simeq (^1(\pi, \pi^*) \ |\hat{H}_{so}|^3 \ (\pi, \pi^*)) \simeq 0$$

$$(^1(n, \pi^*) \ |\hat{H}_{so}|^3 \ (\pi, \pi^*)) \neq 0 \tag{48}$$

In chlorophyll the main perturbation is due to the magnesium atom which distorts the spatical symmetry of the nitrogen of the pyrrole, and hence the above selection rules may not apply.

12. Emission or Quantum Yields

Emission or quantum yields are very important quantities which denote the amounts of excitaton energy which are obtained from the various de-excitation processes. The quantum yield is defined.

$$\Phi = \frac{\text{number of quanta emitted}}{\text{number of quanta absorbed}} \tag{49}$$

The *total instrinsic* quantum yield (if no external deactivation occurs) for a molecule may then be described by

$$\Phi_F^o + \Phi_P^o + \Phi_{int} = 1 \tag{50}$$

where Φ_F^o is the intrinsic quantum yield of fluorescence, Φ_P^o is the intrinsic quantum yield of phosphorescence, and Φ_{int} is the quantum yield for internal thermal degradation. For most rigid molecules, or at least molecules in *rigid* solvents, $\Phi_{int} \simeq 0$, so that

$$\Phi_F^o + \Phi_P^o \simeq 1 \tag{51}$$

It is important to recognize that Φ_F^o and Φ_P^o are complementary in nature. For example, if $\Phi_F^o = 0.3$, then $\Phi_P^o = 0.7$. Even in *fluid* solutions where collisional deactivation is prominent the corresponding triplet has a probability of being excited of 0.7. Therefore if the triplet state is the photochemically reactive species the limiting quantum yield of the reaction is 0.7.

Finally, the relationship between the observed and intrinsic lifetimes of the various emission processes is given by

$$\tau = \left(\frac{\Phi}{\Phi^o} \right) \tau^o \tag{52}$$

Where τ is the observed lifetime and Φ is the observed quantum yield of the particular process. Φ^o and τ^o, as previously defined, represent the *intrinsic* quantities. The fact that τ differs from τ^o is due to bimolecular quenching from the solvent molecules. This is greatly reduced in rigid solvents.

G. M. Maggiora and L. L. Ingraham

III. Spectral Considerations

1. $\pi \to \pi^*$ and n $\to \pi^*$ Transitions

The spectral bands of chlorophyll [Fig. 9] arise from transitions very similar to those in other porphyrin molecules. The intense Soret bands located between 400 mμ and 500 mμ and the long wave bands in the red and near infrared both result from $\pi \to \pi^*$ transitions in the porphyrin ring in which the optical electrons move toward the periphery. The satellite bands of lower intensity represent transitions from vibrational substates (12). $\pi \to \pi^*$ transitions in which electrons located on the carbon atoms of the pyrrole move towards the periphery of the porphyrin ring give rise to the Soret band, while $\pi \to \pi^*$ transitions in which electrons flow from the pyrrole nitrogens to the peripheral carbon atoms lead to the long wavelength transitions [see Figs. (9) and (10)] (9).

n $\to \pi^*$ transitions play a most important role in the fluorescence and phosphorescence of chlorophyll. These transitions give rise to $^1(n, \pi)$ states that are important in the radiationless intersystem crossing processes which lead to the formation of triplets and concomitant quenching of fluorescence. These triplets are lost either by phosphorescence, radiationless deactivation, or reaction. Recall that in most systems measured in rigid glass solvents, where chemical reactions are not important,

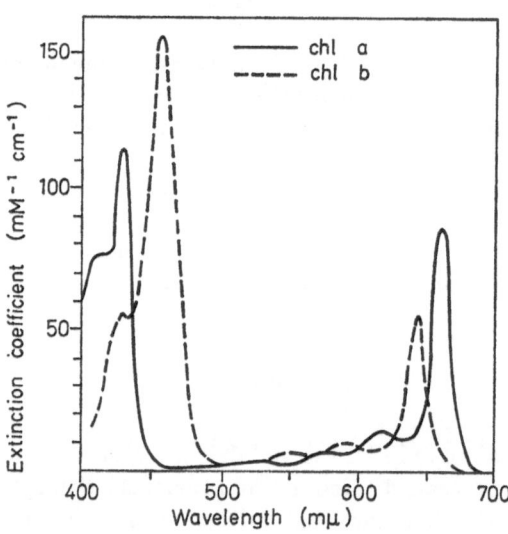

Fig. 9. Absorption Spectra of chlorophyll *a* and *b* in ether.

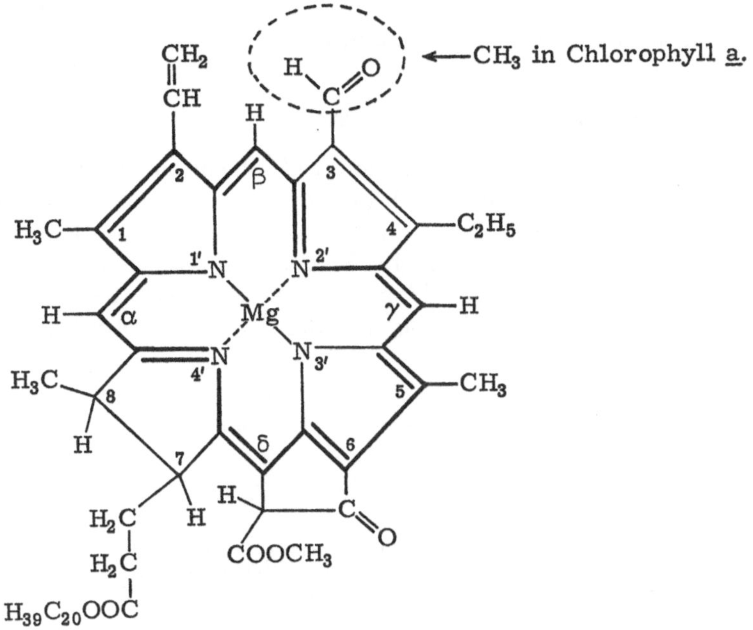

Fig. 10a and b. Two mesomeric forms of chlorophyll *b*. A methyl group replaces the carbonyl at position 3 in chlorophyll *a*.

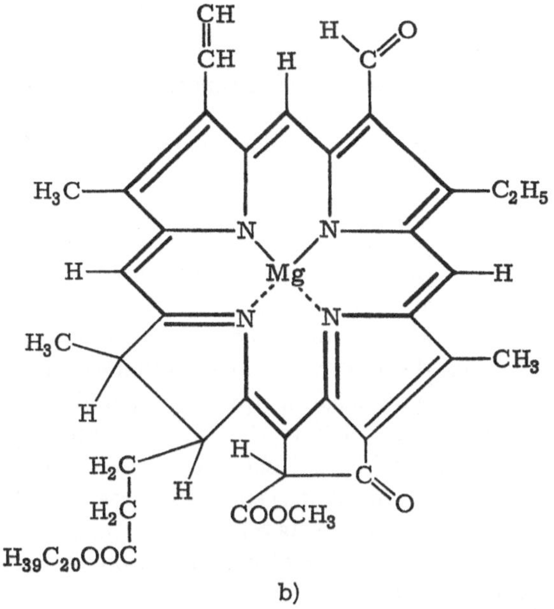

153

$\Phi_F^\circ + \Phi_P^\circ \simeq 1$, so that by measuring Φ_F° or Φ_P° we can determine the amount of intersystem crossing occurring.

2. Solvent Effects — Inversion of States

Many authors have discussed the effect of solvent on the lowest $^1(n, \pi^*)$ and $^1(\pi, \pi^*)$ states in chlorophyll (3, 10, 23). By increasing the polarity of the solvent [see Table 2] the $\pi \rightarrow \pi^*$ and $n \rightarrow \pi^*$ transitions undergo red and blue shifts, respectively. Therefore if the solvent polarity is changed sufficiently, it should be possible to invert the order of the $^1(n, \pi^*)$ and $^1(\pi, \pi^*)$ states [Fig. 11] and hence control Φ_F° and Φ_P°. A

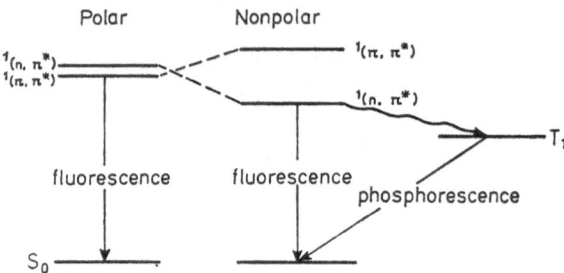

Fig. 11. Effect of polar and non-polar solvents on the energies of electronic transitions of molecules in which the $^1(n, \pi^*)$ and $^1(\pi, \pi^*)$ states are close together.

certain amount of spin-orbit coupling is necessary, however, in order that intersystem crossing processes may occur, i. e. we will not get complete quenching of fluorescence in non-polar solvents unless spin-orbit coupling is present.

Fernandez and *Becker* have made some interesting observations concerning $n \rightarrow \pi^*$ transitions and phosphorescence in chlorophyll a and b (10). They found that in non-polar hydrocarbon solvents the long wavelength side of the intense red ($\pi \rightarrow \pi^*$) band exhibited a shoulder. Phosphorescent emission was observed. When ethanol was added to the solvent the shoulder disappeared, and phosphorescent emission was no longer observed. These facts can be quite easily understood if we appeal to Fig. 11. In non-polar solvents the $^1(n, \pi^*)$ state is lower than the $^1(\pi, \pi^*)$ state as evidenced by the shoulder on the long wavelength side of the main red band. Phosphorescence would be expected in this case because intersystem crossing is favored by a $^1(n, \pi^*)$ state. When the polarity of the solvent is increased by the addition of ethanol the $^1(\pi, \pi^*)$ and $^1(n, \pi^*)$ states invert. The $^1(\pi, \pi^*)$ state is now lower in energy. This correlates with the loss of the shoulder as well as the quenching of phosphorescence in that the intersystem crossing process necessary to form the phosphorescent triplets does not usually occur from a $^1(\pi, \pi^*)$ state.

3. Non-bonding Electrons

Several authors feel that the above mentioned n $\rightarrow \pi^*$ transitions of chlorophyll result from non-bonding electrons residing on nitrogens 2' and 4' or 3' and 4' of the mesomeric forms shown in Fig. 10, rather than from non-bonding electrons residing on the carbonyl groups.

It may be possible to rule out carbonyl n $\rightarrow \pi^*$ transitions on the following grounds. Electron donating substituents cause n $\rightarrow \pi^*$ transitions on carbonyls to undergo blue shifts (15). In the $\pi \rightarrow \pi^*$ transition, which gives rise to the intense red band that is *almost* isoenergetic with the observed n $\rightarrow \pi^*$ transition, the electrons move towards the periphery of the ring in the direction of the carbonyl group. This is analogous to a powerful electron donating substitutent and would cause a large blue shift that would shift the carbonyl n $\rightarrow \pi^*$ transition considerably out of the energy range of interest. The opposite would be true for nitrogen n $\rightarrow \pi^*$ transitions. In fact, carbonyl groups conjugated to the nitrogen atoms would tend to lower the energy of these transitions. Thus, these two effects could be sufficient to place such transitions in the proper energy range.

It is interesting to note that Mg (II) etioporphyrins, which contain no carbonyl groups, exhibit phosphorescence (7). If the $S_1 \leadsto T_1$ process, necessary for phosphorescence originates from a S_1 (n, π^*) state the non-bonding electrons must be located on the pyrrole nitrogens. This, however, does not constitute definitive proof that non-bonding electrons on the pyrrole nitrogen exist, because it is possible for the $S_1 \leadsto T_1$ process to originate from an S_1 (π, π^*) state if strong spin-orbit coupling is present.

Franck has offered a possible mechanism which would explain the blue shift in polar solvents in terms of an n $\rightarrow \pi^*$ transition originating on the pyrrole nitrogens. He bases his mechanism on the following observations: (1) The sensitivity of the fluorescence (in benzene) to water is very high. A 1:1 stoichiometry between water and chlorophyll is sufficient for fluorescence to reappear. Furthermore, water must have a direct or indirect influence on the magnesium atom of chlorophyll because removal of the magnesium destroys the effect. (2) Small concentrations of water change the absorption spectrum of chlorophyll. Other polar molecules cause similar effects but only if they are present at much higher concentrations.

Franck interprets the high sensitivity to water as being a result of a water molecule becoming attached to two of the nitrogens by hydrogen bonding, which binds the water molecule relatively strongly. This

155

enhances the force by which the non-bonding electrons are held in their ground state, thus raising the energy between the ground and $^1(n, \pi^*)$ state— causing a blue shift. This has little effect on the (π, π^*) state due to the highly delocalized nature of the electrons. This also explains the fact that weaker hydrogen bonding molecules are required in greater concentration than water. Furthermore, the hydrogen bonding reduces the mesomerism of the magnesium bonds [Fig. 10a, b]. These bonds will now stay with one pair of nitrogen atoms which will attract the magnesium atom, and draw it away from the center of the porphyrin. Hence the charge distribution of the ring is changed. The excited electron in such an (n, π^*) state would therefore have less chance of penetrating the field of the magnesium atom and spin-orbit interaction should be concomittantly reduced and fluorescence increased.

Another suggestion has been made by *Clayton* that the water molecule bonds to the magnesium directly which in turn influences the non-bonding nitrogen electrons (5). This suggestion seems to be confirmed by nuclear magnetic resonance studies (6). It was found that in nonpolar solvents chlorophylls *a* and *b* exist as phosphorescent dimers. The bonding is between the carbonyl oxygens and the central magnesium atom. Chlorophyll *b* which possesses two carbonyl groups has been reported to be bound more tightly than chlorophyll *a*. Upon addition of methanol the dimerization is broken down and fluorescence is again observed. It was also shown that the methanol is bound to the magnesium atom. Infrared studies confirmed this finding. These findings can be consistently interpreted by assuming, as did *Franck*, that it is the non-bonding electrons of the pyrrole nitrogens, and not those of the carbonyl group, which undergo the n $\rightarrow \pi^*$ transition. In fact, the necessity of the magnesium is more logically explained.

This data, however, admits an alternative explanation in which the carbonyl group gives rise to the important n $\rightarrow \pi^*$ transition. The fact that in the phosphorescent dimer the carbonyl oxygen is bound tightly to the magnesium is significant. The proximity of the electron in the $^1(n, \pi^*)$ state "localized" in the non-bonding orbital of the carbonyl oxygen is such that spin-orbit coupling due to the magnesium atom could occur. When polar solvents are added two important effects take place: (1) The dimer is no longer formed so that spin-orbit coupling between electrons on the carbonyl oxygen and the magnesium is not possible, and (2) the carbonyl n $\rightarrow \pi^*$ transition undergoes a blue shift [see Table 2] due to the polarity increase of the solvent. The inversion of states plus the elimination of spin-orbit coupling due to the magnesium would cause an increase in fluorescence.

The definite assignment of which electrons undergo the n $\rightarrow \pi^*$ transition is still unresolved. The solution of this problem awaits more detailed experimental and theoretical analysis. Especially valuable would be further studies of the bonding between the magnesium atom and the pyrrole nitrogens. If the bonding between magnesium and all four nitrogens is largely covalent, non-bonding orbitals on the nitrogens would no longer be possible (*22*). This would give strong support to the postulate that the non-bonding electrons which undergo transition are, indeed, on the carbonyl oxygen.

IV. Speculations on the "Fitness" of Chlorophyll

Before proceeding it would be well to restate that our speculations on the "fitness" of chlorophyll relate principally to its triplet state reactions in solution. In determining its fitness, we should then ask what physical properties it possesses that make it such a good producer of long-lived triplets, which can act as intermediates in photochemical chlorophyll-sensitized reactions. From a cursory glance at the molecule [Fig. 10] four things are readily apparent:

1. The extensive conjugation of the ring allows electronic transitions in the range of absorbed visible radiation to take place.

2. The conjugation also helps to rigidify the ring.

3. The magnesium atom can promote spin-orbit coupling.

4. The carbonyl oxygen and pyrrole nitrogens contain nonbonding electrons.

It is of primary importance if chlorophyll is to function *in vivo* as an energy trapping source, that the energy of its excited electronic states should be within the energy range of visible radiation. Chlorophyll would hardly be able to play such a universal role in photosynthesis if its primary energy absorption were in a much less prevalent energy range. In chlorophyll-sensitized reactions visible radiation is the most important source of sensitizing radiation. The extensive conjugation of the ring allows e ectronic transitions to fall within this essential energy range.

The rigidity of the ring is also important. From previous discussions we have shown that in rigid molecules $\Phi_{int} \simeq 0$. This is quite important if we are to build up a large, long-lived triplet population. If $\Phi_{int} \neq 0$ the processes $S_0 \rightsquigarrow S_1$ and $S_0 \rightsquigarrow T_1$ would gain in importance, and the population of S_1 and T_1 would then be dissipated by radiationless as well as fluorescent and phosphorescent processes. This would be deleterious to the production of a large triplet population.

The spin-orbit coupling produced by the magnesium atom is essential, for without it multiplicity forbidden intersystem crossing processes would not readily occur. Experimentally this has been demonstrated by the fact

G. M. Maggiora and L. L. Ingraham

that porphyrins which do not possess a coordinated metal ion show only fluorescent emission, while metalloporphyrins show some phosphorescent emission (1). Therefore in the case in which no metal is present population of the triplet state by $S_1 \leadsto T_1$ would be forbidden and chlorophyll-sensitized reactions would not take place.

The question may then arise that if spin-orbit coupling is so important why not use a metal with a larger Z such as Zinc ($Mg:Z = 12; Zn:Z = 30$) or a paramagnetic metal such as Copper which is known to produce a much stronger spin-orbit coupling? The answer lies in the fact that even though increased spin-orbit coupling greatly enhances $S_1 \leadsto T_1$ it also enhances $S_0 \leftarrow T_1$. If the lifetime of the $S_0 \leftarrow T_1$ process is shortened, the triplet state may become depopulated faster than it can react.

Allison and *Becker* have elegantly shown the effect of different metals on the luminescent spectra of porphyrins (1). Their results plainly show that as the metal increases in Z Φ_F^o decreases as Φ_P^o increases, and the lifetime of the triplet state decreases. The same is true for paramagnetic metals but to a much greater degree. Therefore it seems that magnesium is perfectly suited for its role in triplet production.

Finally, the non-bonding electrons can form $^1(n, \pi^*)$ states advantageous for the $S_1 \leadsto T_1$ process necessary for triplet formation. Whether the non-bonding electrons on the pyrrole nitrogens, as opposed to those on the carbonyl oxygens, are primarily responsible for the formation of the $^1(n, \pi^*)$ state has not been definitively shown at this time.

From the above conclusions, it should become apparent that the "design" of chlorophyll is based on a series of physico-chemical compromises. The delicate balance of the various physical processes combine to make chlorophyll a highly efficient photosensitizing molecule. In this paper, we have tried to discuss some of the background necessary for an understanding of the formation of triplet states in chlorophyll, and how the various physical factors work together to promote the production of these triplets.

Acknowledgment. We wish to thank *Linda Maggiora* and *Charles Bowen* for their many helpful suggestions and for their patient reading of the manuscript.

V. Bibliography

1. *Allison, J. B.*, and *R. S. Becher:* Effect of Metal Atom Perturbations on the Luminescent Spectra of Porphyrins. J. Chem. Phys. *32*, 1410 (1960).
2. *Ballhausen, C. J.*, and *H. B. Gray:* Molecular Orbital Theory. New York: W. A. Benjamin, Inc. 1964.

3. *Becker, R. S.,* and *M. Kasha:* The Luminescence of Biological Systems, p. 25. Washington: American Association for the Advancement of Science 1955.
4. *Bethe, H. A.:* Intermediate Quantum Mechanics. New York: W. A. Benjamin, Inc. 1964.
5. *Clayton, R. K.:* Molecular Physics in Photosynthesis. New York: Blaisdell Publishing Company 1965.
6. *Closs, G. L., J. J. Katz, F. C. Pennington, M. R. Thomas,* and *H. H. Strain:* Nuclear Magnetic Resonance Spectra and Molecular Association of Chlorophylls *a* and *b*, Methyl Chlorophyllides, Pheophytins, and Methyl Pheophorbides. J. Am. Chem. Soc. *85*, 3809 (1963).
7. *El-Sayed, M. A.:* Spin-orbit Coupling and the Radiationless Process in Nitrogen Hetercyclics. J. Chem. Phys. *38*, 2834 (1963).
8. *Eyring, H., J. Walter,* and *G. E. Kimball:* Quantum Chemistry. New York: John Wiley and Sons 1944.
9. *Falk, J. E.:* Porphyrins and Metalloporphyrins, Vol. 2. B. B. A. Library. New York: Elsevier Publishing Co. 1964.
10. *Fernandez, J.,* and *R. S. Becker:* Unique Luminescence in Dry Chlorophylls. J. Chem. Phys. *31*, 467 (1959).
11. *Franck, J.:* Remarks on the Long Wavelength Limits of Photosynthesis and Chlorophyll Fluorescence. Proc. Nat. Acad. Sci. *44*, 941 (1958).
12. *Gouterman, M.:* Spectra of Porphyrin. J. Mol. Spectr. *6*, 138 (1961).
13. *Hanna, M.:* Quantum Mechanics in Chemistry. New York: W. A. Benjamin, Inc. 1965.
14. *Hochstrasser, R. M.:* Molecular Aspects of Symmetry. New York: W. A. Benjamin, Inc. 1966.
15. *Jaffe, H. H.,* and *M. Orchin:* Theory and Applications of Ultraviolet Spectroscopy. New York: John Wiley and Sons, Inc. 1962.
16. *Kauzman, W.:* Quantum Chemistry. New York: Academic Press 1957.
17. *Livingston, R.,* and *K. E. Owens:* A Diffusion-controlled Step in Chlorophyll-sensitized Photochemical Auto-oxidations. J. Am. Chem. Soc. *78*, 3301 (1956).
18. *McGlynn, S. P., F. J. Smith,* and *G. Cilento:* Some Aspects of the Triplet State. Photochemistry and Photobiology *3*, 269 (1964).
19. *Salem, L.:* Molecular Orbital Theory of Conjugated Systems. New York: W. A. Benjamin, Inc. 1966.
20. *Schonland, D. S.:* Molecular Symmetry. New York: D. Van Nostrand Company, Ltd. 1965.
21. *Sidman, J.:* Electronic Transitions Due to Non-bonding Electrons in Carbonyl, Aza-Aromatic, and Other Compounds. Chem Rev. *58*, 689 (1958).
22. *Storm, C. B., A. H. Corwin, R. Arellano, M. Martz,* and *R. Weintraub:* Stability Constants of Magnesium-Porphyrin-Pyridine Complexes: Solvent and Substitutent Effects. J. Am. Chem. Soc. *88*, 2525 (1966).
23. *Turro, N. J.:* Molecular Photochemistry. New York: W. A. Benjamin, Inc. 1966.

(Received January 24, 1967)

Chemistry and Structure of some Borate Polyol Compounds of Biochemical Interest

Dr. U. Weser[1]

Chemistry Department, Indiana University, Bloomington, Indiana, USA

Table of Contents

I. Introduction

The biological significance of boron in plant metabolism has been established over half a century ago and there is also a high probability that boron is strongly involved in mammalian metabolism. More information about the biological role of borate may be obtained from the review by *Zittle* (77). On the other hand, borate buffers are quite extensively employed in preparative and analytical biochemistry especially in the chromatography and electrophoresis of saccharides, nucleosides, nucleotides and mucoproteins. However, the molecular structure and chemistry of all these borate compounds has not always been studied carefully and were often neglected. The purpose of this review is therefore to place

---

[1] Present address: Physiologisch-Chemisches Institut der Universität Tübingen, Germany

emphasis on the molecular structure of boric acid and various poly-
borates and to discuss their ability to react with some polyols of biochemical significance.

II. Boric Acid: Structure and Reactivity

Although there are intensive studies on the structure and chemistry of
phosphoric and silicic acids, a comprehensive scope of the monomer and
polymer boric acids is still lacking (49). There is a tendency to compare
the chemistry of borate with the chemical behaviour of phosphate and
silicate. However, the only similarity which may be considered is the
formation of isopolyacids.

Monomer boric acid itself is stable as the free acid as well as in solution
and may be obtained as monomer even from concentrated solutions after
adding strong mineral acids. In the slightly hydrophobic boric acid
crystal the $B(OH)_3$ molecules are arranged in layers and tied together
over hydrogen bondings (Fig. 1). The distance between consecutive layers
is 3.18 Å. The layers themselves may be randomly displaced with respect
to each other, indicating that only van der Waals forces bind the layers
together (74, 75, 58).

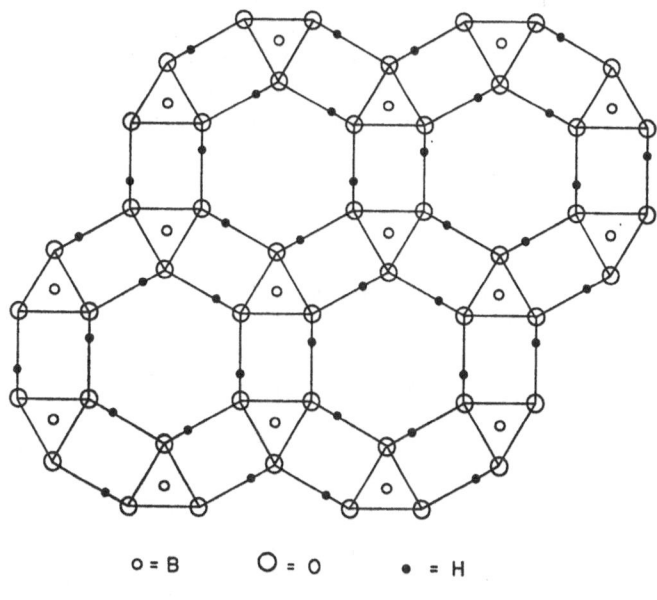

o = B   O = O   • = H

Fig. 1 (Ref. 75)

In aqueous solutions boric acid acts exclusively as a monobasic acid (34). Hydrated salts with more than one cationic charge per boron atom are unknown. This indicates that boric acid acts, not as a proton donor, but as a Lewis acid, accepting the electron pair of OH^- to form a tetrahedral anion (16, 20, 34). The weakness of boric acid is attributed to its reluctance to change from the stabilized mesomeric form, which is arranged trigonal and planar (B-O distance 1.37 Å (10)) into the tetrahedral structure where no mesomerism is possible (B-O distance 1.48 Å (10)) (Fig. 2). The low energy differences between structures with trigonal and tetrahedral boron and their ready interconversion explains the practically instantaneous establishment of equilibria in condensation and hydrolysis of boric acid and borates (21) as well as in the formation and hydrolysis of boric esters.

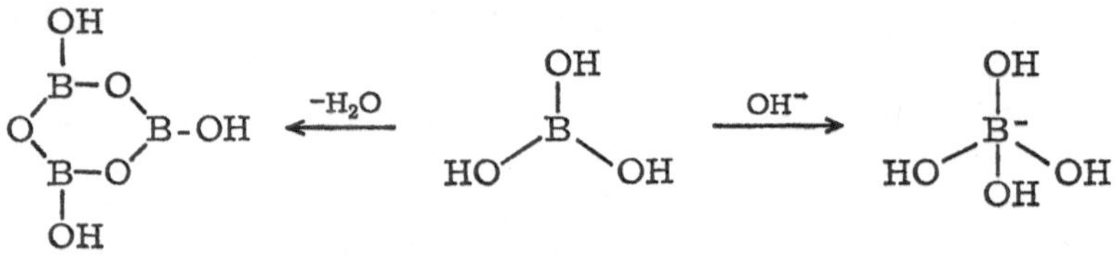

Fig. 2 (Ref. 16)

With increasing concentration of boric acid (> 0.025 M) the hydration-dissociation — equilibrium is shifted towards free $B(OH)_3$ molecules, which associate progressively. These macromolecules can be regarded as polyborate anions (34). Some polynuclear species, $B_3O_3(OH)_4^-$, $B_3O_3(OH)_5^{2-}$ and the very stable tetraborate ion, which is the most predominant dinegative ion in sodium borate solutions, have been detected (Fig. 3).

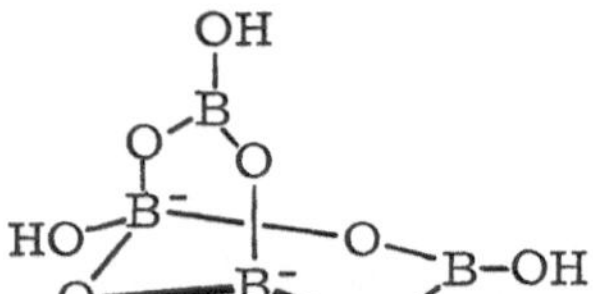

Fig. 3 (Ref. 16)

A. Boric Esters

Hermans (28) and later other workers (61) showed that boric esters are formed more readily from 1,3-diols than from 1,2-diols. Early stereochemical considerations (29) suggested that the different stabilities should

B—O: $B\langle\begin{smallmatrix}O\\O\end{smallmatrix}$ 2.36—2.39 Å   B—O: $B\langle\begin{smallmatrix}O\\O\end{smallmatrix}$ 2.40—2.44 Å

1.36—1.38 Å   1.47—1.41 Å

            trigonal   tetrahedral

be attributed to differences in the O—O distances for trigonal and tetrahedral boron, the first of which fits better with 1,3-diols, the latter better with 1,2-diols. Newer investigations (10) show that the two O—O distances are much closer and the difference most likely too small to account for such a strong effect. Studies of the relative stabilities of cyclic boric esters of 1,2-diols and 1,3-diols by means of determining the heat of reaction with amines indicate that the five membered ring of the cyclic boric ester of 1,2-diols must be strained with trigonal boron (32). The BOC angle of 120° in the planar boric acid system is mainly responsible for the strain (51). 1,3-diols, however, allow much better attachment of the planar boric acid to form an unstrained six membered ring. In biochemical reactions which principally take place in aqueous systems boric esters of 1,2-diols are expected to be very unlikely.

B. Borate Complexes

It has been shown that boric acid behaves exclusively like a monobasic acid and the anion has a tetrahedral structure. Chelate complexes with cis 1,2-diols are readily formed since the resulting 5-membered ring does not contain any strain (Fig. 4). In the presence of excessive ligands the predominant compound is the 1:2 complex.

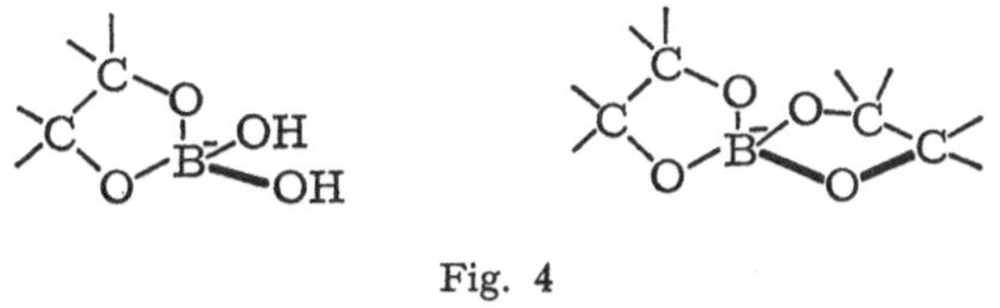

Fig. 4

1,3-diols form according to refs. (1, 2, 16, 32) borate complexes having six-membered non planar cyclohexane-like ring systems (Fig. 5). Proof of the puckered chair form with two types of substituents, axial and

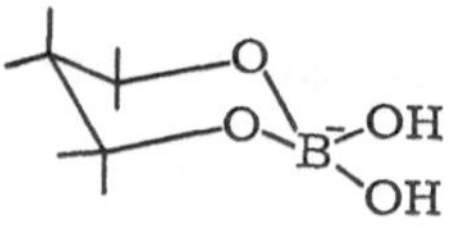

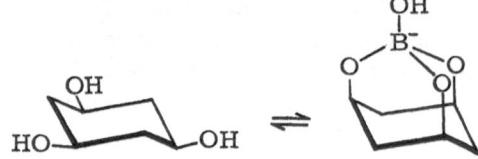

Fig. 5 (Ref. *16*)

equatorial, was obtained from the fact that racemic pentane-2,4-diol with one axial group forms a slightly less stable complex than the diequatorial mesomer. When the axial group is phenyl no complex is formed at all with the racemate.

Evidence for a cage like structure of a 1,3,5-triol boric acid complex gave the strong electrophoretic mobility of cyclitols containing three cis oriented hydroxyl groups (*2*). The formation of such a tridentate complex occurs by converting the more stable chair form (Fig. 6) (with equatorial hydroxyl groups) into the less stable chair form where the hydroxyl groups are arranged in axial positions.

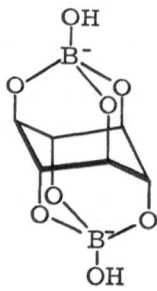

Fig. 6 (Ref. *2*)

A scyllitol diborate complex of tridentate structure has been reported by *Weissbach* (*70*), the structure is presented in Fig. 7.

Fig. 7 (Ref. *70*)

C. Determination Methods

Numerous procedures have been described to study the reaction of boric acid with polyols. Titrimetric methods employing conductometry (*8, 19*) and potentiometry (*2—6, 42a*) are applied to measure the stability and coordination number of the various boric complexes. Stereochemical considerations and differences in molecular size were investigated by means of chromatography or electrophoresis (7, 12, 13, 25, 26, 52, 69). For preparative work paper-, plate-, and column chromatography is extensively employed. Absorption spectroscopy has been successfully applied for the detection of boric complexes with aromatic ligands. Optical rotation (33, 37) and NMR studies (50) give also valuable information on polyol-borate interactions.

1. Potentiometry

Earlier studies (2, 8, 19, 65) appeared to be rather erratic. However, the differential potentiometric procedure (3—6, 42a) seemed to eliminate quite a number of errors.

The reactions studied are of the type

$$\text{HB} + n\text{D} \rightleftharpoons \text{H}^+ + \text{BD}_n^- \qquad (a)$$

HB; boric acid, D; saccharide;
The equilibrium constant is expressed by

$$K_n = [\text{H}^+] \, [\text{BD}_n^-] \, / \, [\text{HB}] \, [\text{D}]^n \qquad (b)$$

while the chelation reaction refers to

$$K_n' = [\text{BD}_n^-] \, / \, [\text{B}^-] \, [\text{D}]^n = K_n / K_1 \qquad (c)$$

K_n' is obtained by dividing K_n over K_1 (the first ionization constant of boric acid).

The numerical values of the constants K_n and K_n' may be calculated from the last equation and

$$\text{p} \, (\text{K}^* - K_1) = \text{p}K_n - n\log C_D \qquad (d)$$

K* is the apparent ionization constant of boric acid in presence of different ligand concentrations and has to be determined. The coordination number can be obtained directly from the slope of the curve p (K* — K₁) against — log C_D. The resulting values of K_n' are apparently independent of the ionic strength. Fig. 8 shows an example of the concentration dependence of the apparent ionization constant of boric acid.

At this relatively high ligand concentrations predominantly one-to-two complexes are formed. This ist not surprising since the tetrahedral borate anion should allow the attachment of even large polyol molecules

165

U. Weser

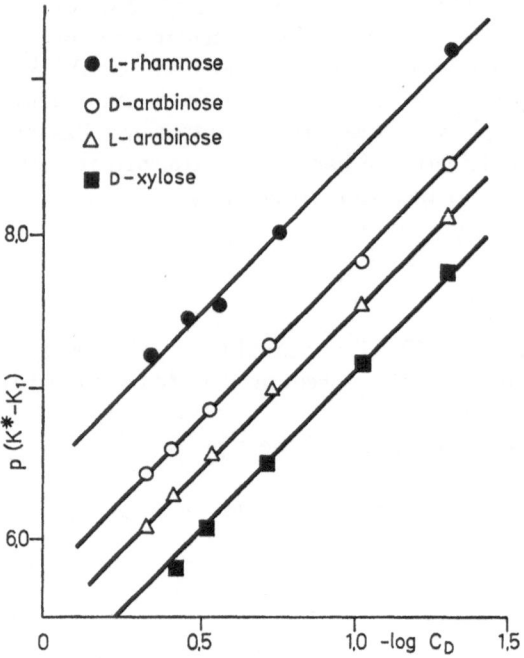

Fig. 8. Concentration dependence of K* in presence of L-rhamnose, D-arabinose, L-arabinose and D-xylose. The measurements were performed in 0.1 M KCl soln. at 25° C. The boric acid concentration was 5 m M; ref. (4).

Table 1. *Formation constants of one-to-two chelates formed by boric acid with some saccharides and alcohols in approx. 0.1 M KCl solns. at 25° C (from refs. 3, 5, 6).*

Ligand	$\log k_2^j$
L-rhamnose	2.61
D-galactose	2.39
D-glucose	2.86
D-arabinose	3.28
L-arabinose	3.55
D-xylose	4.01
D-mannose	4.52
D-fructose	5.04
L-sorbose	5.80
D-mannitol	4.92
D-dulcitol	5.23
D-sorbitol	5.65

166

without causing strong steric hindrance (see also Fig. 4). The numerical values of the stability constants of some hexoses and pentoses are all roughly in the same orders of magnitude (Table 1).

The stability constants show that the ketohexose chelates are more stable than the corresponding aldose chelates. This may be attributed to the fact that the carbon atoms 1 and 2 of aldoses contribute to the heterocyclic lactol ring, while the carbon atom 1 of the ketoses is outside the ring. This agrees also with the observation that the boric aldose complexes are less stable than the corresponding boric alcohol-chelates. This may be displayed by comparing the stability constants of boric-D-mannitol and boric-D-mannose, respectively. However, it is also obvious that groups outside the lactol rings are affecting the stability of the chelates only to a certain extent. The steric configuration of the hydroxyl groups within the lactol rings is by far more important for the stability of the complexes.

2. Electrophoresis

It has been reported in 1952 that quite a number of neutral carbohydrates, dissolved in borate buffers, migrate towards the anode when they are subjected to zone electrophoresis (*14, 27, 38, 39, 53*). The maximum mobilities are generally in alkaline pH regions (pH 9—10) (*15*). However, the relative mobilities of various carbohydrates may be pH dependent. Thus, at pH 7—8 D-fructose migrates faster than D-glucose while at pH 9—10 the reverse can be observed. It has to be recalled that at lower pH boric acid starts to form isopolyacids which might react in a different way with the ligands, whereas at higher pH predominantly the monomer borate tetrahedron is present (*34*). The presence of presumably different complexes may also be attributed to the fact that at high pH values, the migrated zones were sharp and circular, while reduced pH values caused elongated zones. For the quantitative detection of the electrophoretic mobility of a substance, a M_G value (*24*) has been suggested, which resembles formally the R_f value employed in chromatography;

$$M_G = \frac{\text{true distance of migration of a substance}}{\text{true distance of migration of D-glucose}}$$

The true distances of migration of a substance in borate buffer are obtained by correcting for movement due to electroendosmotic flow toward the cathode; i. e. the true distances of migration are usually greater than the apparent distances.

In the following Table the M_G and R_f values of some monosaccharides are compared.

Table 2. *Comparative M_G and R_f values of some monosaccharides (25).*

Sugar	$M_G{}^{(9)}$	R_f in solvent system*			
		1	2	3	4
L-arabinose	0.96	0.43	0.21	0.51	0.12
D-ribose	0.77	0.56	0.31	—	0.21
D-xylose	1.00	0.50	0.28	0.34	0.15
L-fucose	0.89	0.44	0.27	0.59	0.21
L-rhamnose	0.52	0.59	0.37	0.56	0.30
D-galactose	0.93	0.34	0.16	0.35	0.06
D-glucose	1.00	0.39	0.18	0.29	0.082
D-mannose	0.72	0.46	0.20	0.35	0.11
D-fructose	0.90	0.42	0.23	0.45	0.12
L-sorbose	0.95	0.40	0.20	0.36	0.10
L-galacto-heptulose	0.89	—	—	—	—
D-manno-heptulose	0.87	—	—	—	0.11

* *1.* Water saturated with 2,4,6-collidine *(60)*; *2.* 1-butanol-acetic acid *(60)*;
3. phenol-acetic-water *(15)*; *4.* 1-butanol-ethanol-water *(30)*.

It can be shown that carbohydrates in zone electrophoresis and in paper chromatography move quite differently. This sometimes enables a more effective separation of sugars, especially of those with close R_f values but wide spreading migration. It is further important to note that saccharides which have identical configuration of the hydroxylgroups in

Table 3. *Mobilities of some sugar derivatives in borate buffers (24, 52).*

Substance	M_G
2-O-methyl-D-glucose	0.23
2-Deoxy-D-glucose	0.29
3-O-methyl-D-glucose	0.80
3-Deoxy-D-glucose	0.85
4-O-methyl-D-glucose	0.24
4-O-benzyl-D-glucose	0.17
5-O-methyl-D-glucose	0.65
6-O-methyl-D-glucose	0.80
2,3-Di-O-methyl-D-glucose	0.12
2-O-methyl-D-galactose	0.43
3-O-methyl-D-galactose	0.63
4-O-methyl-D-galactose	0.30
6-O-methyl-D-galactose	0.86
2,6-di-O-methyl-D-galactose	—
2-O-methyl-D-xylose	0.39
3-O-methyl-D-xylose	0.66
4-O-methyl-D-xylose	0.21

the pyran ring also have a very similar migration. This can be seen in the pairs L-arabinose— D-galactose and D-xylose— D-glucose, respectively.

Zone electrophoresis employing borate buffers is a very appropriate means to separate active polyol compounds from those species whose hydroxylgroups have been substituted. Thus mixtures of 2,4- and 3,4-di-O-methyl-L-rhamnose, which are difficult to resolve by paper chromatography, are easily separated (Table 3) (23, 24, 52).

3. Optical Rotation

Extensive investigations have been carried out to study the effect of borate on the specific optical rotations of a great number of carbohydrates (37, 33). Table 4 presents the results for a series of sugars at pH 10 using

Table 4. *Optical rotations of sugars in water and in 0.6M borate buffer (33).*

Sugar	$[\alpha]_D^{20}$ in water (A)	$[\alpha]_D^{20}$ in water (B)	$\Delta [M]_D$ (A—B)	M_G value
L-arabinose	+104°	+ 3.1°	+151	0.96
D-fructose	− 92	− 14	+140	0.90
L-fucose	− 76	+ 60	−223	0.82
D-galactose	+ 80	− 44	+233	0.93
D-glucose	+ 53	− 5.2	+104	1.00
D-glucoheptose	− 20	− 2.5	+ 37	—
D-mannose	+ 14	− 2.5	+ 30	0.72
L-rhamnose, H_2O	+ 8.2	+ 16	+ 14	0.52
D-ribose	− 24	+ 40	+ 96	0.77
2-deoxy-D-glucose	+ 90	+ 39	+ 84	0.29
D-glucosamine hydrochloride	+ 73	+ 1.9	+153	—
N-acetyl-D-glucosamine	+ 41	+ 5.0	+ 81	0.23
2-O-methyl-D-glucose	+ 65	+ 48	+ 33	0.23
3-O-methyl-D-glucose	+ 56	− 31	+169	0.80
2:3-di-O-methyl-D-glucose	+ 50	+ 40	+ 21	0.12
2:3:6-tri-O-methyl-D-glucose	+ 70	+ 70	0	0.00
Methyl-α-L-fucofuranoside	−108	−108	0	—
Methyl-α-L-fucopyranoside	−196	−196	0	—
Methyl-α-D-glucopyranoside	+159	+159	0	0.11
D-glucitol	− 2.0	+ 15	− 31	—
Cellobiose	+ 35	+ 25	+ 34	0.23
Isomaltose	+122	+ 88	+166	0.69
Laminaribose	+ 19	− 44	+215	0.69
Leucrose	− 7.5	0	− 26	—
Maltose, H_2O	+130	+130	0	0.32
Melibiose, $2H_2O$	+129	+129	0	0.80
Raffinose, $5H_2O$	+105	+ 98	+ 42	—
Sucrose	+ 67	+ 51	+ 55	0.17
αα-trehalose, $2H_2O$	+178	+178	0	0.19

a molar excess of borate of at least 45:1 to achieve maximal rotational change (37).

There is no direct correlation between the velocity of the electrophoretic migration of sugars and the change in molecular rotation (Δ [M]$_D$). Compounds with equal M$_G$ values show different Δ [M]$_D$ values, cf. isomaltose and laminaribose. It has to be pointed out that, with the exception of melobiose, the greatest change in molecular rotation also results in a higher M$_G$ value. The absence of a change in rotation cannot therefore be attributed to a non occuring complex formation between borate and the polyol.

4. Nuclear Magnetic Resonance Spectroscopy

To study boric complexes of carbohydrates in solution the saccharides are dissolved in saturated deuterium oxide solutions of borax and then subjected to NMR spectroscopy (50). The proton spectra of glucose and xylose in presence and absence of borax are shown in Fig. 9.

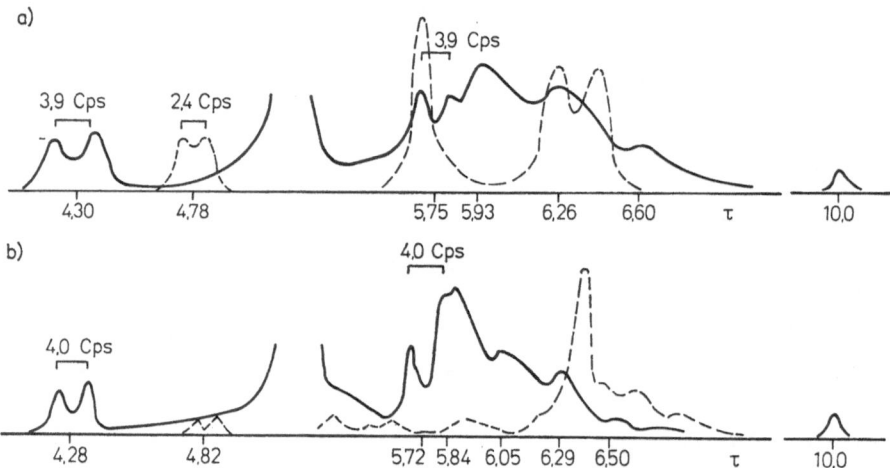

Fig. 9a and b. NMR spectra of deuterium oxide solutions (dotted line) and deuterium oxide solutions saturated with borax (solid line); a) glucose; b) xylose (50).

Profound changes in the spectra upon formation of the boric complex as well as strong similarities, both between the free carbohydrates and their complexes, can be observed. There was only one type of anomeric proton for the complexes, suggesting either one exclusive molecular structure in the complex or an extremely fast exchange among two or more conformations. The spectra of the borate complexes were exactly reproducible even after several weeks, thus apparently indicated no slow equilibration or decomposition. It is believed that the changes in the spectra of glucose and xylose in the presence of borax may be attributed to structural alterations of these carbohydrates on conversion to the complexes rather than to changes in the borate or tetraborate molecules. This is supported by the increased H_1—H_2 coupling constants of both complexes, which indicate a significant decrease in the C_1H_1—C_2H_2 dihedral angle. The reduced angle was expected to form a five membered ring with the borate molecule and the 1,2- cis hydroxyl groups. Actually, the magnitude of the coupling constant and the absence of a peak attributable to the C_1 proton of the other anomer β-glucose, both support the hypothesis that only the α-anomer was able to form a borate complex between the OH groups on the C_1 and C_2 atoms of the pyranoside ring.

5. Chromatography

Borate polyol complexes with their relatively strong ionization constants may easily be separated when subjected to ion exchange chromatography. This procedure originally has been employed (44, 45) to separate a mixture of carbohydrates on a borate-anion exchanger with borate eluting solutions. Similar separations have been demonstrated for specific groups of monosaccharides (45), for di-tri-, and tetrasaccharides (59) and for sugar alcohols (76).

Mixtures of nucleosides have been separated by this method (40) (Fig. 10). Weakly sorbed nucleosides as cytidine and adensoine become

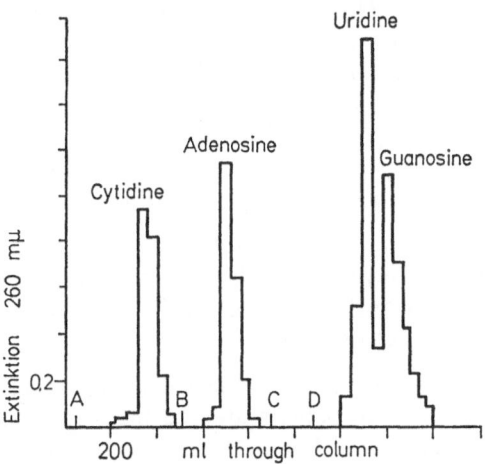

Fig. 10. Dowex-1-borate, loaded with 10^{-2} mM Cytidine, Adenosine, Guanosine and Uridine, respectively. Elution; A; 10^{-2} N borate buffer, pH 9.2; B; 5×10^{-2} N borate buffer, pH 9.2; C; 5×10^{-2} N borate buffer, pH 9.2 + 10^{-2} N NaCl; D; 5×10^{-2} N borate buffer, pH 9.2 + 0.1 N NaCl (ref. 40).

171

more strongly sorbed and more easily separable. This is further a powerful method to separate deoxynucleosides and ribonucleosides since the deoxynucleosides do not form any complexes with borate (43a).

The plate theory of ion exchange chromatography was applied to the separation of some borate-glycol complexes. A theoretical elution equation was derived and tested. Equilibrium constants for the formation of the complexes were determined. The position of the peak of an eluted glycol could be predicted for any concentration of Na-borate (63).

Examples for separating boric complexes by means of paper chromatography have been given for different sugar phosphates (11) and for various nucleosides (64); also o-dihydroxy phenols have been successfully separated (41).

6. Absorption Spectroscopy

Boric complexes or esters with aromatic polyols can easily be detected by means of absorption spectroscopy either in the visible or in the ultraviolet region. It has to be pointed out, however, that only vicinal hydroxyl groups of an aromatic ring may react with the borate tetrahedron or the trigonal $B(OH)_3$ molecule since such a ring is strongly planar and there is no puckered ring or chair form possible which would facilitate the approach of 1,3-diols. The atomic distance between these hydroxyl-groups is too far to allow a reaction with borate. If, however, there is a flexible hydroxyl group like the methylenehydroxy group at the 4 position of pyridoxol, a reaction of the two hydroxyl groups in 1,3-position with borate is possible (66). Also the formation of boric esters with 1,3-OH groups in peri position of condensed aromatic rings, such as in anthraquinones, has been discussed (22, 47).

In most cases there is a considerable increase of the absorption of the boric compound, with respect to the ligand, and the absorption maximum is shifted to a longer wavelength. Fig. 11 (71) illustrates the UV spectra of adrenaline in presence of borate.

The increased absorption, as well as the bathochromic shift, of boric compounds with aromatic polyols are the basis for a great variety of analytical determinations of boron. Thus, the reaction of $B(OH)_3$ with alizarin in conc. H_2SO_4 to form the stable ester is probably the most common procedure (22, 57). The orange-red colored curcumin-borate ester has a very high molar extinction coefficient which makes this compound a powerful means to detect even traces of boric acid (68).

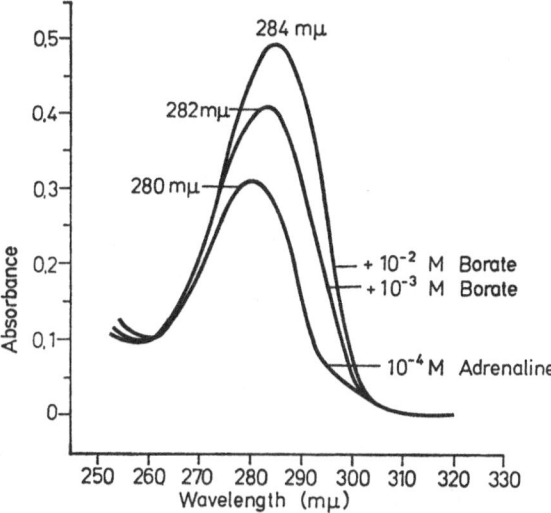

Fig. 11. Ultraviolet-absorption spectra of 10^{-4} M adrenaline and additional borate (10^{-3} M and 10^{-2} M) at pH 7.5 in 3×10^{-3} M Tris buffer.

III. Reaction of Borate with Saccharides

In the previous chapters reactions of borate with monomer or oligomer saccharides have been extensively described. Also very valuable information may be obtained from the comprehensive review by *Böeseken (8)*. In this section interactions of borate with polysaccharides will be discussed.

Highly polymerized polysaccharides with adjacent OH-groups in cis position, or even 1,3-diols, such as in polyvinyl alcohol react with borax (*17, 18, 55, 77*). The reaction was studied by measuring the increase in viscosity of the respective polymer. For the molecular structure a cross linkage of the borate tetrahedron to the polysaccharides has been suggested (*17*) (Fig. 12).

Each borate molecule should form the 1:2 complex. However, the viscosity of the polymers decreases at high borax concentrations. The authors attributed this fact to the increasing pH, while *Zittle (77)* stated that the decrease must be due to the formation of 1:1-complexes. On the other hand it has to be realized that at high borax concentrations tetraborate anions are present to a large extent which might react in a completely different manner with the polysaccharides.

173

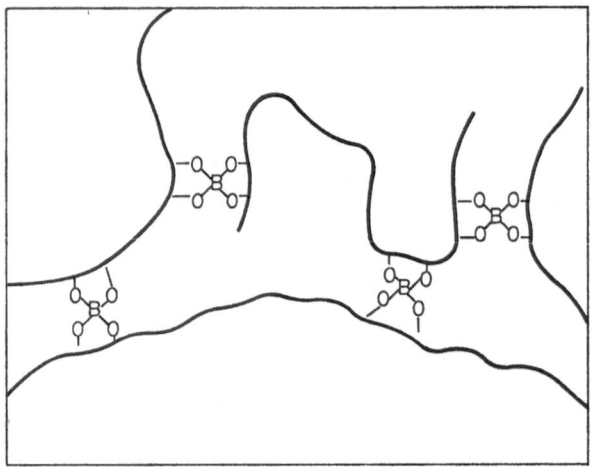

Fig. 12. Coupling of linear polysaccharides in the borate-didiol complex (17).

The stability of the boric-polysaccharides is of a much lower order of magnitude than that of the monomer complexes. This can be shown by ligand exchange reactions employing fructose, glucose, mannitol, glycerol, glycol or glyoxal.

Very important for the gel formation seems to be the molecular structure of the polymer. Thus a reaction with borate is more likely with a linear molecule like that of salep mannan than with a molecule of branched structure like in the yeast mannan. With increasing molecular weight, the better is the gel formation. The reverse takes place when the polymer borate complexes are subjected to enzymatic hydrolysis (17).

IV. Nucleotide and Nucleoside Interactions with Borate

As early as 1932 it was reported that an increase in the acidity of AMP was observed when borate was present (46). Twenty years later the complex reactions of borate with nucleosides were employed to separate mixtures of nucleosides and even nucleotides (see also sections on chromatography and electrophoresis). Only very little is known about whether or not borate interferes with biochemical reactions where nucleosides or nucleotides are involved. Thus, the stability constants of some boric nucleosides (72) are roughly in the same order of magnitude as the formation constants of earth alkaline or transition metal nucleotide complexes (62). It is therefore not unlikely that borate could influence to a certain extent some biochemical reactions by blocking the 2'3'-hydroxyl groups of the ribose residue.

The phosphorylation of nucleosides with phosphorylchloride in dimethylformamid succeeds when the 2'3'-hydroxylgroups of the ribose moiety are masked with boric acid (35, 36). The biological activity of the resulting 5'-nucleotides was tested by the enzymatic digestion with bull semen 5'-nucleotidase.

V. Borate and Hydroxycarbonic Acids

α-Hydroxy acids (lactic acid etc) react quite strongly with boric acid (8). Even long chain fatty acids such as hydroxystearic acid form stable complexes in alcoholic solutions (6a). The necessary OH groups are presumably supplied by the hydration of the COOH group. Cryoscopic measurments of borotartrates suggest the complex ions $(BO_2 \cdot C_4H_4O_6)^{3-}$ and $(BO_2 \cdot 2\ C_4H_4O_6)^{5-}$ (67). The latter is stable only in concentrations > 0.01 M.

In alkaline solutions the oxidation of 5-keto-D-gluconic acid, dehydro-L-ascorbic acid, and 2,3-diketo-L-gulonic acid are completely inhibited by borate which would indicate the reaction of these ligands with borate (54).

VI. Borate and Aromatic Polyols

In section C6 the application of absorption spectroscopy for the determination of boric-aromatic polyol complexes has been described. Some more specific examples shall be given here.

The effect of boric acid in alcoholic sodium acetate solutions on the spectra of a variety of flavonoid compounds has been investigated (31,42). The boric acid reacts specifically with o-dihydroxy groups, and this reaction results in a bathochromic shift of the λ max (Table 5).

Boric complexes with pyrocatechol and gallic acid have been reported (43, 48). The stability constant of boric-catechol-3,5-disulphonic acid has been determined potentiometrically (56). Borate inhibits tyrosinase activity by forming 1:1 or 1:2 complexes with the substrates catechol or dopa. The stability constants of these complexes have been determined spectrophotometrically and are in the order of 5.3×10^{-5} and 6.7×10^{-5}, respectively (73).

The primary purpose of this review was to emphasize the chemistry and structure of boric acid and some borate polyol compounds of biochemical interest. However, much work remains to establish the exact

Table 5. *The influence of boric acid in ethanolic sodium acetate solution on the spectra of flavonoid compounds (42).*

Compound	EtOH		B(OH)$_3$/NaOAc		Δ λ*
	λ max mμ	log ε	λ max mμ	log ε	mμ
2'-Hydroxychalcone	316	4.36	316	4.53	0
	221	4.12			
3,4-Dihydoxychalcone	368	4.34	403.5	4.40	35.5
	264.5	4.20	273	4.23	8.5
2',4-Dihydroxy-3-methoxy-chalcone	383	4.42	383	4.34	0
	266	4.07	266	3.98	0
2',3,4-Trihydroxy-chalcone	388	4.46	420	4.48	32
	269	4.14	277	4.09	8
			249	3.96	
5-Hydroxyflavone	337	3.88	337	3.66	0
	272	4.35	271	4.30	− 1
Apigenin	336	4.32	340	4.11	4
	269	4.31	269	4.15	0
3',4'-Dihydroxyflavone	342.5	4.52	371	4.47	28.5
	310	4.38	304	4.24	− 6
	244	4.46	244	4.41	0
Luteolin	350	4.17	373.5	4.21	23.5
	255	4.13	261	4.21	6
Flavonol	344	4.30	344	4.28	0
	305	4.15	306	4.12	1
	239	4.32			
Kaempferol	367.5	4.32	367.5	4.32	0
	326	4.03	326	3.93	0
	266	4.22	268	4.25	2
Quercetin	373	4.32	389	4.32	16
	301	3.87	305 **	3.76	4
	257	4.31	260	4.30	3
3,3',4'-Trihydroxyflavone	369	4.40	393	4.32	24
	250	4.33	253.5	4.21	3.5
p-Coumaric acid	312	4.36	285	4.37	−27
	227	4.09			
Methyl p-coumarate	313	4.40	312	4.33	− 1
	228	4.06			
Ferulic acid	323	4.22	310	4.13	−13
	296 **	4.09	285.5	4.14	−10.5
	233.5	4.10			
Caffeic acid	326	4.27	331	4.27	5
	299	4.18	295	4.09	− 4
	244	4.05			
Ethyl caffeate	328	4.35	352	4.36	24
	299	4.25	305	4.10	6
	244.5	4.20	255	4.18	10.5
	217	4.28			

* Δ λ = λ max (B(OH)$_3$ − λ max (EtOH).
** Inflection.

reaction mechanism of boron-oxygen compounds in plant or mammalian metabolism. The literature cited here may be of value for further investigations on this subject.

I am greatly indepted to Profs. *H. G. Day* (Bloomington) and *J. B. Neilands* (Univ. Calif. Berkeley) for financial support and valuable discussions. This work was aided by a PHS Grant (No. AMO 8 209—02 to *H. G. Day*).

VII. References

1. *Angyal, S. J.*, and *D. J. McHugh:* Interaction energies of axial hydroxyl groups. Chem. Ind. (London) 1147 (1956).
2. — — Cyclitols, Part V, paper ionophoresis, complex formation with borate, and the rate of periodic acid oxidation. J. Chem. Soc. 1423 (1957).
3. *Antikainen, P. J.:* On the chelation of boric acid with hexoses. Suomen Kemistilehti B *31*, 255 (1958).
4. — A comparative study on the chelate formation between germanic acid and some glycols and polyalcohols in aqueous solution. Acta Chem. Scand. *13*, 312 (1959).
5. —, and *V. M. K. Rossi:* The influence of the ligand configuration on the stability of oxyanion chelates. I. Chelation of boric acid and germanic acid with hexols. Suomen Kemistilehti B *32*, 182 (1959).
6. —, and *K. Tevanen:* The influence of ligand configuration on the stability of oxyanion chelates. II. Chelation of boric and germanic acid with saccharides. Suomen Kemistilehti B *32*, 214 (1959).
6a. *Azarova, E. J.*, and *V. A. Zarinskii:* Formation of borohydroxystearic acid in alcoholic solutions. Khim. i Geol. Nauk. *5*, 39 (1965).
7. *Barker, S. A., E. J. Bourne, A. B. Foster*, and *R. M. Pinkard:* Steric effects in the ionophoresis of carbohydrates. Chem. Ind. (London) 226 (1959).
8. *Böeseken, J.:* The use of boric acid for the determination of the configuration of carbohydrates. Adv. Carbohydrate Chem. *4*, 189 (1949).
9. *Bourne, E. J., A. B. Foster*, and *P. M. Grant:* Ionophoresis of carbohydrates on glass fiber sheets. J. Chem. Soc. 4311 (1956).
10. *Christ, C. L., J. P. Clark*, and *H. T. Evans jr.:* Studies of borate minerals (III): The crystal structure of colemannite, $CaB_3O_4(OH)_3 \cdot H_2O$. Acta Cryst. *11*, 761 (1958).
11. *Cohen, S. S.*, and *D. B. Mc N. Scott:* Formation of pentose phosphate from 6-phosphogluconate. Science *111*, 543 (1950).
12. *Cohn, W. E.:* The nucleic acids, chemistry and biology, ed. by *E. Chargaff*, and *J. N. Davidson*, New York (1955), p. 235—241.
13. *Consden, R.*, and *M. N. Powell:* Use of borate buffer for a better separation of the β-globulin components. J. Clin. Pathol. *8*, 150 (1955).
14. —, and *W. M. Stanier:* A simple paper electrophoresis apparatus. Nature *170*, 1069 (1952).
15. *Counsell, J. N., L. Hough*, and *W. H. Wadman:* Distribution chromatography at elevated temperatures. Research (London) *4*, 143 (1951).

16. *Dale, J.:* The stereochemistry of polyborate anions and of borate complexes of diols and certain polyols. J. Chem. Soc. 922 (1961).
17. *Deuel, H.,* and *H. Neukom:* The reaction of boric acid and borax with poly-saccharides and other high-molecular compounds. Makromol. Chem. *3,* 13 (1949).
18. — —, and *F. Weber:* Reaction of boric acid with polysaccharides. Nature *161,* 96 (1948).
19. *Deutsch, A.,* and *S. Osoling:* Conductometric and potentiometric studies of the stoichiometry and equilibria of the boric acid mannitol complexes. J. Amer. Chem. Soc. *71,* 1637 (1949).
20. *Edwards, J. O., G. C. Morrison, V. F. Ross,* and *J. W. Schultz:* The structure of the aqeous borate ion. J. Amer. Chem. Soc. *77,* 266 (1955).
21. — Detection of anionic complexes by pH measurements. I. polymeric borates. J. Amer. Chem. Soc. *75,* 6151 (1953).
22. *Feigel, F.:* Spot tests in inorganic analysis, Elsevier publishing company, Amsterdam (1958, p. 339—341.
23. *Foster, A. B.:* Separation of the dimethyl-L-rhamnopyranoses by ionophoresis. Chem. Ind. (London) 828 (1952).
24. — Ionophoresis of some disaccharides. J. Chem. Soc. 982 (1953).
25. — Zone electrophoresis of carbohydrates. Adv. Carbohydrate Chem. *12,* 81 (1957).
26. *Frahn, J. L.,* and *J. A. Mills:* Paper ionophoresis of carbohydrates. Australian J. Chem. *12,* 65 (1959).
27. *Hashimoto, Y., I. Mori,* and *M. Kimura:* Paper electromigration of flavonoids and sugars using a high constant voltage current. Nature *170,* 975 (1952).
28. *Hermans, P. H.:* Structure of boric acids and some of their derivatives. Z. anorg. Chem. *142,* 83 (1925).
29. — On the structure of the Böeseken's boric acid compounds with diols. Rec. Trav. Chim. *57,* 333 (1938).
30. *Hirst, E. L.,* and *J. K. N. Jones:* The application of partition chromatography to the separation of the sugars and their derivatives. Discussions Faraday Soc. *7,* 268 (1949).
31. *Hörhammer, L.,* and *R. Hänsel:* Analysis of flavones VIII. Further properties of halochromic boron complexes. Arch. Pharm. *288,* 315 (1955).
32. *Hubert, A. J., B. Hargitay,* and *J. Dale:* The structure and relative stabilities of boric esters of 1,2- and 1,3-diols. J. Chem. Soc. 931 (1961).
33. *Hughes, R. C.,* and *W. J. Whelan:* Effect of sodium borate on the optical rotation of sugars. Chem. Ind. (London) 50 (1959).
34. *Ingri, N., G. Lagerström, M. Frydman,* and *L. G. Sillén:* Equilibrium studies of polyanions. Acta Chem. Scand. *11,* 1034 (1957).
35. *Ikehara, M., E. Ohtsuka,* and *Y. Kodama:* Nucleosides and nucleotides: XXII phosphorylation of adenosine using borate complex as protecting group for 2′,3′-hydroxylgroups. Chem. Pharm. Bull. (Tokyo) *12,* 145 (1964).
36. *Imai, K., T. Hirata,* and *M. Honjo:* Phosphorylation of nucleosides at the 5′-position through their borate complexes. Takeda Kenkyusho Nempo *23,* 1 (1964), Chem. Abstracts *63,* 4553 (1965).
37. *Isbell, H. S., J. F. Brewster, N. B. Holt,* and *H. L. Frush:* Behavior of certain sugars and sugar alcohols in presence of tetraborates; correlation of optical rotation and compound formation. Res. nat. Bur. Stand. *40,* 129 (1948).
38. *Jaenicke, L.:* Paper electrophoresis of sugars and sugar derivatives. Natur-wissenschaften *39,* 86 (1952).
39. —, and *I. Vollbrechtshausen:* Electrophoretic separation of nucleosides as boric complexes. Naturwissenschaften *39,* 86 (1952).

40. —, and *K. von Dahl:* Separation of nucleosides as complex boric acids on ex-changers. Naturwissenschaften *39*, 87 (1952).
41. *Jurd, L.:* Chromatography of phenolic compounds on borate impregnated paper. J. Chromatog. *4*, 369 (1960).
42. — A spectrophotometric method for the detection of o-dihyrdoxyl groups in flavonoid compounds. Arch. Biochem. Biophys. *63*, 376 (1956).
42a. *Kilipi, S.:* A differential potentiometric method of measuring acid and base dissociation constants. J. Amer. Chem. Soc. *74*, 5296 (1952).
43. *Khan, J. A.,* and *D. Sen:* Studies on the complex formation of boron with aromatic hydroxy compounds. Proc. Indian Acad. Sci. *49a*, 226 (1959).
43a. *Khym, J. X.,* and *W. E. Cohn:* The ion-exchange separation of the 5'-ribo-nucleotides and deoxyribonucleotides. Biochim. Biophys. Acta. *15*, 139 (1954).
44. —, and *L. P. Zill:* The separation of monosaccharides by ion exchange. J. Amer. Chem. Soc. *73*, 2399 (1951).
45. — — The separation of sugars by ion exchange. J. Amer. Chem. Soc. *74*, 2090 (1952).
46. *Klimek, R.,* and *J. K. Parnas:* Adenylic acid and adeninenucleotide. Biochem. Z. *252*, 392 (1932).
47. *Koremann, I. M.,* and *N. V. Kurina:* Anthraquinone derivatives as reagent for the boric acid. Trudy Khim i Khim. Tekhnol. *1*, 573 (1958).
48. *Kümmel, D. F.,* and *M. G. Mellow:* Structure and composition of pyrocatechol-boric acid-pyridine complexes. J. Amer. Chem. Soc. *78*, 4572 (1956).
49. *Lehmann, H. A.:* Recent chemistry of boric acids salts. Z. Chem. *3*, 284 (1963).
50. *Lenz, R. W.,* and *J. P. Heeschen:* The application of NMR to structural studies of carbohydrates in aqueous solutions. J. Polym. Sci. *51*, 247 (1961).
51. *Lewis, G. L.,* and *C. P. Smyth:* The dipole moments and structures of the esters of some fatty and some inorganic acids. J. Amer. Chem. Soc. *62*, 1529 (1940).
52. *Lindberg, B.,* and *B. Swan:* Paper electrophoresis of carbohydrates in germanate buffer. Acta Chem. Scand. *14*, 1043 (1960).
53. *Michl, H.:* Paper ionophoresis at potential gradients of 50 V/cm. II. Organic borate complexes. Monatsh. *83*, 737 (1952).
54. *Militzer, W. E.:* The inhibition of carbohydrate oxidations by borate. J. Biol. Chem. *158*, 247 (1945).
55. *Moe, O. A., S. E. Miller,* and *M. H. Iwen:* Investigation of the reserve carbo-hydrates of leguminous seeds. I. Periodate separation. J. Amer. Chem. Soc. *69*, 2621 (1947).
56. *Näsänen, R.:* Complex formation between boric and catechol-3,5-disulphonic acid. Suomen Kemistilehti B *33*, 1 (1960).
57. *Nazarchuck, T. N.:* Compounds of boric acid with hydroxyanthraquinones. Ukr. Khim. Zh. *28*, 233 (1962).
58. *Nies, N. P.,* and *G. W. Campbell:* Boron, Metallo boron compounds and boranes; ed. by *R. M. Adams;* Interscience publ., New York 1964, p. 53.
59. *Noggle, C. R.,* and *L. P. Zill:* The quantitative analysis of sugars in plant extracts by ion-exchange chromatography. Arch. Biochem. Biophys. *41*, 21 (1952).
60. *Partridge, S. M.,* and *R. G. Westall:* Filter paper partition chromatography of sugars. Biochem. J. *42*, 238 (1948).
61. *Pastureau, P.,* and *M. Veiler:* Boric acid esters of tetrasubstituted glycerols. Compt. Rend. *202*, 1683 (1936).
62. *Phillips, R. S. J.:* Adenosine and the adenine nucleotides, ionization, metal complex formation, and conformation in solution. Chem. Revs. *66*, 502 (1966).

63. *Rieman, W.:* Application of the plate theory to the anion exchange chromatography of glycols. J. Phys. Chem. *60*, 1370 (1956).
64. *Rose, J. A.*, and *B. S. Schweigert:* Use of borate in paper chromatography of nucleosides. J. Amer. Chem. Soc. *73*, 5903 (1951).
65. *Roy, G. L., A. L. Laferriere,* and *J. O. Edwards:* A comparative study of polyol complexes of arsenite, borate, and tellurate ions. J. Inorg. Nucl. Chem. *4*, 106 (1957).
66. *Scudi, J. V., W. A. Bastedo,* and *T. J. Webb:* The formation of a vitamin B_6-borate complex. J. Biol. Chem. *136*, 399 (1940).
67. *Shvarcs, E.,* and *A. F. Ievinš:* The complex nature of borotartrate ions in aqueous solutions. Zhur. Neorg. Khym *4*, 1835 (1959).
68. *Spicer, G. S.,* and *J. D. H. Strickland:* Determination of microgram amounts of boron. I. Absorptiometric determination with curcumin. Analyt. Chim. Acta *18*, 231 (1958).
69. *Weigel, H.:* Paper electrophoresis of carbohydrates. Adv. Carbohydrate Chem. *18*, 61 (1963).
70. *Weissbach, A.:* Scyllitol diborate. J. Org. Chem. *23*, 329 (1958).
71. *Weser, U.:* Inhibition of adrenaline action on liver dephosphophosphorylase by borate. Biochim. Biophys. Acta *121*, 413 (1966).
72. — Chelation of boric acid with some nucleosides. Z. Naturforschg. 22B, (1967).
73. *Yasunobu, K. T.,* and *E. R. Norris:* Mechanism of borate inhibition of diphenoloxidation by tyrosinase. J. Biol. Chem. *227*, 473 (1957).
74. *Zachariasen, W. H.:* The crystal structure of boric acid. Z. Krist. *88*, 150 (1934).
75. — The precise structure of orthoboric acid. Acta Crystallogr. (Copenhagen) *7*, 305 (1954).
76. *Zill, L. P., J. X. Khym,* and *G. M. Cheniae:* Further studies on the separation of the borate complexes of sugars and related compounds by ion-exchange chromatography. J. Amer. Chem. Soc. *75*, 1339 (1953).
77. *Zittle, C. A.:* Reaction of borate with substances of biological interest. Adv. Enzymol. *12*, 493 (1951).

(Received January 24, 1967)

Reversible Oxygenierung von Metallkomplexen

Prof. Dr. E. Bayer und Dipl.-Chem. P. Schretzmann

Lehrstuhl für Organische Chemie der Universität Tübingen

Inhaltsverzeichnis

I. Einleitung

Eines der interessantesten ungelösten Probleme der Biochemie ist die Frage, warum die Atmungspigmente molekularen Sauerstoff reversibel fixieren können, ohne dabei irreversibel oxidiert zu werden.

Hämoglobine und Myoglobine (1, 2, 3, 4) enthalten als Aktivzentrum der Sauerstoffbindung das Protohäm IX, einen Porphyrinkomplex des zweiwertigen Eisens (Abb. 1a). Chlorocruorin (5), das grüne Atmungspigment von Polychaeten-Würmern, trägt das Chlorocruorohäm (Spirographis-Häm) als prosthetische Gruppe (Abb. 1b). Diese Hämoproteine kombinieren mit molekularem Sauerstoff im Verhältnis Eisen : $O_2 = 1:1$.

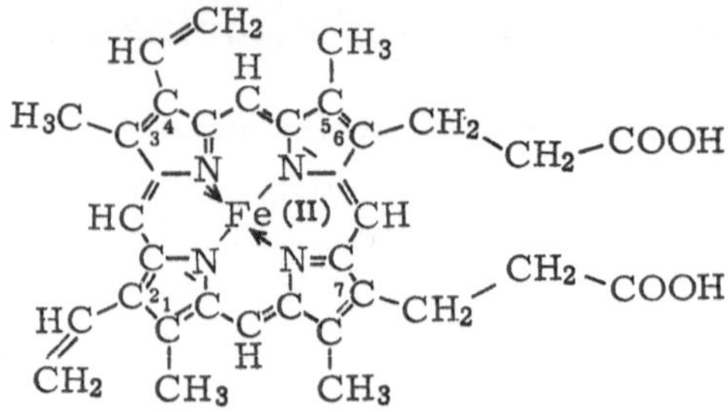

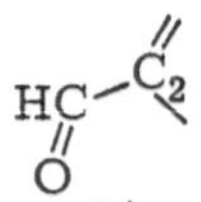

Abb. 1a und b. a) Protohäm IX, b) Chlorocruorohäm (der Vinylrest in Position 2 im Protohäm ist durch einen Formylrest ersetzt).

In manchen Würmern (Sipunculiden, Polychaeten) und in Brachiopoden kommt ein reversibel oxygenierbares Nicht-Häm-Eisenprotein vor, das Hämerythrin (5, 6, 7). Hämocyanine sind Kupferproteine aus Wirbellosen (Mollusken, Arthropoden, Crustaceen) (5, 6, 7). Diese Nicht-Häm-Metallproteine enthalten Eisen(II) bzw. Kupfer(I) direkt am Protein gebunden. Sie binden Sauerstoff im Verhältnis Metall : $O_2 = 2 : 1$.

Über die besonderen Strukturfaktoren in reversibel oxygenierbaren Metallkomplexen ist bisher wenig bekannt; das gleiche gilt für die Art der Bindung des Sauerstoffs, seine Orientierung relativ zu den Metallatomen und den Valenzzustand der Metalle in den oxygenierten Komplexen. Die Lösung dieser Fragen an den Blutfarbstoffen wird erschwert durch den hochmolekularen Proteinteil. Es besteht daher großes Interesse an niedermolekularen synthetischen Metallkomplexen, die reversibel mit molekularem Sauerstoff kombinieren können (8, 9).

Solche Komplexe sind neben ihrer Bedeutung als Modellsubstanzen für die Aktivzentren in den Atmungspigmenten als potentielle Rohstoffe für die Gewinnung von reinem Sauerstoff aus der Atmosphäre von technischem Interesse. Die Zahl der reversibel oxygenierbaren Verbindungen ist klein.

Besonders eingehend untersucht ist der Chelatkomplex Bis(salicyliden)-äthylendiaminkobalt(II) (Salcomin, Abb. 2) (8, 9, 10). Das „Salcomin"-Verfahren zur Sauerstoffgewinnung aus der Luft ist zwar teurer als der „Linde"-Prozeß, es liegt aber nur knapp unter der Grenze der Wirtschaftlichkeit. Einer technischen Anwendung steht noch die irreversible Oxidation des Chelatkomplexes hindernd im Wege, die nach einigen hundert Oxygenierungs-Deoxygenierungs-Kreisläufen die Sauerstoffkapazität vermindert.

Es soll versucht werden, aus der Reaktivität des Sauerstoffmoleküls und den Eigenschaften der Modellkomplexe Hinweise für die Oxygenierung der Atmungspigmente zu gewinnen.

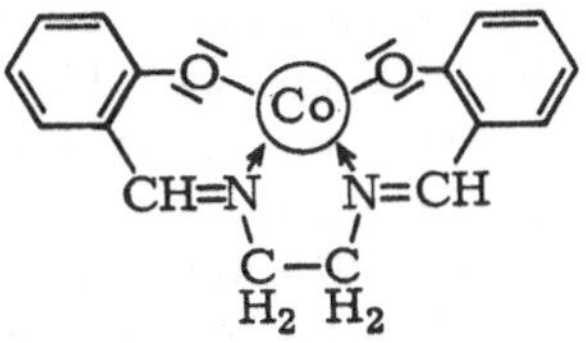

Abb. 2. Bis(salicyliden)-äthylendiaminkobalt(II).

II. Das Sauerstoffmolekül

1. Elektronische Struktur

Sauerstoff ist im elektronischen Grundzustand ein Biradikal. Die Theorie der Molekülorbitale liefert dafür eine plausible Erklärung (11, 12). In Abb. 3 ist das Orbitalschema für ein Molekül AB aus Atomen der 2. Periode des Periodensystems angegeben. Abb. 4 zeigt schematisch die räumliche Anordnung dieser Molekülorbitale und ihre Besetzung im elektronischen Grundzustand des Sauerstoffmoleküls.

Atom-orbitale A	Molekülorbitale A B	Atom-orbitale B	Besetzung der Molekülorbitale in					
A	A B	B	N_2	NO	$(^3\Sigma)\,O_2$	$(^1\Delta)\,O_2$	O_2^-	O_2^{2-}
	σ_{Py}^* —		—	—	—	—	—	—
	π_x^* — π_z^* —							
2	antibindende MO	2	—	↓	↓ ↓	↓↑	↓↑ ↓	↓↑ ↓↑
Px Py Pz	π_x π_z	Px Py Pz						
	σ_{Py}		↓↑ ↓↑	↓↑ ↓↑	↓↑ ↓↑	↓↑ ↓↑	↓↑ ↓↑	↓↑ ↓↑
	bindende MO		↓↑	↓↑	↓↑	↓↑	↓↑	↓↑
2 s — 1 s	lone pair Orbitale	2 s — 1 s	↓↑ ↓↑	↓↑ ↓↑	↓↑ ↓↑	↓↑ ↓↑	↓↑ ↓↑	↓↑ ↓↑
—	innere Schalen der Atome	—	↓↑ ↓↑	↓↑ ↓↑	↓↑ ↓↑	↓↑ ↓↑	↓↑ ↓↑	↓↑ ↓↑

Abb. 3. Orbitalschema für Moleküle AB mit Atomen der zweiten Periode des Periodensystems.

Die inneren Schalen beider Atome beeinflussen sich nicht. Die 2s-Atomorbitale bilden im Molekülverband zwei lone pair-Orbitale. Wenn die y-Achse des Koordinatensystems als Kernverbindungslinie gewählt wird, so ergeben die beiden $2p_y$-Atomorbitale den σ-bindenden Molekülorbital σp_y und den antibindenden Molekülorbital σp_y^*. Diese vier Molekülorbitale sind achsensymmetrisch zur Kernverbindungslinie angeordnet.

Aus den Atomorbitalen $2p_x$ und $2p_z$ resultiert je ein Paar π-bindender und π-antibindender Molekülorbitale. Während im Stickstoffmolekül alle 10 Valenzelektronen in bindenden Orbitalen untergebracht werden können oder als einsame Elektronenpaare auftreten, sind im Sauerstoffmolekül 2 Elektronen in antibindende Orbitale einzubauen. Im elektronischen Grundzustand (spektroskop. Symbol $^3\Sigma$) des Sauerstoffs werden sie unter Beachtung der Hund'schen Regel mit parallelem Spin in die energiegleichen Orbitale π_x^* und π_z^* eingebaut. Sauerstoff enthält demnach das gleiche σ—π—π-Bindungsgerüst wie Stickstoff, zusätzlich aber 2 Elektronen in antibindenden Molekülorbitalen mit parallelem Spin.

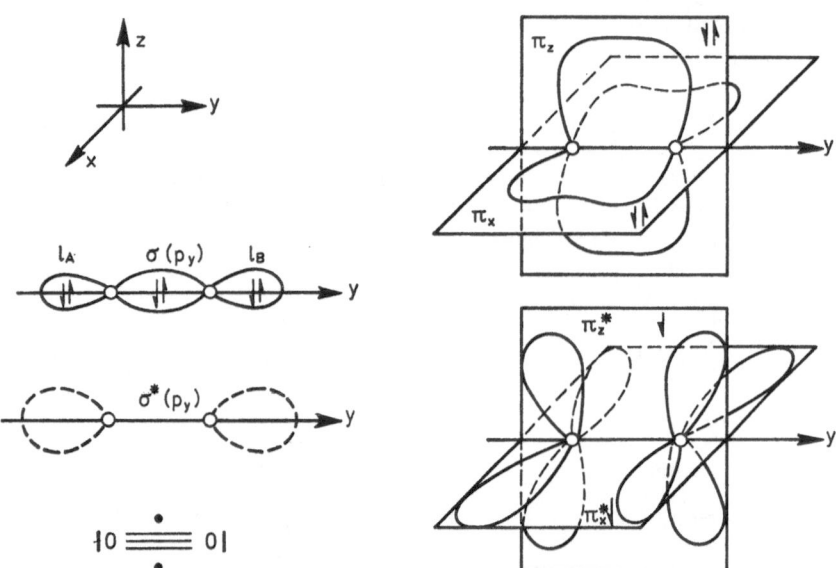

Abb. 4. Räumliche Anordnung und Besetzung der Molekülorbitale im elektronischen Grundzustand des Sauerstoff-Moleküls.

2. Das Sauerstoffmolekül in Redoxreaktionen

Die Chemie des Sauerstoffs ist geprägt durch seine stark oxidierende Wirkung bei gleichzeitiger kinetischer Trägheit. Reduzierend wirkt Sauerstoff nur gegenüber den stärksten elektronenaffinen Verbindungen wie Fluor oder PtF_6 (13, 14). Das ist aus der Ionisationsenergie und der Elektronenaffinität des O_2-Moleküls unmittelbar zu entnehmen. Die niedrigste Ionisationsenergie — die eines π^*-Orbitals — beträgt für $(^3\Sigma)O_2 = 12,1$ eV (15, 16).

Dagegen ist die Zuführung eines Elektrons mit 0,6—0,9 eV, entsprechend 15—21 Kcal/mol, eine exotherme Reaktion (17, 18). Nach George (19) ist die oxidierende Wirkung des Sauerstoffs sehr unterschiedlich, je nachdem, ob die Reaktion

$$O_2 \xrightarrow{e^-} O_2^- \xrightarrow{e^-} O_2^{2-} \xrightarrow{2\,e^-} 2\,O^{2-} \tag{1}$$

auf der Oxidationsstufe des Superoxids, Peroxids oder Oxids stehen bleibt, wie Tabelle 1 zeigt.

Tabelle 1. *Standardredoxpotentiale für die Reduktion von* O_2 (pH = 0).

Reaktion	E° (Volt)	Literatur
$O_2 + e^- \rightleftharpoons O_2^{\bullet-}$	—0,4	(20)
$O_2 + H^+ + e^- \rightleftharpoons HO_2^{\bullet}$	—0,1	(20)
$1/2\,O_2 + H^+ + e^- \rightleftharpoons 1/2\,H_2O_2$	+0,68	(19)
$1/4\,O_2 + H^+ + e^- \rightleftharpoons 1/2\,H_2O$	+1,23	(19)
$Fe^{3+} + e^- \rightleftharpoons Fe^{2+}$	+0,77	(20)
$1/2\,Cl_2 + e^- \rightleftharpoons Cl^-$	+1,36	(19)

Führt die Reduktion des molekularen Sauerstoffs nur zum Superoxidanion, und kann dieses nicht weiter reagieren, so ist Sauerstoff ein sehr schwaches Oxidationsmittel. Das Bild wird günstiger, wenn sich das Superoxidanion mit einem Proton vereinigen kann (20):

$$O_2^{\bullet-} + H^+ \rightleftharpoons HO_2^{\bullet} \tag{2}$$

Bei der Reduktion zum Hydrogenperoxid verhält sich Sauerstoff wie ein mittelstarkes Oxidationsmittel, etwa wie Fe^{3+}. Das ist bei der Besprechung der Sauerstoffaddukte von Metallkomplexen im Verhältnis $Me:O_2 = 1:1$ und $Me:O_2 = 2:1$ im Auge zu behalten. Allerdings sind diese Eigenschaften des freien Sauerstoffmoleküls nur mit Vorsicht auf den Sauerstoff im Ligandenfeld von Metallkomplexen übertragbar, für den eine elektronische Struktur postuliert wurde, die dem ersten Anregungszustand des O_2-Moleküls entspricht.

3. Chemie des Sauerstoffmoleküls im ersten elektronischen Anregungszustand: $(^1\Delta)O_2$, Singlettsauerstoff

Werden die beiden antibindenden Elektronen des Sauerstoffmoleküls mit antiparallelem Spin in einem der beiden energieentarteten π^*-Orbitale untergebracht, so erhält man den diamagnetischen Singlettsauerstoff, der um 23 Kcal/mol energiereicher ist als der $(^3\Sigma)$-Sauerstoff (27).

Flüssiger Sauerstoff ist blau. Diese Farbe beruht auf einer Lichtabsorption, die nach

$$(\downarrow\downarrow)\,O_2 + (\uparrow\uparrow)\,O_2 \rightleftharpoons [(\downarrow\downarrow)\,O_2 \cdot (\uparrow\uparrow)\,O_2] \underset{-h\nu\,(\text{Lumineszenz})}{\overset{+h\nu\,\text{Absorption}}{\rightleftharpoons}} [(\uparrow\downarrow)\,O_2]_2 \rightleftharpoons 2\,(\uparrow\downarrow)\,O_2 \tag{3}$$

zu einer photochemischen Bildung des Singlettsauerstoffs führt (20, 21, 22). Der Zerfall des Singlettsauerstoffs kann unter orangeroter Lumineszenz im gleichen Spektralbereich (600—800 mμ) erfolgen (20, 26).

$(^1\Delta)\,O_2$ ist metastabil, da er bei Abwesenheit von freien Radikalen oder Molekülen mit Triplettzuständen niederer Energie nur in dimeren Stoßkomplexen $[(^1\Delta)\,O_2]_2$ strahlen kann. Eine Reaktion

$$(^3\Sigma)\,O_2 + h\nu \rightleftharpoons (^1\Delta)\,O_2 \tag{4}$$

ist wegen der Änderung der Spinmultiplizität nicht möglich. Dieser Singlettsauerstoff besitzt einen leeren π^*-Orbital, der ihn Elektronendonoren gegenüber zu einem starken Elektronenacceptor macht (20).

Besonders charakteristisch sind die Cycloadditionen, die $(^1\Delta)\,O_2$ mit Dienen eingeht (23, 24, 25, 26). Seine Dienophilie legt einen Vergleich mit elektronenaffinen Olefinen wie Tetracyanäthylen nahe.

4. Valenzzustand des Sauerstoffmoleküls

Im gasförmigen Sauerstoff sind die π^*-Orbitale energiegleich. Tritt das Sauerstoffmolekül aber in unsymmetrische Wechselwirkung mit anderen Molekülen, so wird diese Energieentartung aufgehoben (Abb. 5).

Bei gelöstem Sauerstoff ist diese Energieaufspaltung klein, beide π^*-Orbitale bleiben einfach besetzt. Der Sauerstoff bleibt paramagnetisch (27).

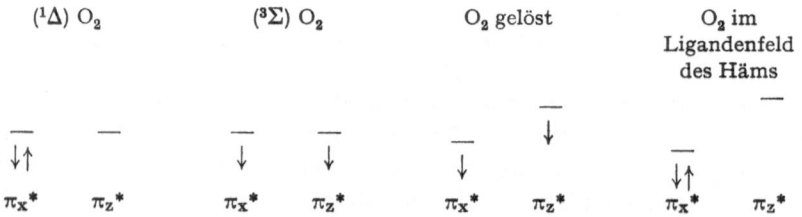

Abb. 5. Ligandenfeldaufspaltung und Besetzung der π^*-Molekülorbitale im Sauerstoff-Molekül.

Griffith (28) hat nun die Hypothese aufgestellt, daß im Ligandenfeld des Häm-Eisens der Hämoproteine diese Energieaufspaltung so groß wird, daß beide Elektronen in den stabilisierten Orbital eintreten. Die Änderung der Spinmultiplizität ist nur möglich, wenn gleichzeitig das Metall seinen magnetischen Zustand ändert. Das ist bei der Oxygenierung von Hämoproteinen der Fall.

Angenommen, das Häm liege in der x—y-Ebene und die Kernverbindungslinie im O_2-Molekül sei parallel zur y-Achse, so wird der π_x^*-Orbital des Sauerstoffs stabilisiert und doppelt besetzt. Der π-bindende Molekülorbital π_x und der antibindende Orbital π_x^*, die beide doppelt besetzt sind, kombinieren unter Ausbildung zweier einsamer Elektronenpaare. Um eine möglichst weitgehende Ladungstrennung zu erreichen, sind die 4 einsamen Elektronenpaare in der x—y-Ebene unter dem Winkel von 120° zum σ-bindenden Molekülorbital σp_y angeordnet.

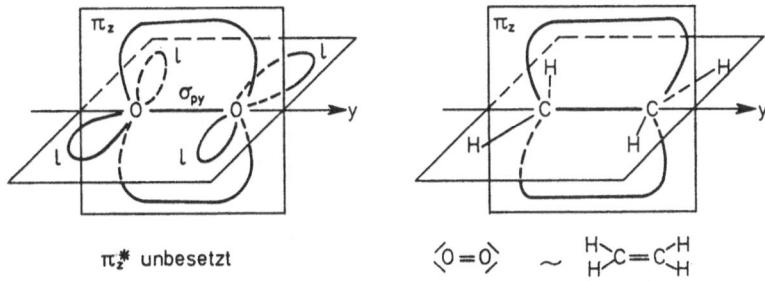

Abb. 6. Valenzzustand des Sauerstoff-Moleküls nach *Griffith (28)*.

Die elektronische Struktur des Sauerstoffmoleküls im Ligandenfeld entspricht somit der des Äthylens (Abb. 6). Seine Bindung an Metallkomplexe im Verhältnis $Me : O_2 = 1 : 1$ ist vergleichbar mit dem Bindungszustand in Olefin-Metallkomplexen, wie er von *Chatt* und *Duncanson* postuliert wurde (29).

III. Ladungszustand und Abstände in O—O-Gruppen

Bei der Reduktion des Sauerstoffmoleküls werden Elektronen in antibindende Orbitale eingebaut. Je größer die Zahl antibindender Elektronen ist, desto lockerer wird die Bindung und desto größer wird der Kernabstand. Tabelle 2 zeigt diese Abhängigkeit des Kernabstandes von der Ladung der O—O-Gruppen.

In Verbindungen wie Kaliumsuperoxid $K^+O_2^-$, Bariumperoxid $Ba^{2+}O_2^{2-}$ oder Magnesiumoxid $Mg^{2+}O^{2-}$ kann von einer Wertigkeit des

Tabelle 2. *Abstände von O–O-Bindungen*

Verbindung	Verbindungstyp	Struktur (schem.)	Dihedralwinkel $\begin{smallmatrix}X\\\\Y\end{smallmatrix}$	O–O-Kernabstand (in Å)	Literatur
O_3	freies Molekül			$1,278 \pm 0,003$	(30)
O_2	freies Molekül			$1,2074 \pm 0,0011$	(31)
				$1,2074 \pm 0,0001$	(30)
O_2^+	freies Molekülion			$1,1227 \pm 0,0001$	(30)
KO_2	Superoxid	O_2^-		$1,28 \pm 0,02$	(30)
BaO_2	Peroxid	O_2^{2-}		$1,49 \pm 0,04$	(30)
H_2O_2	Peroxid (homöopolar)	H H	$93° 51'$	$1,49 \pm 0,01$	(32)
$C_6H_5{-}CO{-}O{-}O{-}CO{-}C_6H_5$	Peroxid (homöopolar)		$91°$	$1,46 \pm 0,015$	(33)
O_2F_2	Dioxygenylverbindung	$O_2^{\delta+}$	$87° 30'$	$1,217 \pm 0,003$	(34)
O_2PtF_6	Dioxygenylverbindung			$1,17 \pm 0,17$	(14)
O_2PF_6	Dioxygenylverbindung				(13)

Komplex	Bezeichnung	Struktur	Winkel	Abstand	Lit.
$[Cr(O_2)_4]^{3-}$	Peroxokomplex		0°	1,405 ± 0,039	(35)
$[U(O)(O_2)_3]^{4-}$	Peroxokomplex		0°	1,51	(64)
$[(O_2)Ir(J)(CO)(P(C_6H_5)_3)_2]$	Peroxokomplex?		0°	1,47 ± 0,02	(108)
$[(O_2)Ir(Cl)(CO)(P(C_6H_5)_3)_2]$	Sauerstoffträger		0°	1,30 ± 0,03	(36)
$[(NH_3)_5CoO_2Co(NH_3)_5]^{4+}(SCN)_4$	μ-Peroxokomplex?		180°	1,65 ± 0,03	(37)
$[(NH_3)_5CoO_2Co(NH_3)_5]^{5+}(SO_4)(HSO_4)_3$	oxidierter μ-Peroxokomplex		180°	1,312	(38)
$[(NH_3)_5CoO_2Co(NH_3)_5]^{5+}(NO_3)_5$	oxidierter Sauerstoffträger?		180°	1,45 ± 0,06	(39)

189

E. Bayer und P. Schretzmann

Metalls und des Liganden gesprochen werden. Ist das Metall aber redoxaktiv, d. h. leicht oxidier- oder reduzierbar, so ist es nach *Jørgensen*
(40a, b) nicht mehr möglich, bei der Kombination mit einem ebenfalls
redoxaktiven Liganden von Wertigkeiten zu sprechen.

Oxygenierte Metallkomplexe enthalten stets redoxaktive Zentralatome

$$Fe(II)-Fe(III); \quad Cu(I)-Cu(II); \quad Co(II)-Co(III); \quad Ir(I)-Ir(III)$$

und den redoxaktiven Sauerstoff als Liganden. Man kann keine Wertigkeiten mehr zuordnen, sondern nur noch von einem vorwiegend auf dem
Metall oder dem Liganden lokalisierten gemeinsamen Molekülorbital
sprechen, mit anderen Worten: von Partialladungen auf den Bindungspartnern. Der O—O-Kernabstand ist ein untrügliches Maß für diese
Partialladungen.

Betrachten wir zum Beispiel den O—O-Abstand im Sauerstoffkomplex

$$[Ir(O_2) (Cl) (CO) (P (C_6H_5)_3)_2].$$

Er beträgt 1,30 Å, was dem Superoxidanion entspricht. Dieses liegt aber
sicher nicht vor, da $O_2^{\cdot-}$ paramagnetisch, das Sauerstoffaddukt aber diamagnetisch ist. Schematisch kann dieser Ladungszustand durch eine
Valenzmesomerie

$$Ir(I) \cdots O_2^0 \longleftrightarrow Ir(II) \cdots O_2^- \longleftrightarrow Ir(III) \cdots O_2^{2-} \tag{5}$$

wiedergegeben werden, wobei jeder Formel mit diskreten Wertigkeits-
angaben die Bedeutung einer mesomeren Grenzstruktur zukommt.

IV. Sauerstoff als Acceptormolekül in Donor-Acceptor-Komplexen

Es ist eine bekannte Erscheinung, daß aromatische Amine wie Dimethyl-
anilin bei Sauerstoffzutritt sofort intensive Färbungen annehmen.

Evans (41) stellte fest, daß allgemein nukleophile Aromaten reversible
Sauerstoffaddukte bilden, wobei starke Lichtabsorption auftritt. Wird
nur kurzzeitig belichtet, so kann im Vakuum oder durch Einleiten von
Stickstoff der Sauerstoff wieder ausgetrieben werden. Wird das Addukt
längere Zeit belichtet, so tritt irreversible Oxidation ein. *Evans* erklärt
das mit der Bildung eines reversiblen Donor-Acceptor-Komplexes als
Vorstufe der lichtinduzierten Oxidation des Aromaten.

Nach *Tsubomura* und *Mulliken (42)* bilden auch σ-Donoren wie aliphatische Amine, z. B. Triäthylamin, und nach *Navon (43)* Halogenanio-

190

nen mit molekularem Sauerstoff Kontakt-charge-transfer-Komplexe. Sogar der schwache σ-Donor H_2O bildet mit O_2 schwache Assoziate (44). Die charge-transfer-Absorption liegt um so langwelliger, je basischer der Donor, d. h. je niedriger die Ionisationsenergie des Donors ist (42, 43, 45). Isolierbar sind solche Sauerstoff-Molekülkomplexe nicht.

V. Molekülkomplexe mit Metallchelaten als Donor-Molekülen

Kupfer-, Nickel- und Palladium-Chelate des 8-Hydroxychinolins bilden mit organischen π-Acceptoren typische Molekülkomplexe (46, 47). Solche Acceptoren sind Chloranil (Tetrachlor-para-benzochinon) und 2.3.5.6-Tetracyanobenzol. In diesen Molekülkomplexen alternieren parallele Schichten planarer Chelatmoleküle mit Schichten der organischen π-Acceptoren. Der Schichtabstand von etwa 3,5 Å entspricht den van der Waals-Abständen. Innerhalb der Moleküle treten bei der Bildung des Molekülkomplexes nur unerhebliche Änderungen auf. Es ist somit gezeigt, daß Metallchelate als reine π-Donoren in Molekülkomplexen auftreten können.

VI. Reversibel oxygenierbare Metallkomplexe (soweit sie unmittelbare Modellsubstanzen für die Atmungspigmente sind, werden sie in den folgenden Kapiteln behandelt).

Eine Zusammenstellung von reversiblen Sauerstoffkomplexen synthetischer Metallchelate ist 1963 von *Vogt, Faigenbaum* und *Wiberley* erschienen (9). An dieser Stelle sollen nur Verbindungen betrachtet werden, die strukturelle Aufschlüsse liefern.

Fallab (48) nimmt an, daß die Bildung reversibler Sauerstoffaddukte als Vorstufe irreversibler Oxidation von Metallkomplexen allgemeiner verbreitet ist. Der Nachweis und die Isolierung dieser Addukte gelingt nur in wenigen Fällen.

Die reversible Oxygenierung wird im wesentlichen bei Elementen der VIII. Nebengruppe des Periodensystems gefunden. Hinweise für eine reversible Oxygenierbarkeit existieren auch für Mangan (49) und Rhenium (9) in der VII. Nebengruppe sowie für Kupfer (50, 51) in der I. Nebengruppe.

A. Sauerstoffkomplexe im Verhältnis Metall : O_2 = 1 : 1

1. Iridium

Der Iridiumkomplex [Ir(Cl)(CO)(P(C₆H₅)₃)₂] nimmt in Benzol reversibel Sauerstoff auf, wie *Vaska* gezeigt hat (*52*). Das 1 : 1-Addukt ist kristallin isolierbar. Abb. 7 gibt die Struktur der Verbindung wieder, wie sie von *Ibers* und *La Placa* (*36*) ermittelt wurde.

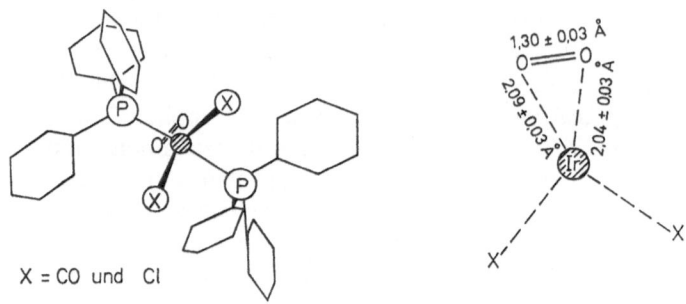

Abb. 7. Struktur von [O₂Ir (CO)(Cl)(P(C₆H₅)₃)₂] nach *Ibers* und *La Placa* (*36*).

Beide Sauerstoffatome haben vom Iridium gleichen Abstand, sie sind also äquivalent gebunden, wie es der Theorie von *Griffith* entspricht. Der Abstand der beiden Sauerstoffatome beträgt 1,30 Å, er ist größer als im molekularen Sauerstoff (1,21 Å), wesentlich kleiner als in Peroxiden und entspricht dem Abstand in Superoxiden (1,28 Å). Trotzdem liegt kein Radikalanion O_2^{-} vor, da der oxygenierte Komplex diamagnetisch ist.

Die reversible Bindung des Äthylens an [Ir(CO)(Cl)(P(C₆H₅)₃)₂] wurde von *Vaska* und *Rhodes* (*53*) nachgewiesen, der reversible Tetrafluoräthylenkomplex von *Jones* und *Parshall* (*54*) isoliert. Ein Vergleich der Bindung des Sauerstoffs mit der Bindung von Olefinen an diesen Komplex ist somit nicht unberechtigt. [Ir(CO)(J)(P(C₆H₅)₃)₂] ist wesentlich O₂-affiner als der Chlorokomplex. Das entstehende dunkelbraune [(O₂) Ir(CO) (J) (P (C₆H₅)₃)₂] enthält den Sauerstoff irreversibel fest gebunden (*108*). Der Ersatz des Chlors durch das elektropositivere, leichter polarisierbare Jod hat demnach eine Verstärkung der Bindung des Sauerstoffs am Iridium zur Folge. Die Röntgenstrukturanalyse nach *McGinnety*, *Doedens* und *Ibers* (*108*) ergab das außerordentlich interessante Resultat, daß der Übergang vom reversiblen zum irreversiblen Sauerstoffkomplex mit einer Vergrößerung des O—O-Kernabstandes von 1,30 Å auf 1,47 Å verbunden ist, wobei jedoch die Äquidistanz der Sauerstoffatome erhalten bleibt.

Dieser Abstand entspricht dem des Peroxidanions (1,49 Å, Tabelle 2). Der Elektronentransfer vom Metall zum O_2-Liganden muß durch den Ersatz des Chlors durch Jod verstärkt worden sein. Auch der Ersatz des Chlors durch einen Phosphin-Liganden im

$$[Ir \ (P(C_6H_5)_2 - (CH_2)_2 - P)C_6H_5)_2)]^+$$

führt zu einer Stabilisierung des O_2-Adduktes (55, 57).

Das O_2-Adukt von $[Co(P \ (C_6H_5)_2 - (CH_2)_2 - P(C_6H_5)_2)_2]^+$ (55) und die von *Wilke* et al. (58) dargestellten O_2-Addukte von Nickel-, Palladium und Platinphosphinkomplexen sind irreversibel fest. Sie gehen leicht eine intramolekulare Redoxreaktion unter Oxidation des Phosphins ein.

Umgekehrt muß in den H_2O_2-Derivaten des Cr(V) und Cr(VI), CrO_5 und $[CrO_8]^{3-}$ ein erheblicher Elektronentransfer vom O_2^{2-}-Liganden zum stark elektronenaffinen Zentralatom stattfinden, wie die O—O-Kernabstände von 1,40 Å ((35), Tabelle 2) ergeben.

Es scheint eine kontinuierliche Zunahme der Bindungsfestigkeit und der O—O-Kernabstände in der Reihe von labilen O_2-Addukten → irreversiblen O_2-Addukten → Peroxokomplexen zu geben.

Tabelle 3 bringt einen Vergleich von O_2- und Olefin-Komplexen von Metallen der VIII. Nebengruppe.

Tabelle 3. *Komplexe, die Sauerstoff im Verhältnis Metall : O_2 = 1 : 1 fixieren.*

Verbindung	Reversibilität der O_2-Bindung	Lit.	Bildung eines Olefinkomplexes mit	Lit.
$[Co((C_6H_5)_2P - (CH_2)_2 - P(C_6H_5)_2)_2]^+X^-$ X = (ClO_4^-), $[B(C_6H_5)_4]^-$	irreversibel	(55)		
$[Rh \ Cl(As(C_6H_5)_3)_3]$	?	(56)		
$[Ir(Cl) \ (CO)(P(C_6H_5)_3)_2]$	reversibel	(52)	$CH_2 = CH_2$ reversibel nicht isolierbar	(53)
			$CF_2 = CF_2$ reversibel isolierbar	(54)
$[Ir \ (CO)(J) \ (P(C_6H_5)_3)_2]$	irreversibel	(108)		
$[Ir((C_6H_5)_2P - (CH_2)_2 - P(C_6H_5)_2)_2]^+X^-$ X = Cl^-, Br^-, J^-, ClO_4^-, $[B \ (C_6H_5)_4]^-$	reversibel	(55, 57)		
$[Ni(PR_3)_2]$ R = C_6H_5	irreversibel	(58)	Äthylen	(58)
$[Pd(PR_3)_2]$ R = C_2H_5	irreversibel	(58)		
$[Pt(PR_3)_2]$	irreversibel	(58, 59)	Olefine	(59, 60)

Diese Komplexe der VIII. Nebengruppe haben folgendes gemeinsam: sie enthalten das Metall in niederer Oxidationsstufe, haben Liganden hoher Polarisierbarkeit, die eine niedere Oxidationsstufe des Metalls stabilisieren, und die Oxygenierung findet in unpolaren, protonenfreien Medien statt. In saurer wässriger Lösung wird $[Ir(O_2)(CO)(Cl)(P(C_6H_5)_3)_2]$ irreversibel autoxidiert, wobei H_2O_2 entsteht (*52*).

Die Bildung eines Sauerstoffadduktes ist kürzlich auch bei dem tetrakoordinierten Eisen(II)-Chelat des N.N'-Diäthylaminotroponimins (= Et$_2$ATI) festgestellt worden (*61*). Da das Di(aminotroponimin)-eisen(II) ein pentakoordiniertes, reversibles Pyridinaddukt (ATI)$_2$Fe(II) · (Pyridin) bildet, ist anzunehmen, daß das O$_2$-Addukt des Diäthylamino-derivates die Zusammensetzung (Et$_2$ATI)$_2$Fe · O$_2$ hat. Dies wäre für die Betrachtung des Oxyhämoglobins von größtem Interesse (s. Kap. VII).

B. Oxygenierung im Verhältnis Metall : $O_2 = 2 : 1$

1. Hexakoordinierte Kobalt(II)-Komplexe

In wässriger Ammoniaklösung ist das Ion $[Co(NH_3)_6]^{2+}$ durch Einleiten von O_2 reversibel oxygenierbar (*48, 62, 63*). Es entsteht ein braunes, dia-magnetisches, binukleares Addukt, das langsam, vornehmlich aber bei erhöhter Temperatur, in mononukleare Kobalt(III)-Komplexe zerfällt (*62*).

Die Bildung des braunen Adduktes ist reversibel durch Erhöhung der Ammoniakkonzentration oder bei Durchleiten von Stickstoff (*62*).

$$[Co(NH_3)_6]^{2+} \underset{\substack{NH_3 \text{ oder} \\ \text{Verminderung} \\ \text{des} \\ O_2\text{-Partialdrucks}}}{\overset{O_2}{\rightleftharpoons}} [(NH_3)_5Co(O_2)Co(NH_3)_5]^{4+} \xrightarrow[\text{langsam}]{H^+}$$

$$\rightarrow \quad [Co(NH_3)_5(H_2O)]^{3+} \tag{6}$$

Außer Ammoniak können auch andere Amine Co^{2+} in hexakoordinier-ten Komplexen zur reversiblen Sauerstoffbindung befähigen (Tabelle 4). Von den 6 Liganden müssen mindestens 4 Stickstoffliganden sein (Rest: H_2O oder Säureanionen), sonst geht die Oxygenierung zu schnell in eine Oxidation zu mononuklearen Kobalt(III)-Komplexen über (*48*).

Tabelle 4. *Stickstoff-Liganden in reversibel oxygenierbaren Kobalt(II)-Komplexen.*

Stickstoff-Ligand	Literatur
NH_3	*(48, 62, 63)*
$H_2N-CH_2-CH_2-NH_2$	*(48)*
$H_2N-CH_2-CH_2-NH-CH_2-CH_2-NH_2$	*(48)*
$H_2N-(CH_2)_2-NH-(CH_2)_2-NH-(CH_2)_2-NH_2$	*(65)*
1.4.8.11-Tetraaza-cyclotetradecan (Cyclam)	*(66)*

Fallab und Mitarb. *(48, 67, 68)* geben aufgrund kinetischer Untersuchungen folgendes Reaktionsschema für die Oxygenierung und Oxidation hexakoordinierter Kobalt(II)-Komplexe an (L = stickstoffhaltiger Ligand):

$$2\ (L)_6\ Co(II) + O_2 \underset{\substack{\text{schnell}\\ \text{reversibel}}}{\rightleftharpoons} (L)_5Co(O_2)Co(L)_5 \xrightarrow{\text{langsam}}$$

$$A \qquad\qquad\qquad B$$

$$\xrightarrow{\text{langsam}} (L)_5Co(III)-O-O-Co(III)(L)_5 \xrightarrow{[H^+]} 2\ Co(III) + H-O-O-H \quad (7)$$

$$ C \qquad\qquad\qquad\qquad D$$

Das braune, diamagnetische Sauerstoffaddukt B soll sich in schneller reversibler Reaktion bilden. Der Kobalt(III)-μ-peroxo-Komplex C bilde sich in einer langsamen intramolekularen Redoxreaktion aus B. Dieser Schritt soll eine Umlagerung des O_2-Liganden im Molekül zur Folge haben, die eine Aktivierungsenergie von 20 Kcal/mol erfordert. C soll zu mononuklearen Kobalt(III)-Komplexen D und H_2O_2 hydrolysierbar sein.

Die Reversibilität der Sauerstoffbindung ist nachgewiesen *(48, 62, 67)*, ebenso H_2O_2 als Hydrolysenprodukt in saurer Lösung *(66, 69)*.

Über die intramolekulare Umlagerung herrscht noch Unklarheit. Darauf soll kurz eingegangen werden.

a) Aus $[(NH_3)_5Co(O_2)Co(NH_3)_5]^{4+}$ kann sich das grüne paramagnetische Kation $[(NH_3)_5Co(O_2)Co(NH_3)_5]^{5+}$ (E) in reversibler Redoxreaktion bilden *(70)*.

Vlček (70) hat für E einen Strukturvorschlag gemacht, den Abb. 8 wiedergibt.

Der O_2-Ligand liegt senkrecht zur Kobalt-Kobalt-Kernverbindungslinie, das paramagnetische Elektron befindet sich frei beweglich in einem 4-Zentren-Molekülorbital, der aus den dyz-Atomorbitalen der Kobaltatome und dem π_y^*-Molekülorbital des Sauerstoffs gebildet werde. Im diamagnetischen Sauerstoffaddukt B soll dieselbe Struktur vorliegen, mit doppelt besetztem 4-Zentren-Molekülorbital.

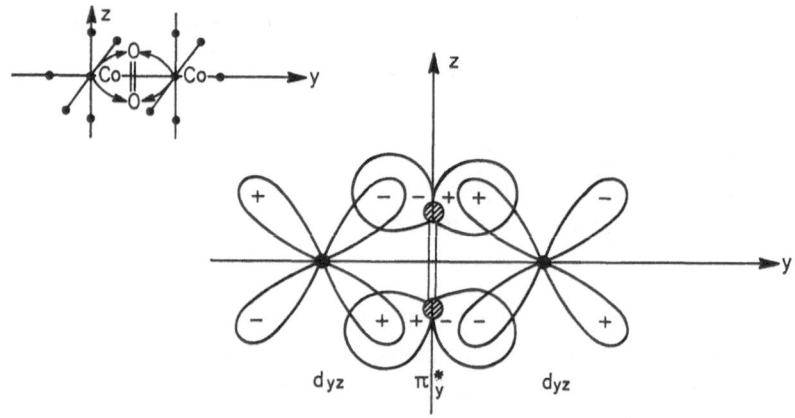

Abb. 8. 4-Zentren-Molekülorbital im Kation [(NH₃)₅ Co O₂ Co (NH₃)₅]⁵⁺ nach *Vlček* (70).

Vlček und *Basolo* (71) haben für diese Hypothese eine ausgezeichnete Stütze gefunden. Das Anion [Co(CN)₅]³⁻ bindet begierig Sauerstoff unter Ausbildung binuklearer Sauerstoffaddukte (69).

Vom p-Benzochinon darf nach dem Vinylogieprinzip der organischen Chemie eine Reaktivität erwartet werden, wie sie dem Singlettsauerstoff entspricht. Tatsächlich bildet sich ein p-Benzochinonaddukt der Zusammensetzung [(CN)₅Co(p-Benzochinon)Co(CN)₅]⁶⁻. Das p-Benzochinon-Addukt soll ein typischer π-Komplex sein (71), ein Analogieschluß zum O₂-Addukt ist nicht unberechtigt.

c) Gestützt wird die These des 4-Zentrenmolekülorbitals auch durch EPR-spektroskopische Untersuchungen am [(NH₃)₅CoO₂Co(NH₃)₅]⁵⁺ und [(CN)₅CoO₂Co(CN)₅]⁵⁻ (72, 73).

Es konnte gezeigt werden, daß beide Kobaltatome gleichwertig sind mit allerdings geringer Aufenthaltswahrscheinlichkeit des paramagnetischen Elektrons. Dieses ist vorwiegend auf der O₂-Gruppe lokalisiert. Das spricht für einen 4-Zentrenmolekülorbital, der jedoch vorwiegend auf der Sauerstoffgruppe lokalisiert ist.

d) Zwei Röntgenstrukturanalysen für das paramagnetische Kation [(NH₃)₅Co(O₂)Co(NH₃)₅]⁵⁺ liegen vor, siehe Abb. 9. Die Analyse von *Vannerberg* und *Brosset* (39) bestätigt den Vlček-Strukturvorschlag, mit der Einschränkung, daß die O₂-Gruppe um 45° um die Co—Co-Achse verdreht ist, während die Analyse von *Schaefer* und *Marsh* (38) eine Struktur beschreibt, bei der je ein O-Atom die Ecke eines Oktaeders um das Kobaltatom bildet. Der O—O-Abstand entspricht dem Abstand im Superoxidanion im Einklang mit den EPR-spektroskopischen Resultaten.

Schaefer und *Marsh* zeigen, daß im Sulfat [(NH₃)₅Co O₂ Co (NH₃)₅]⁵⁺ (SO₄) (HSO₄)₃ die O₂-Gruppe Wasserstoffbrücken ausbildet. Es wäre möglich, daß dies eine Verzerrung der Vlček-Struktur im Kristall zur Folge hat, oder die von *Fallab* postulierte Umlagerung beim Auskristallisieren eintritt.

Mit der Analyse von *Vannerberg* und *Brosset* hat die von *Schaefer* und *Marsh* ermittelte Struktur die Ebene gemeinsam, in der beide Kobalt- und Sauerstoffatome liegen.

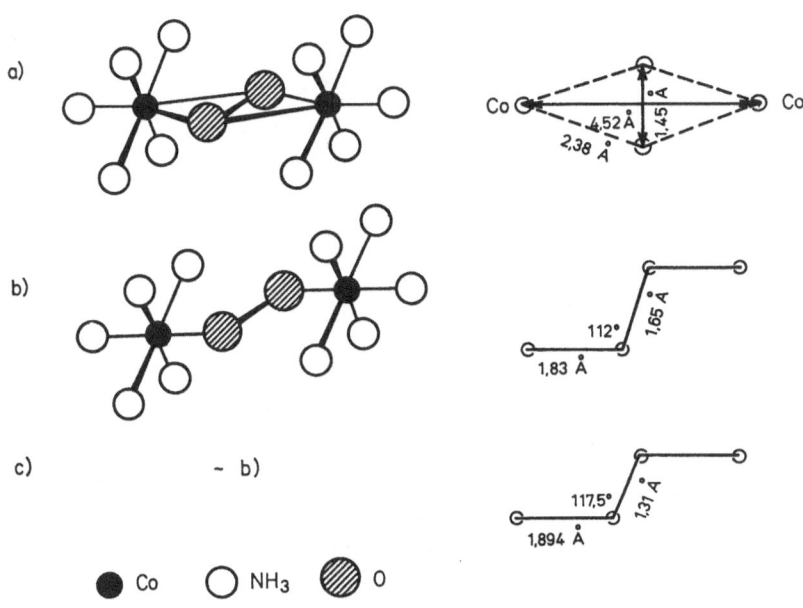

Abb. 9a — c. Strukturen von

a) $[(NH_3)_5 Co\ O_2\ Co\ (NH_3)_5]^{5+}$ $(NO_3)_5$ nach *Vannerberg* und *Brosset* (*39*)

b) $[(NH_3)_5 Co\ O_2\ Co\ (NH_3)_5]^{4+}$ $(SCN)_4$ nach *Vannerberg* (*37*)

c) $[(NH_3)_5 Co\ O_2\ Co\ (NH_3)_5]^{5+}$ $(SO_4)\ (HSO_4)_3$ nach *Schaefer* und *Marsh* (*38*)

e) Die Struktur von $[(NH_3)_5Co\ O_2\ (NH_3)_5]^{4+}$ $(SCN)_4$ ist in Abb. 9b wiedergegeben (*37*). Auch sie zeigt eine gewinkelte Co—O—O—Co-Brücke mit allen 4 Brückenatomen in einer Ebene. *Vannerberg* beschreibt die Verbindung als μ-Peroxo-Kobalt(III)-Komplex.

f) *Bosnich, Poon* und *Tobe* (*66*) finden in den Reflexionsspektren von festen Salzen des $[X(Cyclam)\ Co\ O_2\ Co\ (Cyclam)X]^{n+}$ Analogien zu den Spektren hexakoordinierter Kobalt(III)-Komplexe, zusätzlich aber intensive charge-transfer-Banden um 450 mμ (s. dazu die Interpretation des Lichtabsorptionsspektrums von Oxyhämocyanin durch *Frieden* et al. (*221*) und *van Holde* (*232*); Kap. VIII).

197

Es existiert heute noch kein einheitliches Bild über diese interessante Verbindungsklasse. Die postulierte Umlagerung der O_2-Gruppe, verbunden mit einem intramolekularen Elektronentransfer, bedarf weiterer experimenteller Bestätigung. Es ist möglich, daß in Lösung und bei bestimmten Voraussetzungen im Kristall eine Struktur der labilen O_2-Addukte vorliegt, wie *Vlček* sie vorgeschlagen hat.

Prinzipiell wäre die Bindung in den reversiblen Addukten auch zu erklären, wie in Abb. 10 wiedergegeben ist (s. auch (*37*)).

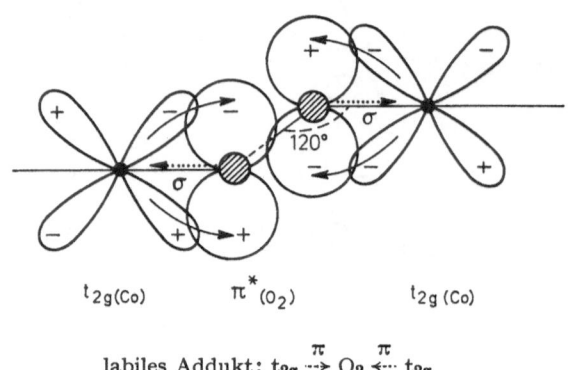

$$t_{2g(Co)} \qquad \pi^*_{(O_2)} \qquad t_{2g\,(Co)}$$

$$\text{labiles Addukt: } t_{2g} \overset{\pi}{\dashrightarrow} O_2 \overset{\pi}{\dashleftarrow} t_{2g}$$

$$\text{irreversibles Peroxid: } t_{2g} \underset{\sigma}{\overset{\pi}{\rightleftarrows}} O_2 \underset{\sigma}{\overset{\pi}{\rightleftarrows}} t_{2g}$$

Abb. 10. Molekülorbitale der Co—O_2-Bindung im Kation $[(NH_3)_5\,Co\,O_2\,Co\,(NH_3)_5]^{4+}$ (eine Alternative zur Hypothese von *Vlček*).

Der Unterschied zwischen reversiblen und irreversiblen binuklearen O_2-Komplexen wäre nach dieser Hypothese nur gradueller Natur, mit dem Vorliegen desselben Grundgerüstes. Tatsache ist, daß von einem binuklearen O_2-Komplex zwei Bindungsisomere noch nicht isoliert wurden.

Man kann sich vorstellen, daß in den reversiblen Addukten nur eine schwache π-Bindung mit geringem Ladungstransfer $t_{2g\,(Co)} \rightarrow \pi^*_{(O_2)}$ in einem 4-Zentren-Molekülorbital vorliegt, während eine σ-Bindung lone pair$_{(O_2)} \rightarrow e_{g(Co)}$ keine Rolle spielen sollte.

Mit zunehmender Donorstärke der t_{2g}-Orbitale des Metalls würde die Ladungsüberführung zur O_2-Gruppe so groß, daß diese in Form von σ-Bindungen einen Teil ihrer negativen Beladung an die Metallatome zurückgeben wird. Die Bindung zwischen Metall und Sauerstoff würde fester, das Addukt irreversibel.

Variierbar ist die Donorstärke der t_{2g}-Orbitale durch die Liganden am Metall. So ist es verständlich, daß zwar $[(NH_3)_5Co(O_2)\,Co(NH_3)_5]^{4+}$

ein reversibles Addukt ist, der Cyanokomplex $[(CN)_5Co\ O_2\ Co\ (CN)_5]^{6-}$ aber keinen Sauerstoff mehr reversibel abgibt (69).

Eine andere Möglichkeit zur Verschiebung der Ladungsverteilung innerhalb des binuklearen Komplexes ist durch das Medium gegeben. Vor allem die Möglichkeit, mit acidem Wasserstoff zu reagieren oder Wasserstoffbrückenbindungen einzugehen, sollte von großem Einfluß auf den Grad der Ladungsüberführung vom Metall auf die Sauerstoffbrücke sein. Damit wäre erklärbar, warum ein und derselbe Komplex in neutralem oder alkalischem Medium reversibel Sauerstoff aufnimmt, während in acidem Medium H_2O_2 entsteht.

2. Aminosäuren- und Peptidkomplexe von Kobalt(II)

a) Oxygenierung von Kobalt(II)-Chelatkomplexen des Histidins und seiner Derivate.

L-Histidin befähigt Kobalt(II) in wässriger Lösung zur reversiblen Sauerstoffbindung (8, 9). Das braune binukleare O_2-Addukt konnte isoliert werden (74, 75), es enthält 2 Histidinreste pro Kobaltatom.

Auch L-Histamin und L-Histidinol bilden reversibel oxygenierbare Kobalt(II)-Komplexe (75). Die Carboxylfunktion des dreizähligen Liganden Histidin ist demnach kein notwendiges Strukturelement für die Oxygenierung.

Fehlt dagegen die primäre Aminogruppe des Histidins, wie es in Imidazolyl(4)-propionsäure und Imidazolyl(4)-milchsäure der Fall ist, oder wird die Basizität der primären Aminogruppe durch Acetylierung herabgesetzt (Nα-Acetylhistidin), so ist reversible Oxygenierung nicht mehr möglich (75).

1-Benzylhistidin bildet reversibel oxygenierbare Kobalt(II)-Chelate. Der Imidazolring ist demnach in den O_2-Addukten als Neutralligand über den tertiären Stickstoff am Kobalt gebunden (75).

Imidazolyl(4)-glycin und 4-Aminomethylimidazol bilden mit Kobalt(II) Fünfring-Chelate, die ebenfalls zur reversiblen Oxygenierung befähigt sind (75).

Es ist nach diesen Ergebnissen anzunehmen, daß 4 Stickstoffliganden: 2 Imidazolringe und 2 primäre aliphatische Aminogruppen die Stabilität dieser binuklearen O_2-Addukte bewirken. Die Hexakoordination des Kobalts wird vervollständigt durch H_2O oder eine Carboxylfunktion als fünftem und der binuklearen O_2-Brücke als sechstem Liganden. Diese Verhältnisse entsprechen der von *Fallab* (48, 67) beschriebenen Oxygenierung von [Co(II) (en)_2 Oxalat].

b) Dipeptide wie Glycylglycin befähigen Kobalt(II) in alkalischem Medium ebenfalls zur reversiblen Sauerstoffbindung (9), jedoch ist N-Glycyclprolin dazu unfähig (76). *Tang* und *Li* (77) sprachen die

Vermutung aus, daß das Fehlen des Amidprotons beim Glycylprolin damit zusammenhängen könne.

Tanford und Mitarbeiter (*78*) fanden, daß die Bildung des braunen Sauerstoffadduktes in alkalischem Medium Hydroxylionen verbraucht. Das ist folgendermaßen zu erklären: Der Stickstoff der Amidgruppierung zeigt nur geringe Basizität. Das Amidproton ist aber in alkalischem Medium abspaltbar und das entstehende Anion B sollte in seiner Basizität

$$
\begin{array}{ccc}
\underset{\substack{|\\N\\ \diagdown\ \diagup}}{H}\diagdown_{\underset{\parallel}{C}}\diagup & \xrightarrow[-H_2O]{+OH^-} & \underset{\substack{\ominus\\N\\ \diagdown\ \diagup}}{}\diagdown_{\underset{\parallel}{C}}\diagup \quad \longleftrightarrow \quad \overline{N}\diagdown_{C}\diagup
\end{array}
\tag{8}
$$

$$
\quad O \qquad\qquad\qquad O \qquad\qquad\qquad |\underline{O}| \ominus
$$

$$
\qquad\qquad\qquad\qquad\qquad B
$$

Ammoniak übertreffen.

Die Bildung des reversiblen Sauerstoffadduktes in alkalischem Medium kann auf folgendem Wege zustande kommen

$$
Co^{2+} + 2\ H_2N-CH_2-\underset{\underset{O}{\parallel}}{\overset{\overset{H}{|}}{C}}-N-CH_2-COO^- \underset{-2\ OH^-}{\overset{+2\ OH^-}{\rightleftharpoons}} 2\ H_2O +
$$

$$
[Co(II)\ (H_2N-CH_2-\underset{\underset{O}{\parallel}}{C}-\overline{N}-CH_2-COO^-)_2]^{2-}
\tag{9}
$$

$$
A
$$

$$
2\ A + O_2 \rightleftharpoons \text{braunes Sauerstoffaddukt C}
$$

Das braune Sauerstoffaddukt geht nach *Tanford* (*78*) in einen roten, nach *Tang* und *Li* einfach negativ geladenen Komplex über (*77*). Dieser ist nach polarographischen Untersuchungen kein Peroxokomplex mehr (*79*). *Gillard, Mason, McKenzie* und *Roberts* haben die Struktur des roten Komplexes röntgenographisch aufgeklärt (*80*).

Die Liganden sind eben gebaut, müssen also sp²-Hybridisierung am Stickstoff aufweisen, was für eine Deprotonisierung des Amidstickstoffes spricht.

Es ist anzunehmen, daß das braune Sauerstoffaddukt die in Abb. 11 angegebene Struktur hat.

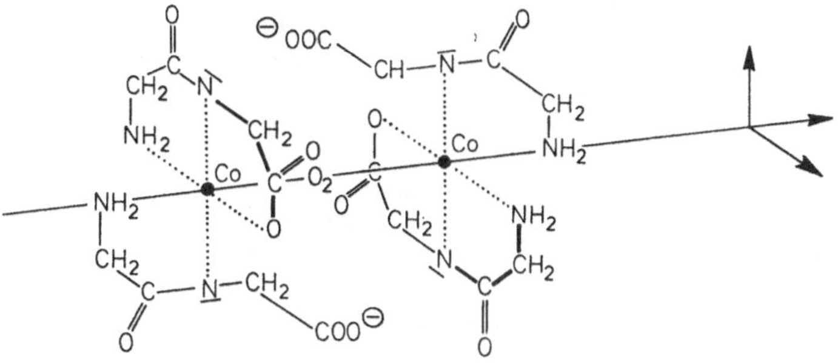

Abb. 11. Strukturvorschlag für den Sauerstoffkomplex des Bis(glycinyl-glycinato)-kobalt (II).

Zusammenfassung:

Hexakoordinierte Kobalt(II)-Komplexe sind reversibel oxygenierbar, wenn genügend Stickstoffliganden ausreichender Basizität am Zentralatom vorhanden sind. Bei der Oxygenierung wird kein Stickstoffligand abgelöst, solange noch andere Liganden wie H_2O oder Carboxylfunktionen das Kobaltzentralatom umgeben. Allen reversiblen Sauerstoffaddukten dieses Typs ist die Säurelabilität gemeinsam.

C. Oxygenierung planarer Kobalt(II)-Chelatkomplexe

Die Schiff'sche Base Bis(salicyliden)-äthylendiamin und ihre Derivate bilden Kobalt(II)-Chelate, die zur reversiblen Oxygenierung befähigt sind (9, 81). Gegenüber den bisher besprochenen Kobalt(II)-komplexen sind sie außergewöhnlich stabil gegen irreversible Oxidation. Die Oxygenierung dieser Chelate im Festzustande wurde von *Calvin* (8), *Diehl* (82), *Stewart* (10) und ihren Mitarbeitern sehr intensiv untersucht.

1. Zur Struktur des Bis(salicyliden)-äthylendiaminkobalt(II)] (= SAD—Co(II), siehe Abb. 2 und Abb. 12)

a) Kristallstruktur

SAD—Co(II) kommt nach Calvin et al. in 3 Modifikationen vor, die im Festzustand mit Sauerstoff im Verhältnis $Co : O_2 = 2 : 0$; $2 : 1$; $3 : 1$ reversibel kombinieren (8). Die kristallographische Untersuchung (83) hat gezeigt, daß der Chelatkomplex eben ist und in den aktiven Modifikationen in Schichtgittern angeordnet ist. Eine lückenlose Packung der Moleküle in den einzelnen Netzebenen ist nicht möglich, vor allem dann, wenn Ringsubstituenten am Komplexbildner vorliegen.

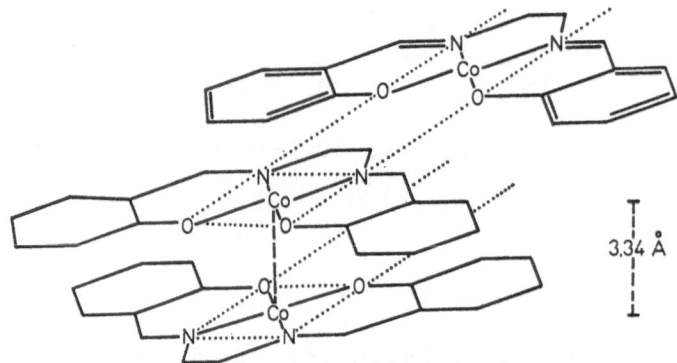

Abb. 12. Struktur des Salcomins, der 2:1 aktiven Modifikation des Bis(salicyliden)-äthylendiaminkobalt (II) nach *Hughes, Barkelew* und *Calvin* (*83*).

Der Sauerstoff hat zwei Möglichkeiten, in den Kristall einzudringen und zu den Aktivzentren der Fixierung zu gelangen: einmal entlang der Netzebenen und zum zweiten durch die Lücken innerhalb der Molekül-lagen. In der 2 : 1 aktiven Modifikation (Abb. 12), liegen jeweils zwei Kobaltatome in benachbarten Gitterlagen senkrecht übereinander, im 3 : 1 aktiven Kristall bleibt ein Drittel der Kobaltatome ohne „Nachbar". Es ist also anzunehmen, daß der Sauerstoff zwischen zwei Kobaltatomen gebunden wird. Dafür spricht auch das Verschwinden des Paramagnetismus bei der Oxygenierung. SAD—Co(II) enthält ein ungepaartes Elektron (*84*) (low spin d^7-Elektronenkonfiguration des Kobalt(II)). Diamagnetismus kann nur auftreten bei gerader Elektronenzahl im Sauerstoffaddukt. Da das Sauerstoffmolekül 16 Elektronen mitbringt, muß es zwangsläufig mit 2 Kobaltatomen in Wechselwirkung treten, wenn ein diamagnetisches Sauerstoffaddukt auftreten soll. Wie genau die Oxygenierung des Salcomins im Verhältnis $Co : O_2 = 2 : 1$ erfolgt, ist aus Tabelle 5 ersichtlich.

Die inaktive Modifikation enthält die SAD—Co(II)-Moleküle in einer für das Sauerstoffmolekül undurchdringlichen Packung. In allen Modifikationen des SAD—Co(II) liegen dieselben Moleküle vor, da sie in Chloroformlösung identische Infrarot-Spektren ergeben (*85*).

b) Da die Auswertung der röntgenographischen Strukturanalyse auf der Annahme eines ebenen Chelatkomplexes beruhte (*83*), ist es von Wichtigkeit, daß diese Planarität des SAD—Co(II) Moleküls durch andere Messungen bestätigt werden konnte. Es gelang, in einem monomolekularen Film von SAD—Co(II) Molekülen die von einem Molekül beanspruchte Fläche zu bestimmen (*86*). Die Fläche von 86 Å^2 spricht für ein planares Molekül.

Planare Kobalt(II)-Komplexe zeigen eine charakteristische Ligandenfeldabsorption bei 1200 mµ im Lichtabsorptionsspektrum (*87*).

Tabelle 5. *Analyse verschiedener Modifikationen und Derivate des Salcomins.*

Nr.	Verbindung	Analyse	C	H	N	O	Reaktionssequenz zur O_2-Aufnahme	Literatur
1	SAD–Co(II) rot, inaktiv	ber.	59,10	4,30	8,63	18,14	1 $\xrightarrow{+ \text{Pyridin}}$ 2	(91)
		gef.	59,09	4,35	8,62	18,15		
2	SAD–Co(II)–Pyridin (Monopyridinat)	ber.	62,4	4,74	10,39	14,6	2 $\xrightarrow{- \text{Pyridin}}$ 3	(92)
		gef.	62,29	4,66	10,3	15	Pyridinabgabe: theor. 19,5%, best. 19,2%	
3	SAD–Co(II) braun, aktiv						3 → 4	(91)
4	SAD–Co · $\frac{1}{2}$ O_2						O_2-Aufnahme: theor. 4,7% O_2, gef. 4,94%	
5	3,3′–Diäthoxy-SAD–Co(II) · H_2O Monohydrat	ber.	55,70	5,57	6,50	13,69	5 $\xrightarrow{- H_2O}$ 6	(91)
		gef.	55,82	5,67	6,44	13,57	H_2O-Verlust theor. 4,2%, gef. 4,32%	
6	3,3′–Diäthoxy-SAD–Co(II) aktiv						6 → 7	
7	3,3′–Diäthoxy-SAD–Co · $\frac{1}{2}$ O_2	ber.	55,95	5,13	6,53	13,75	O_2-Aufnahme: theor. 3,87%, gef. 3,87%	(91)
		gef.	55,85	5,20	6,51	13,6		
8	(SAD–Co)$_2$ · O_2 · 2 DMF	ber.	55,08	5,11	10,14		O_2-Abgabe in Chloroform: ber. 3,86%, gef. 3,74%	(93)
		gef.	55,25	5,08	10,11			
9	(SAD–Co)$_2$ · O_2 · 2 Pyridin	ber.	60,01	4,56	10,00		O_2-Abgabe in Chloroform: ber. 3,81%, gef. 0,85%	(93)
		gef.	60,06	4,51	9,83			

(Ligandenfeldabsorptionen treten auf bei Elektronenübergängen zwischen d-Orbitalen des Kobalts, die im Ligandenfeld energetisch aufgespalten werden). Bei SAD—Co(II) wird diese Absorption beobachtet (88).

Die Messung der magnetischen Suszeptibilität spricht für ein ungepaartes Elektron im SAD—Co(II). Nach *Figgis* und *Nyholm* (89) und *West* (90) sind low spin-Kobalt(II)-Komplexe eben gebaut.

c) *Diehl* et al. haben das aktive Salcomin als binuklearen Komplex μ-Aquodi(bis(salicyliden)-äthylendiamin)-kobalt(II) angesehen (82). Diese Struktur, in der zwei Moleküle SAD—Co(II) über eine Wasserbrücke verbunden sein sollen, wird auch von *Vogt* et al. (9) und *Stewart* et al. (10) wiedergegeben. Der Strukturvorschlag kann aber nicht zutreffen, denn im IR-Spektrum der aktiven wie der inaktiven Modifikationen wird kein Wasser nachgewiesen (85, in KBr). Die Darstellung der 2 : 1-aktiven Modifikation aus der hochreinen inaktiven Modifikation über das genau definierte Monopyridinat (91) und dessen anschließende Aktivierung unter Entweichen von 19,2% Pyridin (Theorie = 19,5%) (92) zeigt, daß die aktive Modifikation kein Wasser enthält (Tabelle 5).

2. Pentakoordination bei SAD—Co(II)-Komplexen

SAD—Co(II) bildet ein Monohydrat (91), Monopyridinat (91) und einen Mononitrosylkomplex (94) definierter Zusammensetzung. Das Kobaltzentralatom hat in diesen Komplexen die Koordinationszahl 5. Leider ist von keinem dieser Komplexe die Struktur röntgenographisch aufgeklärt. Eine Strukturanalyse liegt für die Komplexe [SAD—Zn(H$_2$O)] und [SAD—Fe(III)(Cl)] vor (95, 96). Sie ergibt eine tetragonal-pyramidale Koordinationssphäre des Zentralatoms. Auch für den Nitrosylkomplex des Bis(dimethyl-dithiocarbaminato)-kobalt(II) ist eine tetragonal-pyramidale Struktur nachgewiesen (97). Es ist anzunehmen, daß auch im Sauerstoffaddukt des SAD—Co(II) im Festzustand eine solche tetragonal-pyramidale Struktur vorliegt, wobei der Sauerstoff die Spitze zweier Pyramiden bildet.

3. Über die Bindung im Sauerstoffaddukt

a) Die Bildungswärme des Sauerstoffadduktes von rd. 20 Kcal/Mol O$_2$ (98) ist für die Annahme einer σ-Valenzbindung zu niedrig. Sie liegt in der Größenordnung der Bildungswärmen von Molekülkomplexen (99). (Zum Vergleich: Bei der Oxygenierung von Hämoglobinen werden maximal 14 Kcal/Mol O$_2$ frei (2)).

b) Reversibilität der Oxygenierung: Der Sauerstoff kann aus dem festen SAD—Co-Addukt wieder ausgetrieben werden: 1. durch Erwärmen auf 80 ° C im Vakuum unter Rückbildung der aktiven Modifikationen; 2. durch Kochen am Rückfluß in Toluol, wobei die rote inaktive Modifikation entsteht; 3. Beim Lösen des O_2-Adduktes in Chloroform wird der Sauerstoff quantitativ ausgetrieben (85). Offensichtlich ist das isolierbare Mono-Chloroform-Addukt stabiler als das O_2-Addukt (100).

c) SAD—Co(II) zeigt im IR-Spektrum die C = N-Valenzschwingung bei 1533 cm⁻¹ (85). Im Sauerstoffaddukt liegt sie bei 1534 cm⁻¹ (88). Die Oxygenierung beeinflußt demnach die Struktur des planaren Chelates nur geringfügig.

d) Die reversible Oxygenierung von Metallkomplexen ist mit dem Auftreten sehr intensiver Lichtabsorptionen verbunden (Tabelle 6).

Tabelle 6. *Farbänderung bei der Oxygenierung von Komplexen.*

Verbindung	vor der Oxygenierung	nach der Oxygenierung	Literatur
SAD—Co(II), fest	zimtbraun	schwarz	(8)
SAD—Co(II), Lösung		665 mμ	(88)
Bis(glycylglycinato)-kobalt(II)	510 mμ; $\varepsilon = 5,5$ rosa	345 mμ; $\varepsilon = 3000$ braun	(79)
Bis(histidin)-kobalt(II)	486 mμ; $\varepsilon = 18,4$ rosa	385 mμ; $\varepsilon = 1620$ braun	(74)
[Ir(Cl)(CO)(P(C$_6$H$_5$)$_3$)$_2$]	gelb	orange	(52)

Nach *Yamada, Nishikawa* und *Yoshida* (88) ist die bei der Oxygenierung von SAD—Co(II) auftretende Lichtabsorption bei 665 mμ senkrecht zur Ebene des planaren Liganden polarisiert.

Die Erhaltung des Bindungsgerüstes der Komponenten, das Neuauftreten intensiver Lichtabsorptionen (charge transfer-Banden) und deren Polarisation sind charakteristisch für Molekülkomplexe (101, 102) wie Chinhydron.

Im Abschnitt III wurde gezeigt, daß molekularer Sauerstoff in der Lage ist, mit geeigneten Elektronendonoren charge transfer-Komplexe zu bilden, die als Vorstufen irreversibler Oxidationen angesehen werden können. Andererseits sind auch planare Chelatkomplexe zur Bildung von π-Molekularkomplexen befähigt (46, 47). Es liegt nahe, das reversible Sauerstoffaddukt des Bis(salicyliden)äthylendiaminkobalt(II) mit π-Molekularkomplexen zu vergleichen.

Formal ist das Sauerstoffaddukt des SAD—Co(II) auch als zweidimensionale Einlagerungsverbindung der Lewissäure O_2 zwischen Schichten der Lewisbase SAD—Co(II) aufzufassen, wie sie etwa bei den Graphiteinlagerungen von Lewissäuren vorliegen (103).

d) Zum Magnetismus im Sauerstoffaddukt von SAD—Co(II)

SAD—Co(II) ist paramagnetisch, jedes Kobaltatom trägt ein ungepaartes Elektron (low spin d^7 Konfiguration). Bei der Einlagerung des Biradikals O_2 wird der Paramagnetismus praktisch vollständig gelöscht, es muß Spinkopplung eingetreten sein. Nimmt man an, daß in der Koordinationssphäre zweier paramagnetischer Kobaltatome Spinkopplung im Sauerstoff induziert wird, wie es für Oxyhämoglobin postuliert wird (28), so ist der Ort größter Lewisacidität der leerstehende π^*-Molekülorbital, der in Richtung der Kobaltatome zeigt. Dies sei die y-Achse. Der π_y^*-Molekülorbital kann mit je einem t_{2g} Atomorbital am Kobalt kombinieren unter Ausbildung eines 4-Zentren-Molekülorbitals nach *Vlček* (70) (Abb. 8 oder nach Abb. 10). *Calvin* (104) schloß aus kinetischen Daten auf die Anwesenheit von „aktiviertem" Sauerstoff im SAD—Co(II)-Kristall vor der eigentlichen Sauerstoffbindung.

In diesem 4-Zentren-Molekülorbital ist es möglich, die beiden einsamen Elektronen der Kobaltatome unter Spinkopplung unterzubringen. Die reversible Bildung binuklearer Sauerstoffaddukte wäre dann folgendermaßen formulierbar:

Aktivierungsschritt

$$2\ Co(II)\downarrow + (\uparrow\uparrow)O_2 \rightleftharpoons Co(II)\downarrow + (\downarrow\uparrow)O_2 + Co(II)\uparrow \tag{10}$$

Reversible Oxygenierung

$$Co(II)\downarrow + (\downarrow\uparrow)O_2 + Co(II)\uparrow \rightleftharpoons Co(t_{2g} \rightarrow \pi^*)\ O_2\ (\pi^* \leftarrow t_{2g})Co \tag{11}$$

π-Molekularkomplex

A   B

Irreversible Oxidation (diese ist im festen $(SAD—Co)_2 \cdot O_2$ unterbunden)

$$B \rightarrow (Co(III)O_2^{2-}\ Co(III)\ ) \rightarrow Co(III) \underset{\pi}{\overset{\sigma}{\underset{\rightarrow}{\leftarrow}}} O \diagdown$$

$$O \underset{\pi}{\overset{\sigma}{\underset{\leftarrow}{\rightarrow}}} Co(III) \tag{12}$$

C   D

Hydrolyse $D \xrightarrow{H^+} Co(III) + H_2O_2$

Der elektronische Zustand im π-Komplex B liegt irgendwo zwischen der "no bond" Grenzstruktur A, bei der antiferromagnetische Spin-Kopplung den Paramagnetismus der Kobaltatome löschen müßte, und der Grenzstruktur mit vollendeter Ladungsüberführung, wie sie in C angegeben ist.

Daß der Übergang B → D eine intramolekulare Redoxreaktion ist, läßt sich daran erkennen, daß im π-Komplex B die Kobaltatome Elektronendonoren sind, während im σ-gebundenen μ-Peroxo-kobalt(III)-komplex D die Kobaltatome Elektronenacceptoren sind.

e) Zur Stabilität gegen irreversible Oxidation

Hexakoordinierte Kobalt(II)-Komplexe zeigen nur geringe Reversibilität der Oxygenierung und große Neigung zur irreversiblen Oxidation. Bei planaren Kobalt(II)-Chelaten im Festzustand ist die Reversibilität sehr groß, wie aus Tabelle 7 zu entnehmen ist.

Tabelle 7. *Cyclische Oxygenierung von Salcomin und dem Fluorderivat im Festzustand.*

Verbindung	Zahl der Oxygenierungs- prozesse	Restaktivität nach angegebener Zahl der Oxy- genierungen	Literatur
SAD—Co(II)	300	70%	(84)
3.3'—Difluor—SAD—Co(II)	1500	60%	(84)
SAD—Co(II)	3000	50%	(105)

Hält man sich vor Augen, daß Kobalt(III)-Komplexe fast ausschließlich die Koordinationszahl 6 haben, so ist es erklärbar, warum SAD—Co(II) im Festzustand so resistent gegen irreversible Oxidation ist. Die O_2-Addukte haben nicht die Möglichkeit in hexakoordinierte Kobalt(III)-Komplexe umzulagern, da der 6. Ligand fehlt.

f) Oxygenierung von SAD—Co(II) in Lösung

SAD—Co(II) ist nur in solvatisierenden Lösungsmitteln wie Chloroform, Pyridin, Chinolin und Dimethylformamid gut löslich (10, 93, 106). Es ist anzunehmen, daß sich pentakoordinierte Solvate bilden, die zum Teil isolierbar sind (81, 106). Aus wasserfreiem Pyridin wurde der Komplex [(Pyridin) (SAD—Co) O_2 (Co—SAD) (Pyridin)] isoliert (97). Unter den Bedingungen der Deoxygenierung von festem (SAD—Co)$_2$ · O_2 ist er stabil. Erst bei 170° tritt im Vakuum völlige Zersetzung des Komplexes ein. Man ist geneigt, ihn als irreversiblen μ-Peroxo-Komplex zu bezeichnen. *Calderazzo* et al. (93) haben jedoch festgestellt, daß die Sauerstoffbindung in diesem Komplex nicht vollständig irreversibel ist. Beim Lösen in Chloroform werden etwa 20% des Sauerstoffs freigesetzt. Aus wasserfreiem Dimethylformamid und Dimethylsulfoxid wurden die

Sauerstoffkomplexe [(X) (SAD—Co) O_2 (Co—SAD) (X)] isoliert (X = DMF oder DMSO, 93, Tabelle 5). Bei 100° verlieren sie den Sauerstoff und die Solvatmoleküle unter Ausbildung der aktiven Modifikation des Salcomins. In Chloroform tritt ebenfalls quantitative Deoxygenierung ein (93).

Der Vergleich der Strukturen im reversiblen Sauerstoffaddukt (SAD—Co)$_2$ · O_2 und im [(Pyridin) (SAD—Co) O_2 (Co—SAD) (Pyridin)] wäre von großem Interesse für die Frage, ob zwei verschiedene Geometrien der Sauerstoffbindung tatsächlich existieren, oder ob der Unterschied zwischen den labilen O_2-Addukten und den „irreversiblen" binuklearen Peroxiden nur gradueller Natur ist, wie es bei den mononuklearen Peroxiden der Fall zu sein scheint (108, Tabelle 2).

Im planaren SAD—Co(II) ist die Donorstärke der t_{2g} Orbitale des Kobalts gerade groß genug, um ein labiles O_2-Addukt zu bilden. Tritt aber ein σ-Donor wie Pyridin als 5. Ligand an das Kobaltatom, so wird die Donorstärke der t_{2g} Orbitale so sehr erhöht, daß die Bindung des Sauerstoffs zu fest wird, um noch reversibel zu sein. Entscheidend für die große Stabilität gegenüber irreversibler Oxidation ist auch die Tatsache, daß im festen (SAD—Co)$_2$ · O_2 und in Medien ohne acide Protonen keine Möglichkeit besteht, den Elektronentransfer vom Metall zur Sauerstoffgruppe durch Wasserstoffbrückenbildung zu erhöhen bzw. H_2O_2 zu bilden.

In wäßriger Lösung entsteht dagegen aus Bis(5-sulfosalicyliden)-äthylendiamin-kobalt(II) und Sauerstoff Hydrogenperoxid, das abdestillierbar ist (107).

In Abb. 13 ist das unterschiedliche Verhalten hexakoordinierter, pentakoordinierter und planarer Kobaltkomplexe schematisch wieder-

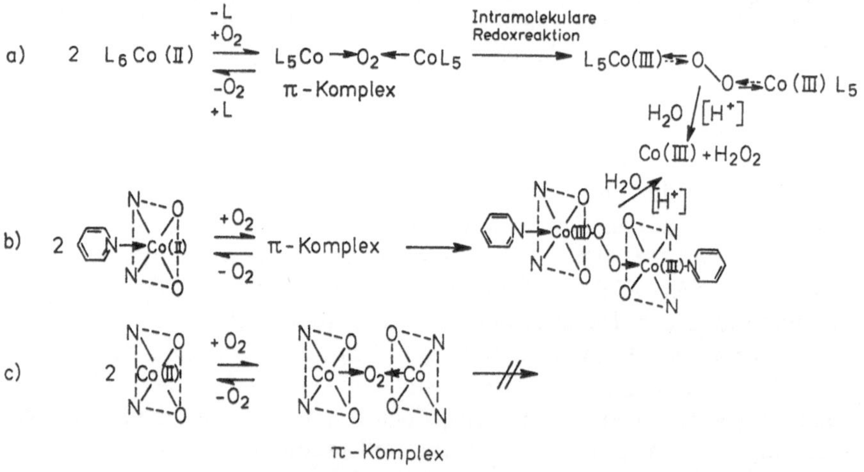

Abb. 13a— c. Schema der Oxygenierung von Kobalt(II)-Komplexen. a) hexakoordiniert, b) pentakoordiniert, c) planar.

gegeben. Die postulierte intramolekulare Redoxreaktion, verbunden mit der Umlagerung zweier strukturisomerer Sauerstoffkomplexe, bedarf weiterer experimenteller Untersuchung.

D. Welche Voraussetzungen muß ein guter Sauerstoffträger erfüllen?

(s. auch *8, 9, 20*).

1. Ein fester O_2-Träger sollte eine lockere, für kleine Moleküle durchdringbare Struktur haben, bzw. eine Schichtgitterstruktur, die eindimensional ausweitbar ist (*8*).
2. Das Reduktionspotential sollte ausreichend nieder sein, um das Metall zu befähigen, mit dem Sauerstoff eine lockere π-Bindung einzugehen. Andererseits darf es nicht zu nieder sein, da sonst der Ladungstransfer zu groß und die Bindung irreversibel fest wird.
3. Die Koordinationsstelle der O_2-Fixierung sollte unbesetzt sein (*20*).
4. Die Koordinationszahl im O_2-Addukt sollte für die niedrige Oxidationsstufe des Metalls nicht ungewöhnlich, für die höhere Oxidationsstufe aber nicht möglich sein.
5. H_2O_2 oder HO_2^- Bildung muß unterbunden werden.

VII. Hämoproteine

Kein Makromolekül ist in seiner Struktur und Funktion eingehender erforscht worden als die Hämoproteine Myoglobin und Hämoglobin. Über die reversible Oxygenierung dieser komplizierten Naturstoffe ist mehr bekannt als bei den niedermolekularen synthetischen Sauerstoff-Trägern. Gegenüber den Metallproteiden ist die Untersuchung der Hämoproteine dadurch erleichtert, daß eine prosthetische Gruppe aus dem Aktivzentrum der O_2-Fixierung abspaltbar ist.

1. Das Häm

In Myoglobinen und Hämoglobinen ist das Protohäm IX die prosthetische Gruppe, der planare Eisen(II)-Chelatkomplex des Protoporphyrins IX (Abb. 1a).

Im Chlorocruorohäm des Chlorocruorins ist die Vinylgruppe in Position 2 des Protohäms IX durch einen Formylrest ersetzt (Abb. 1b, (*109*)).

E. Bayer und P. Schretzmann

2. Assoziation der Untereinheiten in Hämoproteinen

Myoglobine bestehen aus einer Proteinkette und einem Häm, sie haben bei höheren Organismen ein Molekulargewicht von etwa 17 000.

Vertebraten-Hämoglobine bestehen aus 4 Molekülen Häm und 4 Proteinketten, die große Ähnlichkeit mit dem Globin der Vertebraten-Myoglobine aufweisen (*110, 111*). Diese 4, jeweils paarweise identischen Untereinheiten der Hämoglobine sind in den Ecken eines Tetraeders so angeordnet, daß der Molekülverband annähernd die Gestalt einer Kugel annimmt (*110, 111*). In den Hämoglobinen von Wirbellosen — auch Erythrocruorine genannt — liegen dagegen lineare (*112*), in Chlorocruorinen scheibenförmige Assoziate vor, die Molekulargewichte bis $2,7 \cdot 10^6$ erreichen können (*113*). Auf die Mechanismen der Wechselwirkung zwischen diesen Untereinheiten bei der Bindung des Sauerstoffs kann hier nicht eingegangen werden. Es sei hingewiesen auf die bahnbrechenden Arbeiten von *Perutz* und Mitarbeitern (*114, 115, 116*) sowie neuere Untersuchungen des Bohr-Effektes, der pH-Abhängigkeit der Oxygenierung bei Hämoglobinen (*117, 118*).

3. Zur Struktur des Myoglobins

Die einfachsten O_2-fixierenden Hämoproteine sind die Myoglobine. Am Pottwal-Myoglobin sind die detailliertesten Angaben über den Bau des Aktivzentrums gewonnen worden. Die dreidimensionale Röntgenstrukturanalyse dieses Makromoleküls mit einem Molekulargewicht um 17 000 ist von *Kendrew* und seinen Mitarbeitern (*119, 120*) derartig vervollkommnet worden, daß es heute möglich ist, mit Ausnahme der Wasserstoffatome praktisch jedes Atom des Makromoleküls räumlich festzulegen (*121*). Für das Eisen sollen nach *Nobbs* (*122*) sogar Lageveränderungen von 0,1 Å feststellbar sein.

Stryer et al. (*123*) gelang es, im Azidkomplex des Metmyoglobins die Lage des N_3^--Anions festzulegen. Verfahrenstechnisch wäre es demnach möglich gewesen, die Geometrie der Sauerstoffbindung im Oxymyoglobin aufzuklären, Oxymyoglobin geht jedoch leicht in Metmyoglobin über und größere Kristalle stehen erst seit kurzem zur Verfügung (*124, 125*).

Kristallines Oxyhämoglobin ist weniger autoxidabel als Oxymyoglobin. Die Röntgenstrukturanalyse von Pferde-Oxyhämoglobin wurde von *Perutz* et al. (*110, 114*) durchgeführt. Da das Molekulargewicht etwa 64 000 beträgt, war die erreichte Auflösung von 5,5 Å nicht ausreichend, um über die Bindung des Sauerstoffs Aussagen machen zu können. Die Erhöhung der Auflösung auf 3 Å ist angekündigt (*110*).

Myoglobin ist annähernd kugelig gebaut, wobei polare Gruppen fast ausschließlich an der "Kugeloberfläche" liegen. In eine periphere Tasche des Globins ist das Häm eingelagert. Seine Umgebung wird weitgehend von unpolaren Proteinbausteinen gebildet (110, 111, 126, 127) wie aus Abb. 14 zu entnehmen ist.

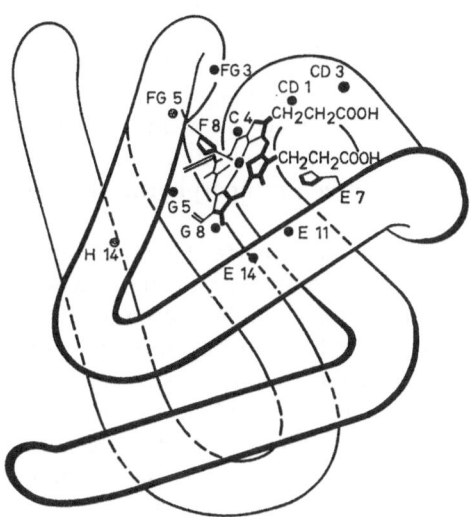

Abb. 14. Struktur des Myoglobins nach *Dickerson* (127), Umgebung des Häms nach *Kendrew* (126). F 8 = proximales Histidin, E 7 = distales Histidin, H 14 = Phe, FG 5 = Ile, FG 3 = His, G 5 = Leu, G 8 = Ile, E 14 = Ala, E 11 = Val, CD 1 = Phe, C 4 = Thr, CD 3 = Arg.

Auch das eisenfreie Protoporphyrin IX reagiert stöchiometrisch mit dem Globin und stabilisiert es gegen Denaturierung (128, 129). Es müssen also erhebliche Wechselwirkungen zwischen Globin und prosthetischer Gruppe stattfinden, die nicht auf eine Koordination des Häm-Eisens zurückzuführen sind. Das Häm ist derart in das Protein eingebaut, daß die hydrophoben Liganden am Porphyrinring, die Vinyl- und Methylgruppen in das hydrophobe Innere des Makromoleküls zeigen. Wie entscheidend die Wechselwirkung der Vinylreste mit dem hydrophoben Innern des Proteins ist, läßt sich daraus erkennen, daß Protohäm IX solche Häme aus der Bindung an das Globin verdrängen kann, die keine Vinylgruppen besitzen (130, 131).

Die Propionsäurereste sind wahrscheinlich salzartig mit den basischen Aminosäureresten Histidin FG 3 und Arginin CD 3 am Rande der Proteintasche verknüpft (110).

Auch das Tetrapyrrolringsystem scheint in Wechselwirkung mit den umgebenden Aminosäureresten zu treten, wobei eine Art π-Komplex zwischen dem Benzolring eines Phenylalanins und einem Pyrrolring zustande kommen soll (132). Daß van der Waals-Kräfte bei der Bindung des Häms eine erhebliche Rolle spielen, ist auch daraus zu ersehen, daß Xenon durch induzierte Dipol-Kräfte zwischen Häm und der Peptidkette des Globins festgehalten werden kann (121).

Die 5. Koordinationsstelle des Häm-Eisens wird vom tertiären Imidazol-Stickstoff des sogenannten proximalen Histidins eingenommen (Histidin F 8 im Pottwal-Myoglobin, Abb. 14 (110, 126, 127)). Diese koordinative σ-Bindung zwischen Häm und Globin ist jedoch für die Stabilität des Hämoproteins von sekundärer Bedeutung, da das Protohäm unfähig ist, das eisenfreie Protoporphyrin aus seiner Bindung an das Globin zu verdrängen (133).

Da Häm leicht hexakoordinierte Hämochrome bildet (Abb. 15) wurde bis vor kurzem angenommen, daß die 6. Koordinationsstelle am Eisen im Ferromyoglobin von einem Wassermolekül besetzt ist, wie es im Metmyoglobin röntgenographisch nachgewiesen wurde (119, 120, 134).

Dieses Wassermolekül fehlt im Ferromyoglobin, wie *Nobbs*, *Watson* und *Kendrew* kürzlich gezeigt haben (120). Das Häm-Eisen ist nur pentakoordiniert. Damit erfüllt es eine Forderung, die bei den synthetischen Sauerstoffträgern aufgestellt wurde (20): O_2 kann gebunden werden, ohne daß ein Ligand substituiert werden muß.

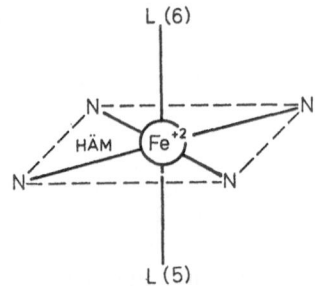

Abb. 15. Hämochrome.

Das Eisen im Deoxymyoglobin hat demnach eine tetragonal-pyramidale Koordinationssphäre, mit dem planaren vierzähligen Makrozyklus des Porphyrinrings als Basis und dem tertiären Stickstoff des proximalen Histidins als Spitze. Tetragonal-pyramidale Struktur wurde von *König* (135), *Hoard*, *Hamor* und *Caughey* (136) bei den Hämichromen [(Cl) Protoporphyrin—Fe(III)] und [(CH_3O) Mesoporphyrin—Fe(III)] nachgewiesen. Bei dieser Koordination wird das Eisen um rd. 0,5 Å aus der Basis in Richtung zum 5. Liganden verschoben.

4. Verhinderung irreversibler Oxidation des Häm-Eisens durch O_2

In unpolaren Medien sind Häme und die meisten Hämochrome unfähig, mit Sauerstoff zu reagieren (137, 138). In Medien hoher Dielektrizitätskonstante mit acidem Wasserstoff tritt dagegen irreversible Oxidation zu Derivaten des Hämatins [(HO) Protoporphyrin—Fe(III)] oder Hämichromen ein (138, 139, 140).

Der Einbau des Häms in das Medium des Globins schützt es vor irreversibler Oxidation durch den Sauerstoff, befähigt es aber gleichzeitig dazu, reversible O_2-Addukte zu bilden. Welche Strukturfaktoren im Aktivzentrum der Hämoproteine sind für dieses Verhalten verantwortlich? *Wang* et al. haben 1958 die Hypothese aufgestellt, daß ein Medium niedriger Dielektrizitätskonstante einen Elektronenübergang vom Eisen zum Sauerstoff nach

$$L\ Fe(II) + O_2 \rightarrow \underset{A}{L\ Fe(III)\ (O_2^-)} \rightarrow L\ Fe(III) + O_2^- \qquad (13)$$

(L = Liganden des Häm-Eisens)

aus elektrostatischen Gründen verhindert, oder zumindest die Dissoziation des Ionendipols A unterbindet, da die bindende Kraft zwischen 2 Ionen gegeben ist durch die Beziehung

$$K = \frac{1}{\varepsilon}\ \frac{e_1 \cdot e_2}{r^2} \qquad (14)$$

Er konnte diese Hypothese durch Modellversuche und kinetische Messungen untermauern (137, 138, 141, 142, 143).

Die kinetische Verfolgung der Oxidationsgeschwindigkeit von Bis(pyridin)-hämochrom zum entsprechenden Hämichrom in Äthanol-Benzol-Gemischen ergab eine Zunahme der Oxidationsgeschwindigkeit mit steigender Polarität des Lösungsmittelgemisches, d. h. mit zunehmender Äthanol-Konzentration. In reinem Benzol ist die Reaktion unmeßbar langsam.

Wang schloß daraus, daß die Einbettung des Häms in die weitgehend unpolare Proteintasche des Globins die Ursache dafür ist, daß das Häm-Eisen durch Sauerstoff nicht zur Ferri-Form oxidiert wird.

Die Fähigkeit, mit Sauerstoff ein reversibles Addukt zu bilden, ist von der Besetzung der 5. Koordinationsstelle am Häm abhängig.

5. Theorien über die Art der Bindung des Sauerstoffs am Häm-Eisen

a) *Die d-Atomorbitale des Eisens* (eine ausführliche Beschreibung ist dem Artikel von Dr. M. Weissbluth in diesem Band zu entnehmen).

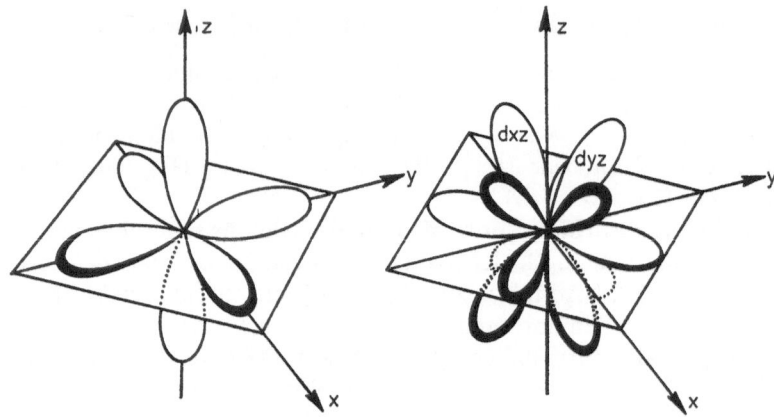

Abb. 16. Räumliche Anordnung der d-Orbitale des Eisens in einem oktaedrischen Ligandenfeld.
Linke Seite: e_g Orbitale dx^2-y^2 und dz^2 bzw. sechs (d^2sp^3) Hybridorbitale.
Rechte Seite: t_{2g} Orbitale dxy, dxz und dyz.

In Abb. 16 sind die 3d-Atomorbitale des Eisens schematisch wiedergegeben. In einem freien Ion oder Atom sind alle fünf Orbitale energiegleich. Wird das Ion aber in Wechselwirkung mit Liganden gebracht, so üben diese einen unterschiedlichen Einfluß auf die d-Atomorbitale aus (Ligandenfeldtheorie). In einem oktaedrischen Ligandenfeld von Elektronendonor-Liganden werden die in Ligandenrichtung zeigenden d-Orbitale aus elektrostatischen Gründen energetisch benachteiligt. Dies sind die e_g-Orbitale $d_{x^2-y^2}$ und d_{z^2}, wenn die Liganden in Richtung der x, y, z-Achsen angeordnet sind.

Im Ligandenfeld des Oxymyoglobins ist die Energieaufspaltung so groß, daß alle sechs d-Elektronen des Eisens(II) unter Spinkopplung in die energetisch weniger beeinflußten t_{2g}-Orbitale d_{xy}, d_{xz} und d_{yz} eingebaut werden.

Die e_g-Orbitale können σ-Bindungen mit den Liganden eingehen, zu deren Ausbildung der Ligand beide Elektronen beisteuert.

Die t_{2g}-Orbitale können aus Symmetriegründen keine σ-Bindungen ausbilden. Hat aber ein Ligand unbesetzte p-Orbitale, so kann das Eisen(II) mit seinen besetzten t_{2g}-Orbitalen dative π-Bindungen zu diesem Liganden ausbilden. Ein solcher Ligand ist das O_2-Molekül im Valenzzustand nach *Griffith* (28).

Zwei Hypothesen über die Geometrie der Bindung des Sauerstoffs stehen im Widerstreit. Beide postulieren eine σ-Bindung, für die vom Sauerstoffmolekül ein Elektronenpaar beigesteuert werden soll und eine π-Bindung, für die das Eisen beide Elektronen liefert.

b) *Theorie von Pauling (144)*

Nach Pauling kommt die σ-Bindung zustande zwischen einem unbesetzten (d²sp³)-Hybridorbital des Eisens und einem einsamen Elektronenpaar des Sauerstoffs. Das Zentralatom und der Ligand schließen einen Winkel von 120° ein (Abb. 17).

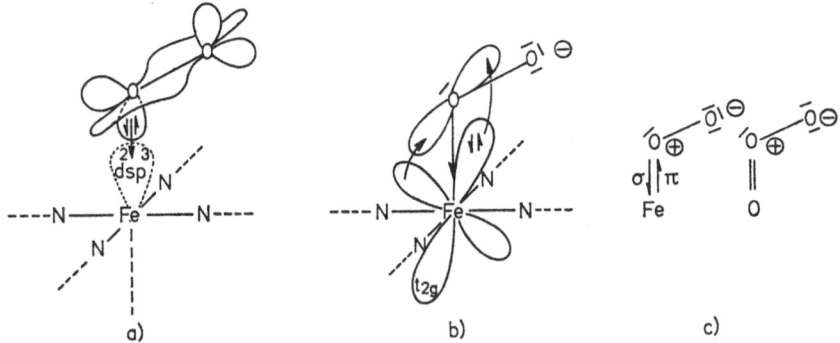

Abb. 17a—c. Bindung des Sauerstoffs am Häm-Eisen nach d. Theorie v. *Pauling (144)*.

Die π-Bindung zwischen dem Eisen und dem benachbarten Sauerstoffatom kann erst nach Aufpolarisierung der O=O Doppelbindung ausgebildet werden (Abb. 17b). Eine Analogie zu diesem Bindungstyp kommt im Ozonmolekül vor (Abb. 17c).

c) *Theorie von Griffith (11, 28)*

Griffith nimmt eine Bindung an, wie sie für Metall-Olefin-Komplexe postuliert (29) und bestätigt (145) wurde (Abb. 18b).

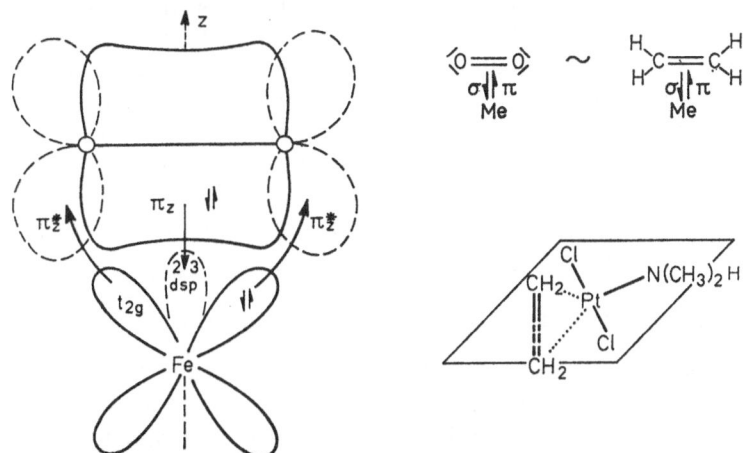

Abb. 18a und b. a) Bindung des Sauerstoffs am Häm-Eisen nach der Theorie von *Griffith (28)*, b) Struktur des [Pt (CH₂ = CH₂) (NH (CH₃)₂) (Cl)₂] nach *Alderman, Owston* und *Rowe (145)*.

Er ging davon aus, daß die Ionisierungsenergie des einsamen Elektronenpaars am Sauerstoff im Valenzzustand höher ist als die des π-bindenden Molekülorbitals π_z. (Einwände gegen diese Hypothese (*142, 143, 146*)). Die σ-Bindung zwischen dem Eisen und dem Sauerstoffmolekül sollte demnach durch Überlappung des π_z-Molekülorbitals des Sauerstoffs mit dem Atomorbital d_{z^2} bzw. dem in Ligandenrichtung zeigenden (d^2sp^3)-Hybridorbital des Eisens zustande kommen. Ein besetzter t_{2g}-Orbital des Eisens sollte eine dative π-Bindung mit dem leerstehenden π_z^*-Orbital des Sauerstoffs eingehen. Nach dieser Theorie haben beide O-Atome gleichen Abstand vom Eisen und sind äquivalent gebunden.

Welche der beiden Anordnungen der Sauerstoffgruppe in den oxygenierten Hämoproteinen vorliegt, ist noch ungeklärt. Im $[Ir(O_2)(CO)(Cl) (P(C_6H_5)_3)_2]$ (*36*) und in Peroxid-Komplexen wie $[Cr(O_2)_4]^{3-}$ (*35*), in denen eine O_2-Gruppe mononuklear an ein Metall gebunden ist, ist röntgenographisch erwiesen, daß beide O-Atome äquidistant gebunden sind. Es ist anzunehmen, daß dies auch im Oxyhämoglobin der Fall ist. Das Eisen würde dann von 7 Nachbaratomen umgeben sein. Heptakoordination bei Eisenkomplexen wurde kürzlich in mehreren Fällen bewiesen (*147, 148, 149*).

6. Zur Ladungsverteilung im Oxymyoglobin und Oxyhämoglobin

Viale, Maggiora und *Ingraham* (*150*) schließen aus MO-Berechnungen, daß der Hauptbeitrag der Sauerstoffbindung von der dativen π-Bindung vom Eisen zum Sauerstoff geleistet wird. Der Sauerstoff trägt demnach eine negative Partial-Ladung. Daß ein erheblicher Ladungstransfer zwischen Eisen und der O_2-Gruppe stattfindet, entnehmen *Lang* und *Marshall* (*151*) dem Mößbauerspektrum von Oxyhämoglobin.

Dagegen ist anzunehmen, daß der Sauerstoff aufgrund seiner hohen Ionisationsenergie (*15*) ein schlechter Elektronendonor ist und die σ-Bindung vom O_2-Liganden zum Eisen nur untergeordnete Bedeutung hat. (Eine analoge Interpretation der NMR- und IR-Spektren von Zeise-Salz $K[Pt\,Cl_3(CH_2 = CH_2)]$ geben *Fritz* et al. (152).)

Die Fähigkeit, O_2 zu binden, hängt also davon ab, wie groß die Donorstärke der t_{2g}-Orbitale des Häm-Eisens ist.

7. Der 5. Ligand am Häm-Eisen: das proximale Histidin

Häm in unpolarem Medium ist unfähig, mit O_2 zu reagieren. Offensichtlich reicht die Donorstärke der t_{2g}-Orbitale nicht aus, um Sauerstoff binden zu können. Es benötigt einen 5. Liganden in trans-Stellung zum Sauerstoff.

Die Bis(imidazol)-hämochrome des Proto- und des Mesohäms lassen sich nach *Corwin* und *Bruck* in kristalliner Form reversibel oxygenieren. Sauerstoff wird dabei angenähert im Verhältnis Fe : $O_2 = 1 : 1$ aufgenommen. Bis(pyridin)-hämochrome zeigen diese Eigenschaft nicht *(153)*. Demnach übt das Imidazol als Ligand des Häm-Eisens und vielleicht auch des fixierten Sauerstoffs einen spezifischen Effekt auf die Oxygenierung aus. Worauf beruht dieser Effekt?

Wang (143, 154) hat am gemischten Hämochrom [(CO) Häm (Pyridin)] eine Hypothese entwickelt, die hier auf das Aktivzentrum der Sauerstoffbindung [(O_2) Häm (Imidazol)] übertragen werden soll.

Sind die 5. und 6. Koordinationsstelle am Häm-Eisen mit 2 gleichen σ-Donorliganden besetzt, so werden im Hämochrom

$$[(L_5) \text{ Häm } (L_6)], \quad L_5 = L_6, \quad \text{z-Achse} = L_5\text{—Fe—}L_6 \tag{15}$$

die beiden t_{2g}-Orbitale d_{xz} und d_{yz} eine symmetrische Verteilung zu beiden Seiten des Häms haben. Ist jedoch L_5 ein stärkerer σ-Donor (high field ligand) als L_6 (low field ligand), so werden aus elektrostatischen Gründen und aufgrund des Pauli-Prinzips die t_{2g}-Orbitale d_{xz} und d_{yz} in Richtung zum low-field-Liganden verschoben. Dieser kann sich deshalb nicht genügend ans Zentralatom annähern und nur eine schwache σ-Bindung ausbilden.

Da Imidazol ein stärkerer Elektronendonor als H_2O ist, erwartet man von einem gemischten Hämochrom [(H_2O) Häm (Imidazol)], daß das Wasser sehr schwach gebunden und leicht abtrennbar ist. Tatsächlich ist dieses gemischte Hämochrom noch nicht dargestellt worden, es ist deshalb auch nicht verwunderlich, daß von *Nobbs* et al. *(120)* röntgenspektroskopisch und *Fabry* et al. *(155)* NMR-spektroskopisch kein H_2O im Deoxymyoglobin bzw. Deoxyhämoglobin gefunden wurde.

Lagert sich aber ein π-Acceptor in trans-Stellung zum Imidazol an das Häm-Eisen an (Abb. 19c), so wird er von einer solchen unsymmetrischen Verteilung der t_{2g}-Orbitale profitieren können, da er dative π-Bindungen des Eisens ermöglicht.

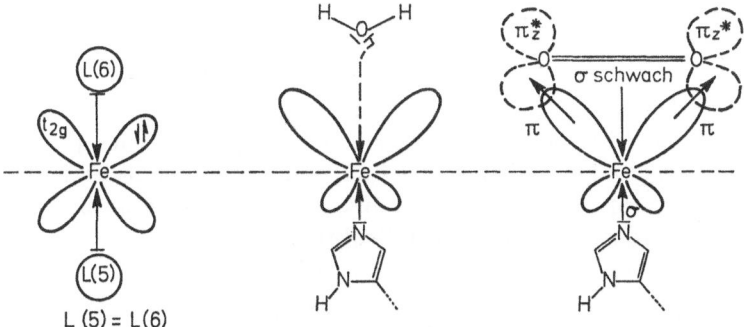

Abb. 19. Trans-Effekt der Liganden des Häms. Ein starker σ-Donor erschwert die Bindung eines zweiten σ-Donors und erleichtert die Bindung eines π-Acceptors in trans-Stellung.

Ein starker σ-Donor erleichtert die Bindung eines π-Acceptors in trans-Stellung am Häm-Eisen. Ein solcher π-Acceptor ist das Sauerstoffmolekül im Valenzzustand.

Imidazol ist ein stärkerer σ-Donor als Pyridin. Das ist daraus ersichtlich, daß ein Hämochrom [(H₂O) Häm (Imidazol)] offensichtlich nicht existent ist, während das entsprechende [(H₂O) Häm (Pyridin)] isoliert werden konnte (156). Worauf beruht dieser Basizitätsunterschied? Bei der Ausbildung einer koordinativen σ-Bindung erscheint zumindest eine positive Partialladung auf dem Stickstoffatom. Diese ist im Pyridiniumsystem weitgehend lokalisiert, während sie im Imidazoliumsystem delokalisierbar und damit mesomeriestabilisiert ist (Abb. 20 b und c, s. auch *Phillips* (157)).

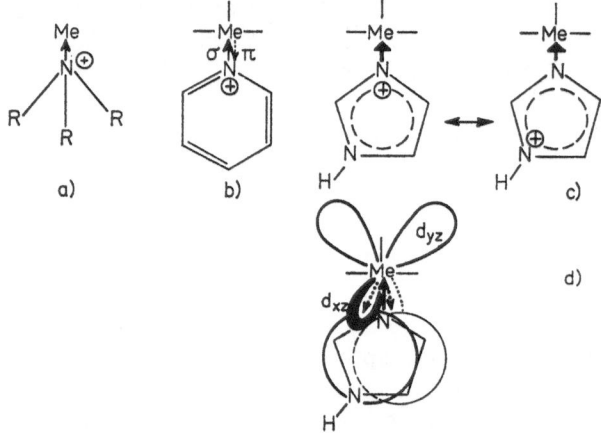

Abb. 20a — d. Stickstoffbasen als Liganden in Metall-Komplexen. a) aliphatisches Amin: σ-Donor, b) Pyridin: σ-Donor und π-Acceptor, c) Imidazol: starker σ-Donor und π-Acceptor.

Die σ-Donorstärke des tertiären Imidazolstickstoffs kann jedoch nicht die alleinige Ursache der funktionellen Überlegenheit des proximalen Histidins gegenüber anderen Stickstoffliganden bei der O₂-Fixierung am Häm-Eisen sein, da aliphatische Amine stärkere σ-Donoren sind. Das ist aus den Basizitätskonstanten der Reaktion

$$\hspace{-2em} \text{>N|} + \text{H}^+ \rightleftharpoons \text{>}\overset{\oplus}{\text{N}}\text{--H} \hspace{4em} (16)$$

zu ersehen (Tabelle 8, (158)).

Tabelle 8. *Basizitätskonstanten von Aminen.*

Base	Hybridisation am Stickstoff	Basizitätskonstante K_B
NH_3	sp^3	$1,62 \cdot 10^{-5}$
$N(CH_3)_3$	sp^3	$5,75 \cdot 10^{-5}$
Pyridin	sp^2	$1,7 \cdot 10^{-9}$
Imidazol	sp^2	$1,07 \cdot 10^{-7}$
4-Aminopyridin	sp^2	$1,48 \cdot 10^{-5}$

$$CH_3-C\underset{\searrow NH}{\overset{\nearrow NH_2}{}} \qquad sp^2 \qquad 2,56 \cdot 10^{-2}$$

(Acetamidin)

$$H_2N-C\underset{\searrow NH}{\overset{\nearrow NH_2}{}} \qquad sp^2 \qquad 5,13 \cdot 10^{-1}$$

(Guanidin)

Stickstoffbasen mit sp^3-Hybridisation am Stickstoff sind nur zur Ausbildung von koordinativen σ-Bindungen befähigt, während Stickstoffbasen mit sp^2-Hybridisation zusätzlich p—d—π-Bindungen mit dem Häm-Eisen eingehen können (Abb. 20b). Liegt das sp^2-Bindungsgerüst in der yz-Ebene, so wird durch diese „back donation" der d_{xz}-Orbital des Häm-Eisens in Richtung zum Stickstoffliganden verlagert (Abb. 20d).

Kein in Proteinen vorkommender Stickstoffligand kann idealere Voraussetzungen zur Bindung des O_2-Moleküls am Häm-Eisen schaffen als das Imidazol. Seinen Eigenschaften kommt das Arginin am nächsten. Liegt das Imidazol in der yz-Ebene, so erhöht es die Basizität des d_{yz}-Orbitals des Eisens und vergrößert seine räumliche Ausdehnung in Richtung zum 6. Liganden. Gleichzeitig erniedrigt es die Basizität des d_{xz}-Orbitals und verkleinert seine räumliche Ausdehnung in Richtung zum 6. Liganden.

Damit sind die Voraussetzungen für eine weitgehende Annäherung des O_2-Moleküls im Valenzzustand mit einem besetzten π_x^*-Orbital und einem unbesetzten, zur π-Bindung befähigten π_z^*-Orbital geschaffen.

8. Das distale Histidin

Von wenigen Ausnahmen abgesehen besitzen Hämo- und Myoglobine 2 Histidinreste in der Umgebung des Aktivzentrums. Die Rolle des proximalen Histidins ist klar ersichtlich: als starker σ-Donor fixiert es das Häm am Globin, erhöht die Basizität des Häm-Eisens und befähigt es damit, dative π-Bindungen zu einem π-Acceptor auszubilden. Weniger durchsichtig sind die Funktionen des sogenannten distalen Histidins.

a) Nach *Pauling* (144) trägt das endständige Atom des am Eisen fixierten Sauerstoffs eine negative Ladung, die es zur Ausbildung einer Wasserstoffbrückenbindung zum distalen Histidin befähigen soll (Abb. 21a). Auch *Perutz* scheint die Bildung einer H-Brücke anzunehmen (116).

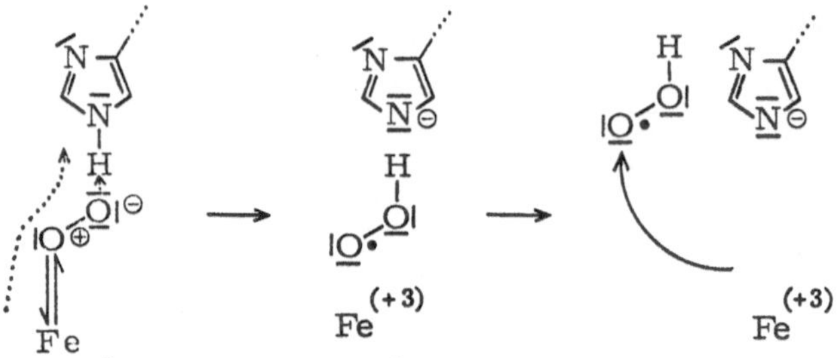

Abb. 21a — c. Effekt einer Wasserstoffbrückenbindung im Oxymyoglobin. a) Verstärkung des Elektronentransfers vom Eisen zum Sauerstoff, b) Bildung des ungeladenen HO_2.

Eine $>$N—H-Gruppe mit einem aciden Proton in unmittelbarer Nähe des Häm-Eisens stünde aber im Widerspruch zur Annahme eines unpolaren, nicht aciden Mediums in der Umgebung der Sauerstoff-Koordinationsstelle (138, 141, 142, 143).

Eine Wasserstoffbrücke würde den Elektronenzug vom Eisen zum Sauerstoff verstärken. Es wäre prinzipiell die Möglichkeit gegeben, aus dem aciden Wasserstoff und der O_2-Gruppe ein elektroneutrales Radikal $HO_2^{\bullet}$ zu bilden, das ohne Rücksicht auf die Dielektrizitätskonstante des Mediums vom oxidierten Häm-Eisen abdissoziieren und die Energiebilanz der Reaktion

$$L\ Fe\ (II) + O_2 \xrightarrow{H^+} L\ Fe(III) + HO_2^{\bullet} \tag{17}$$

(L = Ligandenfeld des Häm-Eisens)

durch Sekundärreaktionen entscheidend mitgestalten könnte (s. dazu *King* (159) und *Weiss* (160)).

Sicherlich liegt im Oxymyoglobin kein O_2^--Ion vor, sondern eine partiell negativierte O_2-Gruppe. Trotzdem ist ein Vergleich mit dem Verhalten von Superoxiden nicht uninteressant. Kaliumsuperoxid KO_2 und

Tetraalkylammoniumsuperoxide $R_4N^+O_2^-$ sind in protonenfreien Medien wie Pyridin, Dimethylformamid und Dimethylsulfoxid stabil (*162, 163, 164*). In Pyridin wird dabei die Bildung eines charge transfer-Komplexes

$$O_2^- \cdot Py \longleftrightarrow O_2 \cdot Py^- \qquad (18)$$

mit einer Absorption bei 450 mμ festgestellt (*163*). Die Superoxidanionen werden jedoch reduziert bzw. disproportionieren zu O_2 und H_2O_2, wenn Substanzen mit acidem Wasserstoff (z.B. Phenol) ins Medium gebracht werden (*164*).

Die polarographische Reduktion des Sauerstoffs in aprotischen Lösungsmitteln nach

$$e_0^- + O_2 \rightarrow O_2^- \qquad (19)$$

erfordert ein negatives Halbstufenpotential von 0,77—0,89 V, während die Reduktion

$$2\,H^+ + 2\,e_0^- + O_2 \rightarrow H_2O_2 \;(pH = 5,\, H_2O) \qquad (20)$$

schon bei einem negativen Halbstufenpotential von 0,1 V erfolgt (*162*).

Die Abwesenheit acider Protonen wirkt demnach außerordentlich stabilisierend auf die Oxidationsstufe des molekularen Sauerstoffs.

Es ist deshalb wahrscheinlicher, daß in oxygenierten Hämoproteinen anstelle einer Wasserstoffbrückenbindung eine Art Donor-Acceptor-Komplex zwischen einem tertiären Stickstoffatom als Donor und der O_2-Gruppe als Acceptor gebildet wird. Solche Donor-Acceptor-Komplexe liegen vor in den Addukten tertiärer Stickstoffbasen, Pyridin $\cdot$ SO₃, Pyridin $\cdot$ Br₂ (*165*) und vor allem in dem schwachen Kontakt-charge transfer-Komplex Triäthylamin $\cdot$ O_2 (*42*).

Der Vorteil des Histidins als distalem Nachbarn des Häm-Eisens gegenüber anderen Aminosäuren ist nach dieser Hypothese darin zu sehen, daß es einen stark basischen Stickstoff enthält — wie er auch im Arginin vorkommt. Im Arginin sitzt jedoch noch ein Wasserstoffatom an diesem Stickstoff. Histidin ist die einzige Aminosäure, die einen stark basischen tertiären Stickstoff besitzt, der dem Sauerstoff gegenüber sowohl als σ- wie als π-Donor wirken kann. Es ist denkbar, daß die Stärke dieser Donor-Acceptor-Wechselwirkung variiert werden kann durch Änderung des Bindungsabstandes oder durch Kontakt des sekundären Stickstoffs $>$N—H des distalen Histidins mit dem Medium, dessen pH-Wert veränderlich ist. Dies ist in Abb. 24 schematisch angedeutet. Die Abstandsänderung könnte eine Folge der Verschiebung der 4 Untereinheiten des Hämoglobins bei den Oxygenierungs-Deoxygenierungsprozessen sein, wie sie von *Perutz* et al. *(116)* festgestellt wurde. Der Kontakt des distalen Histidins mit dem Medium wurde von *Stryer* et al. am Metmyoglobinazid demonstriert *(123)*.

b) Es gibt Hämoproteine, die reversibel oxygenierbar sind, aber nur 1 Histidin enthalten, die Aplysia-Hämoglobine, wie *Wittenberg* et al. *(166)* und *Rossi-Fanelli* et al. *(167)* gezeigt haben (Tabelle 9).

Sie enthalten größere Mengen Arginin. Es ist denkbar, daß ein Histidinrest im Aktivzentrum durch Arginin ersetzt ist. Zu beachten ist die große Ähnlichkeit der Lichtabsorptionsspektren mit den Spektren von Hämoglobin. In dem abnormalen menschlichen Hämoglobin $M_{Zürich}$ *(170)* ist das distale Histidin in Position 63 einer ß-Kette durch Arginin ersetzt *(171)*. Nach *Bachmann* und *Marti* *(172)* hat das im Vergleich zu Hämoglobin A keine wesentlichen Veränderungen im Lichtabsorptionsspektrum des Deoxy-, Oxy- und Methämoglobins $M_{Zürich}$ zur Folge. Es wird aber leichter autoxidiert als Hämoglobin A.

Murawski et al. *(173)* fanden, daß im abnormalen Hämoglobin M_{Radom} das distale Histidin 63 einer ß-Kette durch Tyrosin ersetzt ist. Hb M_{Radom} ist ebenfalls oxygenierbar ($p_{1/2}$ = 10 mm Hg). Das Oxyhämoglobin geht jedoch außerordentlich leicht in das Methämoglobin über.

c) Die Bildung von Methämoglobin im Blut ist nicht vollständig auszuschließen. Das Methämoglobin ist aber enzymatisch reduzierbar. Abnormale Hämoglobine M mit Tyrosin bilden jedoch sehr stabile Methämoglobine, die im Blut nur schwierig oder gar nicht reduzierbar sind. Der Grund ist wohl darin zu sehen, daß in normalen Methämoglobinen die 6. Koordinationsstelle am Eisen mit einem H_2O-Molekül besetzt ist *(120, 134)*, das bei der Reduktion spontan abdissoziiert, unter Rückbildung des pentakoordinierten Häm-Eisens, während in abnormalen Methämoglobinen das Tyrosinanion den 6. Liganden bildet *(174)*.

Definitiv läßt sich die Funktion des distalen Histidins noch nicht angeben. Es ist aber aus den Beispielen, in denen es fehlt, ersichtlich, daß es bei der eigentlichen O_2-Fixierung am Eisen keine entscheidende Rolle spielen sollte. Allen Aminosäuren ist es jedoch bei der Stabilisierung des O_2-Adduktes gegen irreversible Oxidation überlegen.

9. Variation des Häms

Die t_{2g}-Orbitale des Häm-Eisens d_{xz} und d_{yz}, die dative π-Bindungen zu Liganden in Richtung der z-Achse ausbilden sollen, stehen in Konjugation zum π-Elektronensystem des Porphyrinrings. Eine Veränderung der π-Elektronendichte durch Substituenten an der Peripherie des Porphyrinrings wird die Basizität des Eisens beeinflussen. Elektronenanziehende Gruppen wie die Formylreste des Chlorocruorins oder die Vinylreste des Protohäms vermindern die Fähigkeit des Häm-Eisens zur Aus-

Tabelle 9. *Vergleich von Aplysia-Hämoglobin mit Vertebraten-Hämoproteinen.*

	Aplysia Nerven Hämoglobin (166)	Hämoproteine höherer Organismen	Literatur
1. Molekulargewicht (aus der Aminosäurenanalyse)	17550	17450 (menschl. Myoglobin)	(3)
2. Aminosäureanalyse	1 His 1 Tyr 5 Arg	12 His (Pottwalmyoglobin) 3 Tyr 4 Arg	(168)
3. Lichtabsorptionsspektren [mμ/ε] Deoxy-	$435/12 \cdot 10^4$, $560/13 \cdot 10^3$	$435/12,1 \cdot 10^4$, $560/14,4 \cdot 10^3$ (Pferdemyoglobin) $430/14,2 \cdot 10^4$, $555/13,4 \cdot 10^3$ (menschl. Hämoglobin)	(3) (169)
Oxy-	$416/13 \cdot 10^4$, $543/15 \cdot 10^3$, $578/14 \cdot 10^3$	$415/13,1 \cdot 10^4$, $542/14,2 \cdot 10^3$, $577/15,2 \cdot 10^3$ (menschl. Hämoglobin)	(169)
4. Halboxygenierung $P_{\frac{1}{2}} = po_2$ für 50%ige Sättigung	4 mm Hg	0,7 mm Hg (Pferdemyoglobin)	(3)

E. Bayer und P. Schretzmann

bildung dativer π-Bindungen gegenüber dem Deuterohäm, das Wasserstoffatome anstelle dieser Gruppen enthält.

Antonini et al. *(175, 176)* haben in ein Globin verschiedene Häme eingebaut und festgestellt, daß die Sauerstoffaffinität dieser Hämoglobine abnimmt in der Reihe vom Mesohäm über Hämatohäm, Deuterohäm, Chlorocruorohäm zum Protohäm (Tabelle 10).

Tabelle 10. *Sauerstoff-Affinität von resynthetisierten Hämoglobinen* [*nach Antonini et al. (175, 176)*].

Häm	Substituent in Position 2	4	$\lg p_{1/2} = \lg p_{O_2}$ für 50%ige Oxygenierung
Mesohäm	$-C_2H_5$	$-C_2H_5$	$-0,35$
Hämatohäm	$-CHOHCH_3$	$-CHOHCH_3$	$+0,03$
Deuterohäm	$-H$	$-H$	$+0,14$
Chlorocruorohäm	$-CHO$	$-CH=CH_2$	$+0,20$
Protohäm	$-CH=CH_2$	$-CH=CH_2$	$+0,36$

Falk, Phillips und *Magnusson (177)* und *Caughey, Alben* und *Beaudreau (178)* haben eine analoge Abnahme der Stabilität von Hämochromen typischer π-Acceptoren wie 4-Cyan-pyridin oder CO festgestellt, wenn der Elektronenzug im Häm zu peripheren Gruppen zunimmt. Das sind chemische Beweise dafür, daß O_2 als π-Acceptor in oxygenierten Hämoproteinen auftritt.

Die Vinylreste am Protohäm der Hämoglobine üben demnach einen Einfluß auf die O_2-Fixierung aus, der im Gegensatz zur Funktion des proximalen Histidins steht. Offenbar soll das Zusammenspiel dieser beiden Antagonisten eine Sauerstoffbindung bewirken, die labil genug ist, um einen schnellen Austausch des Sauerstoffs mit dem Medium zu ermöglichen.

10. Oxygenierung planarer Eisen(II)-Chelate

Daß Porphyrine als planare Liganden des zweiwertigen Eisens nicht die Voraussetzung für die Oxygenierbarkeit sind, haben *Williams* et al. gezeigt *(179, 180)*. Bis(pyridin)-(diacetyldioxim)eisen(II) und Bis(imidazol)-(diacetyldioxim)eisen(II) bilden ebenfalls reversible Sauerstoffkomplexe.

Benson und *McClellan* haben mitgeteilt, daß der Eisen(II)-Chelatkomplex des N.N'-Diäthylaminotroponimins in protonenfreiem Medium einen Sauerstoffkomplex bildet *(67)*. Wenn es sich dabei um ein Addukt mit dem Verhältnis Metall : $O_2 = 1:1$ handelt, und tetragonal-pyramidale Struktur des O_2-Adduktes nachgewiesen wird, so wäre damit das einfachste Modell für das Aktivzentrum der O_2-Fixierung in Hämoproteinen gegeben.

224

Das Ligandenfeld im Häm sollte mit dem Ligandenfeld eines planaren N.N'-Diäthylaminotroponimin-eisen(II) vergleichbar sein (ohne 5. Liganden liegt in diesem Komplex Tetraederstruktur vor (*67*)).

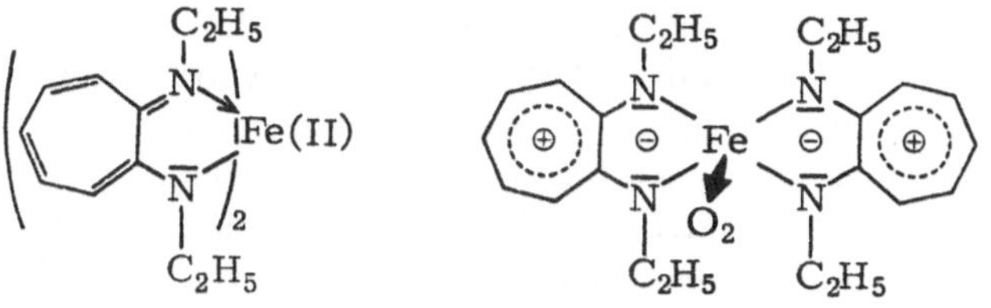

Abb. 22a und b. a) Eisen(II)-Chelatkomplex des N.N'-Diäthylaminotroponimins nach *McClellan* und *Benson* (*67*), b) Hypothetische Struktur des Sauerstoff-Adduktes.

Beim Protohäm ist zur Erhöhung der Basizität der t_{2g}-Orbitale des Eisens das proximale Histidin nötig. Im Eisen(II)-Chelat $(Et_2ATi)_2 \cdot Fe(II)$ sollte die Mesomerie in den Liganden mit der Tendenz zur Ausbildung von Tropyliumkationen eine solche Elektronendichte in der Umgebung des Eisens hervorrufen, daß zur Fixierung des π-Acceptors O_2 kein 6. Ligand nötig ist. Einen analogen Effekt würden Aminogruppen am Porphyrinring bewirken (Substituenten 1. Ordnung).

Der Eisen(II)-Komplex der Phthalocyanintetrasulfonsäure enthält wie das Häm ein ebenes vierzähliges Tetrapyrrol-System als Liganden. Er ist nach *Fallab* et al. bei 20° C in wässrigem Medium oxygenierbar (*48*, *181*). Bei 70° C ist die Oxygenierung im Stickstoffstrom vollständig reversibel. Es bildet sich ein binukleares Addukt, bei dem eine O_2-Gruppe „sandwich"-artig von zwei Chelatkomplexen eingeschlossen und damit dem Einfluß des Wassers entzogen wird. Nach *Fallab* (*48*) ist die Reaktion zu formulieren

$$2\ Fe^{+2}\ (PTS) + O_2 \rightleftharpoons [(PTS)Fe^{+2}\ (O_2)\ Fe^{+2}\ (PTS)] \qquad (21)$$

Das Sauerstoffaddukt unterscheidet sich im Lichtabsorptionsspektrum vom $[Fe^{+3}\ (PTS)]^+$ Kation. Daß keine Autoxidation

$$2\ Fe^{+2}\ (PTS) + O_2 \rightarrow 2[Fe^{3+}\ (PTS)]^+ + O_2^{2-} \qquad (22)$$

stattfindet, ist auch daraus zu ersehen, daß aus $[Fe^{+3}\ (PTS)]^+$ und H_2O_2 das binukleare Addukt nicht dargestellt werden kann. Das Sauerstoffaddukt ist also kein μ-Peroxo-Eisen(III)-Komplex.

11. Hämoglobin-Modelle

Es wurden systematische Versuche unternommen, die Koordinationssphäre des Häm-Eisens und das Medium in den Aktivzentren von Myoglobin und Hämoglobin nachzubauen.

a) *Wang* et al. *(141, 142, 143)* stellten aus einer Lösung des gemischten Hämochroms [(CO) Hämdiäthylester) (1- (2-Phenyläthyl)-imidazol)], Polystyrol und überschüssigem 1-(2-Phenyläthyl)-imidazol (PI) in Benzol unter Kohlenmonoxid Folien her. Diese Folien enthalten das gemischte Hämochrom in einer Matrix aus Polystyrol eingebettet (Abb. 23). Mit Stickstoff kann das CO ausgetrieben werden. Es entsteht ein hellrotes reaktives Hämochrom, dessen Lichtabsorptionsspektrum für ein locker gebundenes PI und ein starr gebundenes PI am Eisen(II) spricht. Dieses reaktive Hämochrom ist in wässrigem Medium reversibel oxygenierbar. Die Einbettung in das hydrophobe Polystyrol unterbindet demnach wirksam die irreversible Oxidation des Häm-Eisens durch molekularen Sauerstoff.

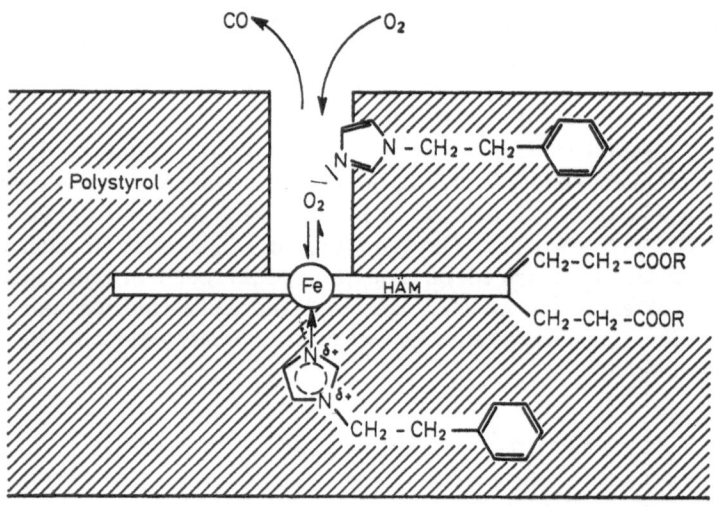

Abb. 23. Hämochrom nach *Wang (142)*.

Wird dieses reaktive Hämochrom einige Stunden bei 80° C getempert, so bildet sich ein echtes Bis-(PI)-Hämochrom. Dieses ist nicht mehr oxygenierbar. Offensichtlich ist das locker gebundene PI in die Lücke diffundiert, die nach Entfernung des CO entstand.

b) *Thojo* und *Shibata* *(182)* gelang es, in einem Modell das Lichtabsorptionsspektrum des Deoxymyoglobins nachzuahmen. Da Lichtabsorptionsspektren Aufschluß über die Elektronenverteilung geben, ist es wahrscheinlich, daß diese Autoren die Koordinationssphäre des Häm-Eisens im Deoxymyoglobin nachgebaut haben. Poly-L-Histidin gibt mit Häm ein typisches Hämochrom-Spektrum mit 2 Absorptionsbanden zwischen 500 und 600 mµ. Wird aber wenig L-Histidin in Poly-L-Glutaminsäure eingebaut, so ist die Chance eines Häms gering, 2 Histidinreste als 5. und 6. Liganden am Häm-Eisen binden zu können. Das Lichtabsorptionsspektrum dieses Komplexes zeigt große Ähnlichkeit mit dem des Deoxyhämoglobins (Tabelle 11).

c) *Hatano (183)* erhielt aus Häm und 1-Polyvinyl-2-methylimidazol Folien, die ein Deoxyhämoglobinspektrum ergeben. Sie scheinen in wasserfreiem Zustand oxygenierbar zu sein. Offensichtlich verhindert die Alkylgruppe am Imidazol die Ausbildung eines hexakoordinierten Hämochroms.

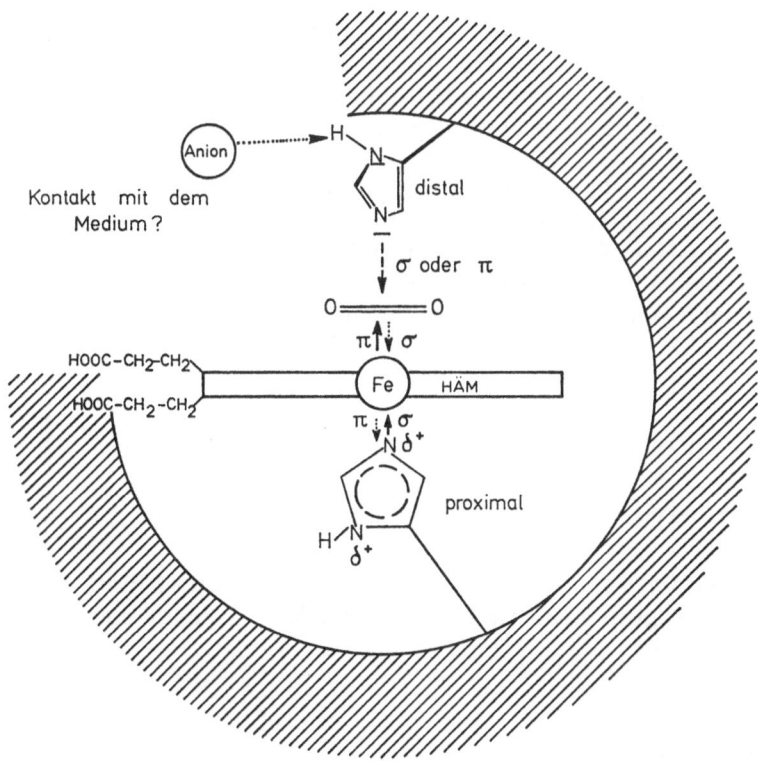

Abb. 24. Hypothetische Darstellung der Bindungsverhältnisse in oxygenierten Hämoproteinen.

Tabelle 11. *Lichtabsorptionsspektren von Hämoglobinmodellen.*

Komplex	Lichtabsorptionsbanden [mµ/ε]		Literatur
	Soret	α	
Deoxymyoglobin	435/144000	555/13500	(183)
Häm in L-Histidin-L-Glutaminsäure-Polyamid	433/ 80000	555/12000	(182)
Häm in 1-Polyvinyl-2-methylimidazol	435/110000	555/11600	(183)

12. Zusammenfassung (hypothetischer Oxygenierungsmechanismus der Hämoproteine).

In einem Medium ohne aciden Wasserstoff wird das Sauerstoffmolekül von 2 Elektronendonoren in die Zange genommen, dem Häm-Eisen und dem tertiären N-Atom des distalen Histidins. Beide ermöglichen eine partielle Ladungsübertragung zum O_2-Molekül, ohne jedoch vollständig oxidiert zu werden.

Das Häm ist in unpolarem Medium unfähig, mit O_2 zu reagieren. Erst die Erhöhung der Basizität des Häm-Eisen durch das proximale Histidin ermöglicht die Bildung eines O_2-Adduktes. Die Geometrie der Bindung ist noch ungelöst, es ist aber anzunehmen, daß eine symmetrische Struktur vorliegt, wie sie in mononuklearen synthetischen O_2-Addukten und mononuklearen Peroxiden röntgenographisch nachgewiesen wurde.

Das distale Histidin hat die Funktion, in Elektronendonor-Konkurrenz zum Häm-Eisen einen zu starken Elektronentransfer vom Metall zum Sauerstoff zu verhindern und damit das O_2-Addukt kinetisch labil, thermodynamisch aber stabil zu halten.

VIII. Nicht-Häm-Metallproteine

A. Hämerythrin

Hämerythrin ist das endocellulare Sauerstofftransport-Protein aus der Coelomflüssigkeit mancher Sipunculiden, Brachiopoden und Anneliden (5, 6, 7, 184).

1. Struktur

Nach Untersuchungen von *Klotz* et al. (185, 186), *Bayer* (187) und *Ghiretti* (184) enthält Hämerythrin keine prosthetische Gruppe, sondern Eisen(II) direkt an das Protein gebunden. Es handelt sich somit um ein

Nicht-Häm-Eisenprotein. Für das Hämerythrin aus Golfingia gouldii wurde ein Molekulargewicht von 107000 (*188*), für das Protein aus Sipunculus nudus ein Molekulargewicht von 66000 gefunden (*189*). Das Golfingia-Hämerythrin ist oktomer (*188, 190*). Nach *Klotz* und *Keresztes-Nagy* (*188, 191*) liegen 8 Untereinheiten mit einem Molekulargewicht von 13500 vor, die sich zu einem Würfel zusammenlagern sollen. Jede Untereinheit enthält 2 Eisenatome und kann 1 Molekül Sauerstoff binden (*191*). Ob eine oder mehrere Proteinketten in der Untereinheit vorliegen, ist noch nicht sicher. Nach *Groskopf* et al. (*192*) ist aber eine Proteinkette mit 112 Aminosäuren anzunehmen.

Deoxyhämerythrin ist paramagnetisch (high spin Fe(II)), dagegen ist im Oxyhämerythrin kein Paramagnetismus feststellbar (*193*).

2. Liganden der Zentralatome

Die Koordinationssphäre des Eisens im Hämerythrin ist noch unbekannt. *Klotz* et al. (*185, 186*) vermuteten ursprünglich, daß Eisen-Schwefelbindungen eine Rolle spielen, während *Manwell* (*5*) der Meinung ist, daß Imidazol und Tyrosin im Aktivzentrum der O_2-Fixierung vorkommen. Der Vergleich mit Ferredoxinen, den redoxaktiven Nicht-Häm-Eisenproteinen (*194, 195, 196*) scheint den Verdacht auf Schwefelliganden zu bestätigen. Um so überraschender ist auf den ersten Blick das Ergebnis der Aminosäureanalyse, die in Tabelle 12 den Ferredoxinwerten gegenübergestellt ist.

Tabelle 12. *Analysenwerte für Alfalfa Ferredoxin und Golfingia Hämerythrin.*

	Chloroplasten-Ferredoxin aus Alfalfa (*197*)	Golfingia-Hämerythrin Untereinheit (*192*)
Molekulargewicht	11500	13500
Eisen	2	2
Cystein	6 (Gesamtschwefel 8)	1 (+1 Methionin)
Histidin	2	7
Arginin	1	3
Lysin	5	11
Tyrosin	4	5

Im Hämerythrin kommt auf 2 Eisenatome 1 Cystein, während im Chloroplasten-Ferredoxin 4 Schwefelliganden pro Eisen anzunehmen sind (*197, 198*). Die Zahl der basischen Aminosäuren ist im Hämerythrin hingegen wesentlich größer als im Ferredoxin. Das ist erklärbar, wenn die physiologische Funktion dieser beiden Nicht-Häm-Eisenproteine verglichen wird. In Redoxenzymen, wie dem Ferredoxin soll das Ligandenfeld Elektronenübertragungen des Zentralatoms (Fe, Mo, Cu) erleichtern (196), während in Sauerstoffträgern eine partielle Ladungsüberführung vom Metall zum Sauerstoff durch das Ligandenfeld zwar erleichtert, eine vollständige Ladungsüberführung (= Oxidation des Metalls) aber unterbunden werden soll.

In vorausgehenden Kapiteln wurde besprochen, daß starke Stickstoffbasen (Histidin) oder Liganden hoher Polarisierbarkeit (P (C_6H_5)$_3$, —$\underline{N}$=) ein redoxaktives Zentralatom in der niederen Oxidationsstufe zur O_2-Fixierung befähigen. Diese Liganden selbst sind nicht redoxaktiv. Das Auftreten redoxaktiver Liganden würde irreversible Reduktion des Sauerstoffs begünstigen. Es ist deshalb nicht überraschend, wenn der O_2-Träger Hämerythrin pro Untereinheit auf 2 Eisenatome nur 1 Cystein enthält und eine Eisen-Schwefel-Bindung im Oxyhämerythrin auszuschließen ist, wie in VIII A 5 gezeigt wird.

Manwells Annahme (5), daß Imidazol im Aktivzentrum des Hämerythrins eine Rolle spielt, dürfte hohen Wahrscheinlichkeitswert haben, obwohl kein Bohr-Effekt bei der Oxygenierung von Hämerythrinen aus Sipunculiden nachgewiesen wurde (7). Nur in einem Fall von Brachiopoden-Hämerythrin wurde ein Bohr-Effekt gefunden (199). Die pH-Abhängigkeit der Oxygenierung ist jedoch auch bei den Myoglobinen in der Regel nicht festzustellen (2). Das Ausbleiben des Bohr-Effektes ist demnach kein Kriterium für eine fehlende Wechselwirkung zwischen dem Metall und dem Imidazol.

3. Oxygenierung

Das farblose Deoxyhämerythrin nimmt nach *Boeri* et al. (200) Sauerstoff im Verhältnis Fe : O_2 = 2 : 1 reversibel auf unter Bildung des burgunderfarbenen Oxyhämerythrins. *Keresztes-Nagy* et al. (191) sind der Meinung, daß bei der Oxygenierung eine Oxidation des Eisens zu Fe(III) O_2^{2-} Fe(III) eintritt.

Tabelle 13. *Oxygenierungswärmen von O_2-Addukten.*

	Addukt Me : O_2	Oxygenierungswärmen in Kcal/Mol O_2 [exoth.]	Literatur
SAD—Co(II)	2 : 1	19	(98)
Hämerythrin	2 : 1	13—20	(5)
Hämocyanin	2 : 1	9—16	(5)
Hämoglobin	1 : 1	8—14	(5)

Da (SAD—Co)$_2$ · O_2 sicher kein Kobalt(III)-Komplex ist, macht ein Vergleich der Oxygenierungswärmen wahrscheinlich, daß auch im Oxyhämerythrin kein Fe(III) O_2^{2-} Fe(III) vorliegt, sondern ein valenz-mesomeres System, wie *Manwell* postulierte (*5, 201*). *Klotz* et al. (*185, 186*) fanden, daß Oxyhämerythrin in saurem Medium einen o-Phenan-trolin-Eisen(III)-Komplex und das Peroxotitanyl-Kation bildet und werten dies als Beweis für das Vorliegen des μ-Peroxo-Komplexes. *Boeri* et al. (*200*) konnten in saurer Lösung aus Oxyhämerythrin keinen Sauer-stoff mehr freisetzen. Es entsteht vermutlich H_2O_2. Daß ein acides Medium und die Möglichkeit, Wasserstoffbrücken auszubilden, eine Oxygenierung in eine irreversible Oxidation überführen können, wurde in den vorangehenden Kapiteln besprochen. Es ist deshalb anzunehmen, daß Oxyhämerythrin ein mesomeres System ist, das mit Protonen aus der mesomeren Grenzstruktur Fe(III) O_2^{2-} Fe(III) heraus reagiert. Diese kann dann durch aciden Wasserstoff „fixiert" werden (*201*). Die Reaktion sagt aber nicht aus, daß im Grundzustand diese Formulierung als Eisen(III)-Komplex das größte Gewicht hat. Es sind vielmehr mindestens die folgenden Strukturen zu berücksichtigen:

$$O_2 + \underset{\underset{\textstyle Hr}{\rule{2.5cm}{0.4pt}}}{Fe(II) \quad Fe(II)} \tag{23}$$

$$\underset{\underset{\textstyle Hr}{\rule{2cm}{0.4pt}}}{Fe(II) \quad O_2 \quad Fe(II)} \leftrightarrow \underset{\underset{\textstyle Hr}{\rule{2cm}{0.4pt}}}{Fe \rightarrow O_2 \leftarrow Fe} \leftrightarrow \underset{\underset{\textstyle Hr}{\rule{2cm}{0.4pt}}}{Fe(III) \quad O_2^{\cdot -} \quad Fe(II)} \leftrightarrow \underset{\underset{\textstyle Hr}{\rule{2cm}{0.4pt}}}{Fe(III) \quad O_2^{2-} \quad Fe(III)}$$

no bond Grenzstruktur	π-Komplex-Schreibweise	ionische Grenzstrukturen

$$[H^+] \downarrow$$
$$H_2O_2$$

(Hr = Ligandenfeld im Hämerythrin)

Eine Formulierung Fe(III) O_2^{2-} Fe(III) wäre demnach zur Beschrei-bung der Ladungsverteilung im Oxyhämerythrin ebenso unvollständig wie eine Kekuléstrukturformel des Benzols.

Manwell (*5*, S. 63) stellte fest, daß Deoxyhämerythrin oxygeniert werden kann, wenn Thiolreagentien in großem Überschuß im Medium vorhanden sind. *Keresztes-Nagy* et al. (*191*) gelang die Oxygenierung der Funktionseinheit des Golfingia-Hämerythrins, dem Deoxymerohäm-erythrin, dessen Mercaptogruppe blockiert war (*202*). Demnach ist offen-bar für die Oxygenierung eine Eisen-Schwefelbindung nicht nötig.

Eine Bindung von Kohlenmonoxid wurde bei Deoxyhämerythrin nicht festgestellt (*5, 203*).

4. Methämerythrin

Oxyhämerythrin geht in Lösung langsam in gelbes Methämerythrin über *(188)*. Beschleunigt wird diese Autoxidation, wenn Anionen im Medium vorliegen, die Eisen(III)-Komplexe bilden *(191)*. Wie bei den Hämoproteinen wird die Oxidation am besten mit K_3FeCN_6 erreicht *(191)*.

Aquomethämerythrin bildet leicht Komplexe mit Anionen wie N_3^-, SCN^-, OCN^-, CN^-, Cl^- und F^-. Für den Azidkomplex wurde ein Bindungsverhältnis $Fe : N_3^- = 2 : 1$ nachgewiesen *(191)*, für die anderen Anionen nicht. (Über einen Komplex mit Azidbrücken siehe *(204)*). Ein Vergleich der Lichtabsorptionsspektren der Komplexe von N_3^-, SCN^-, CN^-, OCN^-, Cl^- mit dem Spektrum von Oxyhämerythrin zeigt eine zunehmende Ähnlichkeit mit zunehmender Polarisierbarkeit des Anions. *Keresztes-Nagy* und *Klotz* *(191)* werten diese Ähnlichkeit als Beweis für das Vorliegen einer Struktur $Fe(III)O_2^{2-}Fe(III)$ im Oxyhämerythrin. Dem ist entgegenzuhalten, daß auch low spin-Methämoglobin- und Metmyoglobinkomplexe ähnliche Spektren wie die entsprechenden Deoxy- und Oxyhämoproteine geben *(5, 205, 206)*.

Bei den Komplexen des Aquomethämerythrins mit leicht polarisierbaren Anionen treten charakteristische Lichtabsorptionen zwischen 650—690 mµ auf. Diese Absorption fehlt beim Mercapto-Komplex des Methämerythrins, sie fehlt auch im Oxyhämerythrin bzw. tritt als Schulter bei 750 mµ auf *(191)*.

5. Reaktionen des Methämerythrins mit redoxaktiven Liganden

Die Bildung von Oxyhämerythrin aus Aquomethämerythrin und O_2^{2-} wäre ein Beweis für die Annahme eines µ-Peroxo-Eisen(III)-Komplexes.

Eine Komplexbildung ist jedoch nicht festgestellt worden. Als Ursache vermuten *Keresztes-Nagy* und *Klotz* *(191)*, daß in dem pH-Bereich, in dem Methämerythrin stabil ist, eine zu geringe O_2^{2-}-Konzentration vorliegt (s. dazu den vergeblichen Versuch *Fallabs* aus H_2O_2 + [Fe^{+3} PTS]⁺ das Sauerstoffaddukt $(Fe\ PTS)_2 \cdot O_2$ zu gewinnen ((*48, 181*), Kap. VI 10).

Mercaptane wie Cysteinmethylester reduzieren Aquomethämerythrin zu Deoxyhämerythrin, wobei Disulfide entstehen *(191)*. Dagegen bilden SH^--Ionen mit Aquomethämerythrin einen roten Komplex, dessen Lichtabsorptionsspektrum mit dem des Oxyhämerythrins erhebliche Übereinstimmungen aufweist *(191)*. Ein Vergleich des Oxyhämerythrins mit dem Mercaptokomplex erscheint uns berechtigt.

Bei der Bildung des Oxyhämerythrins tritt das redoxaktive Fe(II) mit dem redoxaktiven O_2-Molekül in Wechselwirkung. Der Valenzzustand beider Reaktionspartner im Oxyhämerythrin ist unbestimmt. Eine ähnliche Ladungsverwischung kann eintreten bei der Bildung des Mercaptokomplexes von Aquomethämerythrin. Diese Hypothese soll anhand des Formelschemas in Abb. 25 näher erläutert werden:

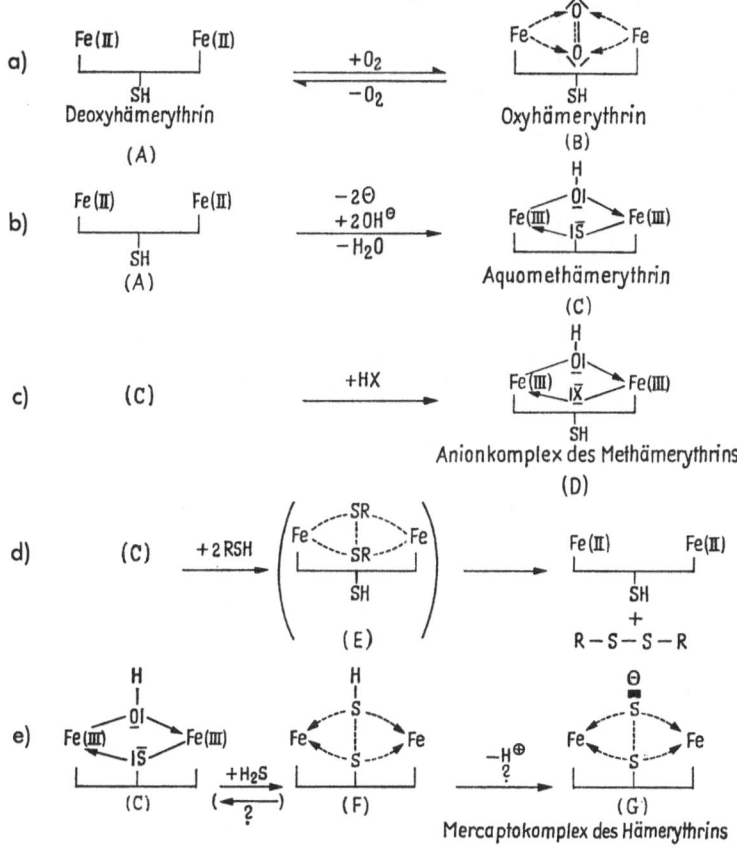

Abb. 25a—e. Reaktionen des Hämerythrins.

Zu Abb. 25a. Deoxyhämerythrin (A) ist auch oxygenierbar, wenn die Mercaptogruppen mit Thiolreagentien blockiert sind ((*191*) S. 924 und (*5*) S. 63). Im Oxyhämerythrin (B) sind die Mercaptogruppen ebenfalls Thiolreagentien zugänglich. Es ist anzunehmen, daß in A und B keine Eisen-Schwefelbindung vorliegt.

Zu Abb. 25b. Im Aquomethämerythrin (C) liegen keine freien Mercaptogruppen vor ((*191*) S. 927). Da Deoxy- und Aquomethämerythrin gleichen isoelektrischen Punkt haben, dürften bei der Oxidation A → C 2 Anionen ins Molekül aufgenommen werden.

Zu Abb. 25c. Mit Anionen reagiert Aquomethämerythrin unter Freisetzung der Mercaptogruppe zum Komplex D. Die Mercaptogruppe ist jetzt wieder Thiolreagentien zugänglich und nachweisbar ((*191*) S. 927).

Zu Abb. 25d. Mit Alkylmercaptanen (Cysteinmethylester) sollte der Prozeß C → E → A ablaufen, wobei E sehr instabil sein muß, da bei der Reaktion die gelbe Farbe des Aquomethämerythrins ohne Zwischenfärbung verblaßt. Bei der Reaktion entsteht Deoxyhämerythrin (A).

Zu Abb. 25e. Mit H_2S wird ein stabiler Komplex gewonnen (pH = 8). Es ist anzunehmen, daß er nach C → F → G entsteht. Das potentielle Hydropersulfid ist am Protein fixiert, das kann der Grund sein, warum kein Deoxyhämerythrin entsteht, wie es mit Alkylmercaptanen der Fall ist. Eine weitere Stabilisierungsmöglichkeit von F wäre die Abgabe eines Protons unter Ausbildung von (G). F würde einen zu B analogen mesomeren Valenzzustand beinhalten mit 2 über 4 Zentren frei beweglichen Elektronen (4-Zentren-Molekülorbital) mit dem Unterschied, daß in B die Metallatome je ein Elektron zu dem 4-Zentren-Molekülorbital beisteuern, während es in F die Thiolanionen sind, von denen beide Elektronen stammen.

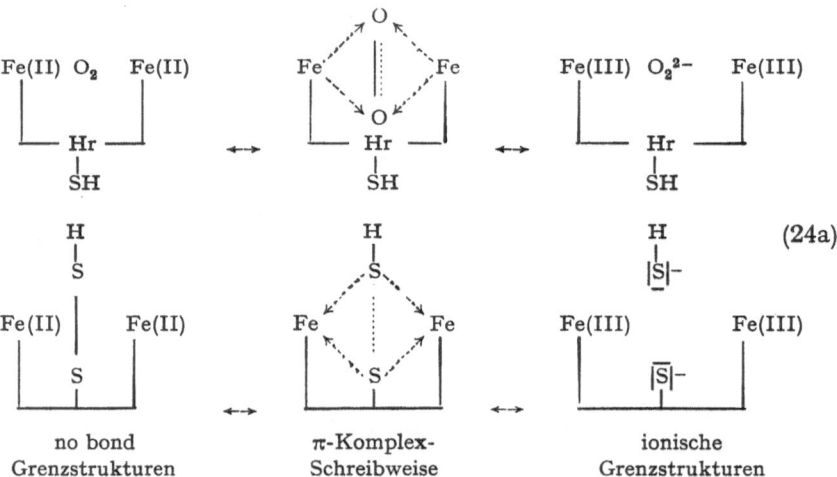

| no bond | π-Komplex- | ionische |
| Grenzstrukturen | Schreibweise | Grenzstrukturen |

(24a)

(Über die Formulierung valenzmesomerer Eisen-Sauerstoffsysteme siehe (*5, 207, 208*), über Eisen-Schwefelsysteme siehe (*196, 209*)).

Folgerungen aus dieser Hypothese

Für das Deoxy-Hämerythrin ist anzunehmen, daß die Koordinationssphäre des Eisen(II) — wie in den Hämoproteinen — von stark basischen, aber nicht redoxaktiven Liganden gebildet wird. Das ist die

Grundvoraussetzung für einen Sauerstoffträger. Eine redoxaktive Koordinationssphäre des Metalls im Hämerythrin hätte keinen O_2-Träger, sondern eine Oxidase bzw. Oxygenase zur Folge.

Wird Deoxyhämerythrin jedoch in Anwesenheit von Mercaptanen oxygeniert, so bildet sich zwar intermediär das rote Oxyhämerythrin; dieses wird jedoch durch die Mercaptane zu Deoxyhämerythrin reduziert. Deoxyhämerythrin zeigt demnach Oxidasenfunktion bei Anwesenheit von Mercaptanen. Es ist in diesem Zusammenhang interessant, daß die Oxygenase 3,4-Dihydroxy-phenylacetat-2,3-Oxygenase aus Pseudomonas aus Untereinheiten besteht, die auf 2 Eisen 2 Mercaptofunktionen enthalten (210).

Den Mercaptokomplex des Aquomethämerythrins kann man mit gleichem Recht als Hydropersulfidkomplex des Deoxyhämerythrins bezeichnen.

Prinzipiell besteht auch die Möglichkeit, daß im Mercapto-Hämerythrin eine Komplexierung des H_2S_2 bzw. des S_2-Moleküls vorliegt:

$$\text{(24b)}$$

oder

$$\text{(24c)}$$

Würden in diesem Komplex die restlichen Liganden des Eisens durch redoxaktive Liganden ersetzt, so wäre ein Redoxenzym, ein Ferredoxin geschaffen. Für das Studium des Aktivzentrums von Chloroplasten-Ferredoxin ist dieser Mercaptokomplex des Aquomethämerythrins von größter Bedeutung.

Möglicherweise liefert dieser Mercaptokomplex den Schlüssel zur endgültigen Lösung des Problems des labilen Schwefels in Ferredoxinen. Dazu muß geprüft werden, ob unter den Bedingungen der H_2S-Eliminierung aus Ferredoxinen die H_2S-Anlagerung an Aquomethämerythrin reversibel ist.

Außerdem ist eine EPR-spektroskopische Untersuchung dieses Mercaptokomplexes und der Vergleich mit den Spektren der Chloroplastenferredoxine in oxidierter und reduzierter Form von großem Interesse.

B. Hämocyanin

Mollusken (z. B. Schnecken, Tintenfische) und Arthropoden (z. B. Hummer) enthalten Hämocyanine als Atmungspigmente im Blut gelöst. Diese Kupferproteine gehören zu den größten einheitlichen Molekülassoziaten (211, 212), die bekannt sind. Sie erreichen Molekulargewichte bis zu 9.10^6 (Helix pomatia (213)).

Gegenüber den Hämoglobinen und Hämerythrinen sind die kleinsten Funktionseinheiten mit Molekulargewichten von 75000 bei Arthropoden und 51000 bei Mollusken außergewöhnlich groß (214). Die Funktionseinheit enthält 2 Atome Kupfer direkt am Protein gebunden (184, 185, 215), sie kann 1 Molekül O_2 reversibel binden (216), wobei mit hoher Wahrscheinlichkeit eine CuO_2Cu-Brücke gebildet wird (217, 218).

Nach Ghiretti-Magaldi et al. (214) besteht die Funktionseinheit des Mollusken-Hämocyanins aus 2 Proteinketten, das gleiche haben Pickett et al. (219) für Arthropoden-Hämocyanin (Hummer) festgestellt. Bei der Oxygenierung wird vermutlich eine O_2-Brücke zwischen den beiden Proteinketten der Funktionseinheit ausgebildet (219).

Da die Biochemie des Kupfers in jüngster Zeit ausführlich beschrieben worden ist (220, 221, 222), sollen an dieser Stelle nur einige wenige Aspekte besprochen werden.

1. Die Bindung des Kupfers in Hämocyaninen

Im Deoxyhämocyanin liegt einwertiges Kupfer vor, wie EPR-spektroskopische Untersuchungen (215, 223) und Resynthesen mit CuCl (215), Cu_2O (214) und $[Cu (N \equiv C—CH_3)_4]^+$ (213, 224) ergaben.

Es ist heute noch nicht zu sagen, ob beide Kupferatome der Funktionseinheit äquivalent gebunden sind. Auch die Koordinationszahl des Kupfers ist unbekannt. Die Koordinationssphäre im Deoxyhämocyanin muß jedoch äußerst Cu(I)-spezifisch sein, da es nach Bayer et al. nicht gelingt, andere Metallionen (Ag^+, Hg^{2+}, Ni^{2+}) in Apohämocyanin einzubauen (215).

Das Ligandenfeld im Deoxyhämocyanin soll zwar das Metall dazu befähigen, O_2 reversibel zu binden, es soll aber eine irreversible Oxidation des Kupfers unterbinden.

Welche Ligandenfelder des Kupfers(I) kommen in Frage? Kupfer(I)-Komplexe treten mit den Koordinationszahlen 2, 3 und 4 auf (225). Das tetrakoordinierte Cu(I)-Kation $[Cu(I) (N \equiv C—CH_3)_4]^+$ ist nach Hemmerich et al. O_2-inert, es ist unfähig, mit O_2 zu reagieren (226). Möglicherweise erhöht der Einbau von Imidazol-Liganden oder Arginin (5) in ein tetraedrisches Ligandenfeld die Basizität des Cu(I) soweit, daß dieses eine dative π-Bindung zu einem O_2-Molekül ausbilden kann. Das Kupfer würde dabei eine trigonal-bipyramidale oder verzerrt tetragonal-pyramidale Koordinationssphäre mit 5 Liganden erhalten, wie sie im tiefblauen $[Cu(NH_3)_5]^{2+}$ (225) vorliegt und vermutlich auch auftritt, wenn der Kupferkomplex des Oxalsäure-bis[äthylidenhydrazides] mit Sauerstoff in Wechselwirkung tritt (51). *Blount* et al. (227) haben im Kupfer(II)-Chelat des Glycyl-histidins röntgenographisch eine tetragonal-pyramidale Struktur nachgewiesen mit einem H_2O Molekül als Pyramidenspitze und einer Basis $\begin{smallmatrix} O & & N \\ & Cu & \\ N & & N \end{smallmatrix}$.

Kupfer(I) bindet jedoch in wäßrigem Medium nur 2 Imidazolliganden unter Ausbildung einer digonalen Koordinationssphäre (226, 228). Bei der Oxygenierung eines digonalen Kupfer(I)-Komplexes würde Kupfer die Koordinationszahl 3 für die nächsten Nachbarn annehmen, wenn die O_2-Gruppe als ein Ligand betrachtet wird. Trigonale Koordination ist Kupfer(I)-spezifisch, eine irreversible Oxidation des Kupfers wäre außerordentlich erschwert. *Zuberbühler* (228) stellte fest, daß die Geschwindigkeit der Oxidation des Kations $[Cu(I) (Imidazol)_2]^+$ mit molekularem Sauerstoff erhöht wird, wenn überschüssiges Imidazol im Medium vorhanden ist, d.h. wenn die Möglichkeit besteht, während der Reaktion mit O_2 die Koordinationszahl des Kupfers zu erhöhen.

Trigonale Koordination der nächsten Nachbarn kommt vor im Anion $[Cu(CN)_3]^{2-}$ (225), vor allem aber im tiefroten labilen Azomethan-Addukt des Kupfer(I)-chlorids (229, 230), dessen Struktur in Abb. 26 wiedergegeben ist.

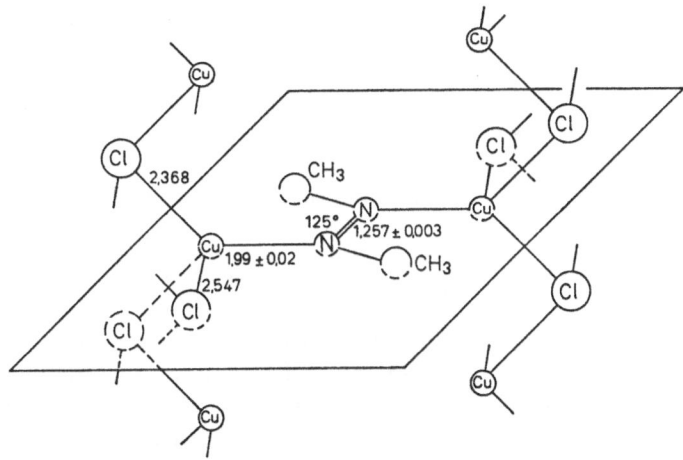

Abb. 26. Struktur des Azomethan-Adduktes von Kupfer(I)-chlorid nach *Brown* und *Dunitz* (230).

Dieses Addukt bildet sich bei Zimmertemperatur aus Azomethan und CuCl in Kaliumchloridlösung. Beim Erwärmen dissoziiert der Komplex. Nach Abkühlung bildet er sich wieder. Die Labilität des Adduktes und die Tatsache, daß Azomethan ein dem Sauerstoff im Valenzzustand nach *Griffith* analoges Bindungsgerüst hat, legen den Vergleich mit Oxyhämocyanin nahe. Es ist aber zu beachten, daß ein erheblicher Unterschied der Ionisationsenergien zwischen dem Azomethan und dem Sauerstoff im Valenzzustand bestehen wird, der sich auf die Geometrie der Chromophore [Cu N $\overset{N\ Cu}{}$] und [Cu O$_2$ Cu] auswirken sollte.

In beiden Fällen ist eine π-Bindung anzunehmen, zu der die Metallatome beide Elektronen liefern (Abb. 9 und Abb. 10, Kap. VI B$_1$).

Diese — vom Metall aus betrachtet — dative π-Bindung dürfte in den binuklearen Sauerstoff-Komplexen den Hauptbeitrag zur Bindung liefern. Dagegen ist anzunehmen, daß eine σ-Bindung, zu der die O$_2$-Gruppe beide Elektronen beisteuern muß, von untergeordneter Bedeutung ist.

Die Benennung des O$_2$-Adduktes als π-Komplex erscheint somit nicht unberechtigt. Ungelöst ist das Problem der Atomanordnung in den binuklearen Sauerstoffaddukten. Im Azomethan-Addukt ist eine koordinative σ-Bindung nicht auszuschließen, zu der die lone pair-Orbitale des Azomethans beide Elektronen beisteuern, wie es bei anderen Azokomplexen erwiesen ist. Für eine Doppelbindung spricht auch der Abstand Cu — N, der mit 1,99 Å nur wenig größer als der Abstand Cu(II) — N$_{tert}$ (Imidazol) im Glycylhistidin-Kupfer(II)-Komplex (*227*) mit 1,98 Å ist.

Solche Stickstoff-Ligandenfelder des Kupfers(I) könnten erklären, warum ein Teil des Kupfers in saurem (*184*) und alkalischem (*219*) Medium aus Hämocyaninen abspaltbar ist. Sie können jedoch die Tatsache nicht erklären, daß ein Teil des Kupfers extrem fest am Protein fixiert ist und nicht einmal mit Kaliumcyanidlösung quantitativ abgetrennt werden kann (*214, 215*). *Klotz* et al. (*185*) vermuteten, daß Kupfer über Schwefelliganden am Protein gebunden ist. *Bayer* (*187*) postulierte eine lineare Koordinationssphäre im Deoxyhämocyanin

$$\text{Protein — S — Cu(I)—N}_{tert}\text{ (Imidazol) — Protein} \qquad (25)$$

Beide Kupferatome der Funktionseinheit können jedoch nicht an Schwefel gebunden sein, da bei der Kupferabspaltung maximal eine Mercaptogruppe pro Funktionseinheit frei wird (*214*). Es ist also nicht ausgeschlossen, daß in der Funktionseinheit des Deoxyhämocyanins ein digonaler Chromophor [—S—Cu(I)—N$_{tert}$] und ein digonaler Chromophor [Cu(I) (N)$_2$] bzw. ein tetraedrisch koordinierter Chromophor [Cu(I) (N)$_4$] vorkommen.

2. Oxygenierung von Hämocyanin

Nimmt man an, daß bei der Oxygenierung des Hämocyanins wie in Hämoproteinen kein Ligand durch O_2 substituiert werden muß, sondern daß eine Erhöhung der Koordinationszahl des Kupfers eintritt, so wäre folgende O_2-Brücke im Oxyhämocyanin denkbar:

$$- S\diagdown \atop \diagup N \diagup \diagdown Cu ----\to O_2 \leftarrow ---- Cu \diagup \diagdown \qquad (26)$$

(trigonale Koordination der nächstliegenden Liganden)

wobei ein (CuCl-Azomethan) oder zwei (Glycyl-histidyl-Kupfer(II)-hydrat) weitere Liganden in größerem Abstande die Koordinationssphäre des Kupfers erweitern könnten.

Nord berichtete, daß die Kinetik der Oxidation von CuCl nur mit der Annahme einer stationären Konzentration eines der Oxidation vorgelagerten O_2-Adduktes erklärbar ist (*231*). Eine analoge Erklärung gibt *Zuberbühler* (*228*) für die Oxidation des Kations [Cu(I) (Imidazol)$_2$]$^+$ in wäßrigem Medium. *Fallab* et al. stellten die reversible Bildung eines O_2-Adduktes als Vorstufe der Oxidation des Cystein-Kupfer(I)-Komplexes fest (*50*).

3. Die Farbe des Oxyhämocyanins

Die Oxygenierung des farblosen Deoxyhämocyanins ist mit dem Auftreten mehrerer intensiver Lichtabsorptionsbanden zwischen 340 mµ und 700 mµ verbunden.

Das Absorptionsspektrum im sichtbaren Spektralbereich zeigt große Ähnlichkeit mit den Spektren von Kupfer(II)-Proteinen, s. Tabelle 14.

Tabelle 14. *Lichtabsorptionsspektren von Kupferproteinen [nach van Holde (232)].*

	Cu/Mol	Wertigkeit	Spektrum ab 300 mµ	mµ/ε
Hämocyanin	2 (Funktionseinheit)	I	—	—
Oxyhämocyanin	2 (Funktionseinheit)	I ↔ II	347/8900	440/<500
			570/500	700/<500
Plastocyanin	2	II		460/590
			597/4900	770/1700
Rhus vernicifera-Kupfer-Protein	1	II		450/970
			608/4030	850/700

Oxyhämocyanin und die Kupfer(II)-Proteine haben 3 Lichtabsorptionsbanden zwischen 400 mμ und 850 mμ gemeinsam. Dagegen fehlt in letzteren die intensive Absorption des Oxyhämocyanins bei 340 mμ.

Frieden et al. (*221*) führen die Absorptionen des Oxyhämocyanins im sichtbaren Spektralbereich auf Ligandenfeldübergänge zwischen d-Orbitalen eines Kupfer(II)-Komplexes niederer Symmetrie zurück, während die Bande bei 340 mμ auf einer Charge-transfer-Absorption beruhen soll (*232*). Diese Deutungen sind mit der Annahme eines valenzmesomeren Systems im Grundzustand des Oxyhämocyanins (*201, 233*)

$$Cu(I) \quad O_2(\downarrow\uparrow) \quad Cu(I) \;\longleftrightarrow\; Cu \cdots\!\rightarrow O_2 \leftarrow\!\cdots Cu \;\longleftrightarrow\; Cu(II)(\downarrow) \quad O_2^{\bullet-}(\uparrow) \quad Cu(I)$$

$$Cu(II)(\downarrow) \quad O_2^{2-} \quad Cu(II)(\uparrow)$$

no bond	π-Komplex	
Grenzstruktur	Schreibweise	ionische Grenzstrukturen
A	B	C

$$(27)$$

nur dann vereinbar, wenn optische Mehrfachanregung postuliert wird. Die charge transfer-Absorption sollte unter Beachtung des Franck-Condon-Prinzips ohne Änderung der Bindungsabstände zu einem kurzlebigen Zustand vollendeter Ladungsüberführung im Oxyhämocyanin führen. Während der Zeit der optischen Anregung liege Cu(II) vor, das die gleichen Ligandenfeldübergänge wie Kupfer(II)-Proteine ergeben sollte.

Da im Oxyhämocyanin kein Paramagnetismus festgestellt (*234*) und kein EPR-Signal erhalten wurde (*215, 223*), sowie durch NMR-spektroskopische Untersuchung der Spin-Gitter-Relaxationszeit des Wassers kein Restparamagnetismus nachgewiesen werden konnte (*5*, S. 119), sollte innerhalb der Gruppe Cu O₂ Cu Spinkopplung stattfinden. Da jedoch angenommen werden darf, daß die Oxygenierung in nicht acidem Medium erfolgt, ist die NMR-spektroskopische Untersuchung kein schlüssiger Beweis für fehlenden Paramagnetismus. Es bleibt noch zu prüfen, welches magnetische Verhalten Oxyhämocyanin bei tiefsten Temperaturen und Belichtung bei 340 mμ zeigt.

Im Zusammenhang mit diesem Problem eines lichtinduzierten Elektronentransfers im Oxyhämocyanin ist es von Interesse, daß der Kohlenmonoxidkomplex von Hämocyaninen, der sich im Verhältnis Cu : CO = = 2 : 1 bildet (*235, 236*), keine Absorptionen im Bereich der „Kupferbanden" liefert (*236*). CO ist kein redoxaktiver Ligand. Ein Elektronentransfer vom Metall zum Liganden kann nicht stattfinden.

4. Methämocyanin, Alterung des Oxyhämocyanins

Oxyhämocyanin geht langsam in Methämocyanin über, dabei verblassen die Absorptionen bei 340 mµ und 550 mµ (*224*), gleichzeitig tritt im EPR-Spektrum ein Cu(II)-Signal auf (*215, 223, 224*).

Nach *Lontie* et al. (*213, 224*) können die Kupferbanden bei 340 mµ und 550 mµ mit H_2O_2 regeneriert werden. Da diese Absorptionen charakteristisch für Oxyhämocyanin sind, führt die Reaktion offensichtlich zur Rückbildung des Oxyhämocyanins. Dies wäre das erste uns bekannte Beispiel der Darstellung eines reversibel oxygenierten Komplexes mit H_2O_2.

Die Reaktion

$$Cu(I) + O_2 + Cu(I) \underset{-2H^+}{\overset{2H^+}{\rightleftharpoons}} 2\,Cu(II) + HO-OH \tag{28}$$

scheint demnach bei Hämocyanin reversibel zu sein.

Bei der Alterung von Oxyhämocyanin verschwinden freie Mercaptogruppen (*213*), es ist nicht ausgeschlossen, daß es dabei zur Bildung von µ-Thiol-Brücken kommt, wie wir sie im Aquomethämerythrin postuliert haben.

$$Cu(II) - - - S - - - Cu(I) \text{ oder } Cu(II) - S \cdots Cu(II) \tag{29}$$
$$\qquad\qquad | \qquad\qquad\qquad\qquad\qquad | $$
$$\qquad\qquad R \qquad\qquad\qquad\qquad\qquad R $$
$$\qquad\qquad A \qquad\qquad\qquad\qquad\qquad B $$

C. Hämovanadin

In den Blutkörperchen mancher Tunikaten ist ein Vanadium-haltiges Nicht-Häm-Metallprotein enthalten, das Hämovanadin (*6, 237*). Nach Untersuchungen von *Bielig* und *Bayer* (*238, 239*) ist dieser Blutfarbstoff als Proteiniumsalz einer Disulfato-vanadin(III)-säure anzusehen. Seine physiologische Funktion ist noch nicht völlig geklärt. *Bielig* et al. (*240*) fanden, daß Hämovanadin Oxidasenfunktion ausüben kann. *Kovalskii* et al. (*241*) erhielten Anhaltspunkte dafür, daß Hämovanadin ein labiles O_2-Addukt bildet. *Kaden* (*242*) erklärt die Kinetik der Autoxidation von Vanadin-Chelaten mit der Annahme eines labilen O_2-Adduktes als Vorstufe der Oxidation.

D. Zusammenfassung

Die Chemie der Nicht-Häm-Metallproteine hat noch nicht den Stand der Kenntnis über die Hämoproteine erreicht. Sie ist jedoch in einer explosionsartigen Entwicklung begriffen. Als wesentliches gemeinsames Bauprinzip zeichnet sich die Tendenz zur Bildung von Mehrzentren-Chromophoren aus redoxaktiven Bindungspartnern ab. Dies ist in Abb. 27 schematisch wiedergegeben.

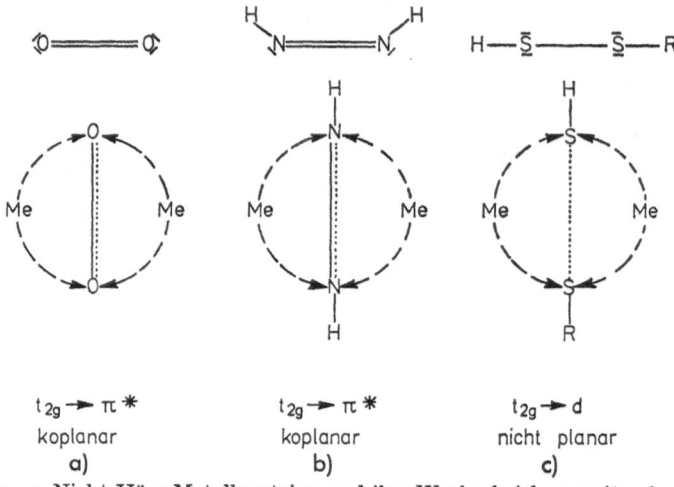

Abb. 27a—c. Nicht-Häm-Metallproteine und ihre Wechselwirkung mit redoxaktiven Liganden der Elemente O, N und S. a) O_2-Träger, Oxygenasen und Oxidasen (Fe, Cu), b) Nitrogenasen (Fe), c) Mercaptokomplex des Methämerythrins (Fe). R = Protein oder —H.

Das Problem der Bindung und der Valenzzustände in den Eisen-Sauerstoff- und Eisen-Schwefel-Systemen erfordert noch eingehende Untersuchungen. Daß dabei Elektronendelokalisierung zwischen Ligand und Zentralatom eine Rolle spielt, zeigten die EPR-spektroskopischen Untersuchungen an paramagnetischen Derivaten dieser Verbindungen (grünes paramagnetisches Kation [$(NH_3)_5$Co O_2 Co$(NH_3)_5$]$^{5+}$ (*72*), partiell reduziertes Nicht-Häm-Eisenprotein aus Azotobacter nach Markierung mit Fe^{57} und S^{33} (*243, 244*)).

IX. Schluß

Es ist heute noch nicht möglich, eine einheitliche Beschreibung des Bindungs- und Valenzzustandes in oxygenierten Metallkomplexen zu geben. Die Sauerstoffaddukte zeigen Merkmale von charge transfer-

Komplexen, Metall-Olefinkomplexen und mesomeren Verbindungstypen. Je nach der — durch das Ligandenfeld variierbaren — Donorstärke des Metalls scheint eine kontinuierliche Ladungsverschiebung vom Metall zur O_2-Gruppe möglich zu sein. Vergrößerung der Donorstärke des Metalls bewirkt erhöhte Affinität zum π-Acceptor Sauerstoff, erhöhte Irreversibilität des O_2-Adduktes, und vermutlich stets eine Vergrößerung des O—O-Kernabstandes. Als Beispiele seien zusammenfassend angeführt (Tabelle 15):

Tabelle 15. *Übergang von labilen zu irreversiblen Sauerstoff-Metallkomplexen.*

Zunahme der Affinität des Komplexes und zunehmende Stabilität des O_2-Adduktes

$[(O_2)Ir(CO)(Cl)(P(C_6H_5)_3)_2] <$ $[(O_2)Ir(CO)(J)(P(C_6H_5)_3)_2]$

              $[(O_2)Ir(P(C_6H_5)_2—(CH_2)_2—P(C_6H_5)_2]^+ <$

       A   C

                  B

                           (irreversibel)

$((SAD)—Co)_2 \cdot O_2 < [DMF(SAD—Co)]_2 \cdot O_2 < [(Pyridin)(SAD—Co)]_2 \cdot O_2$

      D   E   F

Oxyhämoglobin	<	Oxymyoglobin	<	Oxyleghämoglobin	<
Schaf (*3*)		Pferd (*3*)		Soya (*245*)	
($p_{1/2}$ = 3 mm Hg)		($p_{1/2}$ = 0,7 mm Hg)		($p_{1/2}$ = 0,04 mm Hg)	

          Oxyhämoprotein aus   <   Oxygenierte Peroxidase

              Ascaris (*246*)   Meerrettich (*247*)

       ($p_{1/2}$ = 0,001 mm Hg)   ($p_{1/2}$ = ?, irreversibel)

Die Bezeichnung als charge transfer-Komplex ist sicher bei den labilsten O_2-Addukten wie $(SAD—Co)_2 \cdot O_2$ nicht unberechtigt, ebenso ein Vergleich der irreversiblen Sauerstoffkomplexe wie $[(O_2)Ir(CO)(J)$ $(P(C_6H_5)_3)_2]$ mit echten Peroxo-Komplexen stark elektronegativer Zentralatome wie $[Cr(O_2)_4]^{3-}$.

Es scheint — zumindest bei den mononuklearen Verbindungen — möglich zu sein, das Kontinuum von den labilen O_2-Molekülkomplexen über die irreversiblen Sauerstoffkomplexe auf die echten Peroxo-Komplexe auszudehnen. Die Geometrie der Bindung in den niedermolekularen Verbindungen dieses Typs ist gelöst. Dagegen ist bei den binuklearen Sauerstoffkomplexen das Problem noch ungelöst, ob — kontinuierlich oder sprunghaft — eine Strukturumwandlung innerhalb des Chromophors MeO_2Me bei Zunahme der Elektronendichte auf der O_2-Gruppe stattfindet. Dazu sind Röntgenstruktur-Analysen innerhalb einer Reihe verwandter Verbindungen wie D, E, F in Tabelle 15 nötig.

E. Bayer und P. Schretzmann

Die Schwierigkeit, Bindungsverhältnisse in Sauerstoffkomplexen zu definieren, hat *Alfred Werner*, der geniale Begründer der Komplexchemie vor nahezu 70 Jahren erspürt und in einer Weise beschrieben, die auch heute noch Gültigkeit hat:

„.... *die Bindung des Sauerstoffs in den hier besprochenen Verbindungen ist eine so eigentümlich labile, daß man auf die Frage, ob Valenzbindung oder Molekular addition vorliegt, insofern diese Frage in dieser Form überhaupt eine Berechtigung hat, schwerlich eine befriedigende Antwort wird geben können."* [1]

Herrn stud. rer. nat. *P. Krauss* danken wir für die Anfertigung der Abbildungen.

X. Literatur

1. *Braunitzer, G., K. Hilse, V. Rudloff*, and *N. Hilschmann:* Adv. in Protein Chemistry *19*, 1 (1964).
2. *Rossi–Fanelli, A., E. Antonini*, and *A. Caputo:* Adv. in Protein Chemistry *19*, 74 (1964).
3. *Antonini, E.:* Physiol. Rev. *45*, 123 (1965).
4. *Lehmann, H.*, and *R. G. Huntsman:* Mens Hemoglobins. Amsterdam: North Holland Publ. Comp. 1966.
5. *Manwell, C.:* In *F. Dickens*, and *E. Neil*, Oxygen in the Animal Organism, S. 49. Oxford: Pergamon Press 1964.
6. *Bielig, H. J.*, u. *E. Bayer:* In *Hoppe-Seyler/Thierfelder*, Handbuch der physiologisch-chemischen Analyse. 10. Aufl., Bd. IV/1, S. 714. Berlin-Göttingen-Heidelberg: Springer 1960.
7. *Boeri, E.:* In *M. Florkin*, and *E. Stotz*, Comprehensive Biochemistry, Vol. 8, S. 38. Amsterdam: Elsevier Publ. Comp. 1963.
8. *Martell, A. E.*, and *M. Calvin:* Chemistry of Metal Chelate Compounds, S. 336. New York: Prentice Hall Inc. 1953.
9. *Vogt, L. H., H. M. Faigenbaum*, and *S. E. Wiberley:* Chem. Rev. *63*, 269 (1963).
10. *Stewart, R. F., P. A. Estep*, and *J. J. S. Sebastian:* U. S. Bureau Mines Inform. Circ. No. 7906 (1959).
11. *Griffith, J. S.:* In *F. Dickens*, and *E. Neil*, Oxygen in the Animal Organism, S. 141. Oxford: Pergamon Press 1964.
12. *Orgel, L.:* In *T. E. King, H. S. Mason*, and *M. Morrison*, Oxidases and Related Redox Systems, S. 110. New York: J. Wiley Inc. 1965.
13. *Solomon, I. J., R. I. Brabets, R. K. Uenishi, J. N. Keith*, and *J. M. McDonough:* Inorg. Chem. *3*, 457 (1964).
14. *Bartlett, N.*, and *D. H. Lohmann:* Proc. Chem. Soc. *1962*, 115.
15. *Al Joboury, M. I., D. W. Turner*, and *D. P. May:* J. Chem. Soc. *1965*, 616.
16. *McNeal, R. J.*, and *G. R. Cook:* J. Chem. Phys. *45*, 3469 (1966).
17. *Pritchard, H. O.:* Chem. Rev. *52*, 552 (1953).
18. *D'Orazio, L. A.*, and *R. H. Wood:* J. Phys. Chem. *69*, 2550 (1965).
19. *George, Ph.:* In *T. E. King, H. S. Mason*, and *M. Morrison*, Oxidases and Related Redox Systems, S. 3. New York: J. Wiley Inc. 1965.

[1] Wörtlich zitiert aus: *Werner, A.*, u. *A. Mylius:* Z. Anorg. Chem. *16*, 245 (1898).

20. *Taube, H.:* J. Gen. Physiology *49, Pt. 2,* 29 (1965).
21. *Arnold, S. J., N. Finlayson,* and *E. A. Ogryzlo:* J. Chem. Phys. *44,* 2529 (1966).
22. *Ogryzlo, E. A.:* J. Chem. Educ. *42,* 647 (1965).
23. *Foote, C. S.,* and *S. Wexler:* J. Amer. Chem. Soc. *86,* 3880 (1964).
24. *Corey, E. J.,* and *W. C. Taylor:* J. Amer. Chem. Soc. *86,* 3881 (1964).
25. *Arbuzov, Y. A.:* Russ. Chem. Rev. *1965,* 558.
26. *McKeown, E.,* and *W. A. Waters:* J. Chem. Soc. B *1966,* 1040.
27. *Griffith, J. S.:* J. Chem. Phys. *40,* 2899 (1964).
28. — Proc. Roy. Soc. A *235,* 23 (1956).
29. *Chatt, J.,* and *L. A. Duncanson:* J. Chem. Soc. *1953,* 2939.
30. Nach *Abrahams, S. C.:* Quart. Rev. *10,* 407 (1956).
31. *Higginbotham, H. K.:* Diss. Abs. *1964,* 2779.
32. *Abrahams, S. C., R. L. Collin,* and *W. N. Lipscomb:* Acta Cryst. *4,* 15 (1951).
33. *Sax, M.,* and *R. K. McMullan:* Acta Cryst. *22,* 281 (1967).
34. *Jackson, R. H.:* J. Chem. Soc. *1962,* 4585.
35. Nach *Connor, J. A.,* and *E. A. V. Ebsworth:* Adv. Inorg. Radiochem. *6,* 279 (1964).
36. *Ibers, J. A.,* and *S. J. La Placa:* J. Amer. Chem. Soc. *87,* 2581 (1965).
37. *Vannerberg, N. G.:* Acta Cryst. *18,* 449 (1965).
38. *Schaefer, W. P.,* and *R. E. Marsh:* Acta Cryst. *21,* 735 (1966).
39. *Vannerberg, N. G.,* and *C. Brosset:* Acta Cryst. *16,* 247 (1963).
40a. *Jørgensen, C. K.:* Seminarwoche über Strukturchemie biologisch aktiver Metallkomplexe, Basel 23. 2. 1965.
40b. — Z. nat.-med. Grundlagenforschung *2,* 248 (1965).
41a. *Evans, D. F.:* J. Chem. Soc. *1953,* 345.
41b. — J. Chem. Soc. *1957,* 3885.
42. *Tsubomura, H.,* and *R. S. Mulliken:* J. Amer. Chem. Soc. *82,* 5966 (1960).
43. *Navon, J.:* J. Phys. Chem. *68,* 969 (1964).
44. *Heijdt, L. J.,* and *A. M. Johnson:* J. Amer. Chem. Soc. *79,* 5587 (1957).
45. *Jortner, J.,* and *K. Sokolow:* J. Phys. Chem. *65,* 1633 (1961).
46. *Bailey, A. S., R. J. P. Williams,* and *J. D. Wright:* J. Chem. Soc. *1965,* 2579.
47a. *Kamenar, B., C. K. Prout,* and *J. D. Wright:* J. Chem. Soc. *1966 A,* 661.
47b. — — — *A. S. Bailey,* and *H. M. Powell:* J. Chem. Soc. *1965,* 4838, 4851, 4867, 4882.
48. *Fallab, S.:* Z. nat.-med. Grundlagenforschung *2,* 220 (1965).
49. *Przywarska–Boniecka, H.:* Roczniki Chem. *39,* 1377 (1965).
50. *Graf, L.,* u. *S. Fallab:* Experientia *XX,* 46 (1964).
51. *Frieden, E., J. A. McDermott,* and *S. Osaki:* In *T. E. King, H. S. Mason,* and *M. Morrison.* Oxidases and Related Redox Systems, S. 257. New York: J. Wiley Inc. 1965.
52. *Vaska, L.:* Science *140,* 809 (1963).
53. —, and *R. E. Rhodes:* J. Amer. Chem. Soc. *87,* 4970 (1965).
54. *Parshall, G. W.,* and *F. N. Jones:* J. Amer. Chem. Soc. *87,* 5356 (1965).
55. *Sacco, A., M. Rossi,* and *C. F. Nobile:* Chem. Comm. *1966,* 589.
56. *Mague, J. T.,* and *G. Wilkinson:* J. Chem. Soc. *1966 A,* 1736.
57. *Vaska, L,* and *D. I. Catone:* J. Amer. Chem. Soc. *88,* 5324 (1966).
58. *Wilke, G., H. Schott* u. *P. Heimbach:* Angew. Chemie *79,* 62 (1967).
59. *Cook, C. D.,* and *G. S. Jauhal:* Inorg. Nuclear Chem. Letters. *3,* 31 (1967).
60. *Baddley, W. H.,* and *L. M. Venanzi:* Inorg. Chem. *5,* 33 (1966).
61. *McClellan, W. R.,* and *R. E. Benson:* J. Amer. Chem. Soc. *88,* 5165 (1966).
62. *Gluud, W., K. Keller* u. *A. Nordt:* Ber. Ges. Kohlentechnik *4,* 210 (1931), nach Gmelins Handbuch der Anorg. Chemie, Bd. 58, Teil B, Erg. Bd. Lfg. 2, S. 615.

245

63. *Werner, A.*, u. *A. Mylius:* Z. Anorg. Chem. *16*, 245 (1898).
64. *Alcock, N. W.:* Chem. Comm. *1966*, 536.
65. *Abrahamson, S. G.:* Diss. Abs. *25*, 4950 (1965).
66. *Bosnich, B., C. K. Poon*, and *M. L. Tobe:* Inorg. Chem. *5*, 1514 (1966).
67. *Erdem, B.*, u. *S. Fallab:* Chimia *19*, 463 (1965).
68. *Fallab, S.:* Chimia *19*, 174 (1965).
69. *Haim, A.*, and *W. K. Wilmarth:* J. Amer. Chem. Soc. *83*, 509 (1961).
70. *Vlček, A. A.:* Trans. Far. Soc. *56*, 1137 (1960).
71. Nach *Vlček, A. A.*, and *F. Basolo:* Inorg. Chem. *5*, 156 (1966).
72. *Ebsworth, E. A. V.*, and *J. A. Weil:* J. Phys. Chem. *63*, 1890 (1959).
73. *Mori, M., J. A. Weil*, and *J. K. Kinnaird:* J. Phys. Chem. *71*, 103 (1967).
74. *Sano, Y.*, and *H. Tanabe:* J. Inorg. Nuclear Chem. *25*, 11 (1963).
75. *Hartwig, G.:* Dissertation Universität Tübingen, 1966.
76. *Crook, E. M.*, and *B. R. Rabin:* Biochem. J. *68*, 177 (1958).
77. *Tang, Ph.*, and *N. C. Li:* J. Amer. Chem. Soc. *86*, 1293 (1964).
78. *Tanford, C., D. C. Kirk*, and *M. K. Chantooni:* J. Amer. Chem. Soc. *76*, 5325 (1954).
79. *Caglioti, V., P. Silvestroni*, and *C. Furlani:* J. Inorg. Nucl. Chem. *13*, 95 (1960).
80. a) *Gillard, R. D., R. Mason, E. D. McKenzie*, and *G. B. Robertson:* Coordination Chem. Rev. *1*, 263 (1966), b) Nature *209*, 1347 (1966).
81. *Tsumaki, T.:* Bull. Chem. Soc. Japan *13*, 252 (1938).
82. *Diehl, H., C. C. Hach, G. C. Harrison, L. M. Ligget*, and *T. S. Chao:* Iowa State Coll. J. Sci. *21*, 283 (1947).
83. *Hughes, E. W., C. H. Barkelew*, and *M. Calvin:* OEMsr - *279*, 1 (1944) nach *A. E. Martell*, and *M. Calvin*, Chemistry of Metal Chelate Compounds, S. 267. New York: Prentice Hall Inc. 1953.
84. *Calvin, M., R. H. Bailes*, and *W. K. Wilmarth:* J. Amer. Chem. Soc. *68*, 2254 (1946).
85. *Ueno, K.*, and *A. E. Martell:* J. Phys. Chem. *60*, 1270 (1956).
86. *Ferroni, E.*, and *A. Ficalbi:* Gazzetta *89*, 750 (1959).
87. *Yamada, S.*, and *H. Nishikawa:* Bull. Chem. Soc. Japan *37*, 8 (1964).
88. — —, and *E. Yoshida:* Proc. Japan Academy *40*, 211 (1964).
89. *Figgis, B. M.*, and *R. S. Nyholm:* J. Chem. Soc. *1954*, 12.
90. *West, B. O.:* J. Chem. Soc. *1954*, 395.
91. *Bailes, R. H.*, and *M. Calvin:* J. Amer. Chem. Soc. *69*, 1886 (1947).
92. *Hewlett, P. C.*, and *L. F. Larkworthy:* J. Chem. Soc. *1965*, 882.
93. *Calderazzo, F., C. Floriani*, and *J. J. Salzmann:* Inorg. Nucl. Chem. Letters *2*, 379 (1966).
94. *Earnshaw, A., P. C. Hewlett*, and *L. F. Larkworthy:* Nature *199*, 483 (1963).
95. *Hall, D.*, and *F. H. Moore:* Proc. Chem. Soc. *1960*, 256.
96. *Gerloch, M., J. Lewis, F. E. Mabbs*, and *A. Richards:* Nature *212*, 810 (1966).
97. *Alderman, P. R. H., P. G. Owston*, and *J. M. Rowe:* J. Chem. Soc. *1962*, 668.
98. *Martell, A. E.*, and *M. Calvin:* Chemistry of Metal Chelate Compounds, S. 347. New York: Prentice Hall Inc. 1953.
99. *Briegleb, G.:* Elektronen-Donator-Acceptor-Komplexe, S. 125. Heidelberg: Springer 1961.
100. *Tsumaki, H.:* J. chem. Soc. Japan *58*, 1288 (1937) nach Gmelins Handbuch der Anorg. Chem., 8. Aufl., Bd. 58 B, Erg. Bd., Lfg. 1, 217.
101. *Briegleb, G.:* Elektronen-Donator-Acceptor Komplexe, S. 71. Heidelberg: Springer 1961.
102. *Staab, H. A.:* Einführung in die theoretische organische Chemie, 2. Aufl. S. 705. Weinheim: Verlag Chemie 1960.

103. *Rüdorff, W., E. Stumpp, W. Spriessler* u. *F. W. Siecke:* Angew. Chemie *75*, 130 (1963).
104. *Martell, A. E.,* and *M. Calvin:* Chemistry of Metal Chelate Compounds, S. 345. New York: Prentice Hall Inc. 1953.
105. *Diehl, H.:* Iowa State Coll. J. Sci. *22*, 271 (1946).
106. Zusammenfassung in Gmelins Handbuch, 8. Aufl., Bd. 58 B, Erg. Bd., Lfg. 1, 214.
107. *Selke, R.,* and *H. W. Krause:* Z. Physiol. Chem. *340*, 181 (1965).
108. *McGinnety, J. A., R. J. Doedens,* and *J. A. Ibers:* Science *155*, 709 (1967).
109. Nach *Falk, J. E.:* Porphyrins and Metalloporphyrins, S. 102. Amsterdam: Elsevier Publ. Comp 1964.
110. *Perutz, M. F.:* J. Mol. Biol. *13*, 646 (1965).
111. —, *Kendrew, J. C.,* and *H. C. Watson:* J. Mol. Biol. *13*, 669 (1965).
112. *Roche, J.:* Studies Comp. Biochem. *1965*, 62.
113. *Guerritore, D., M. L. Bonacci, M. Brunori, E. Antonini, J. Wymann,* and *A. Rossi-Fanelli:* J. Mol. Biol. *13*, 234 (1965).
114. *Cullis, A. F., H. Muirhead, M. F. Perutz, M. G. Rossmann,* and *A. C. T. North:* Proc. Roy. Soc. *A 265*, 161 (1962).
115. Zusammenfassung *Muirhead, H.:* Fortschr. Chem. Forschung *6*, 41 (1966).
116. *Perutz, M. F.:* Umschau *66*, 597 (1966).
117. *Antonini, E., J. Wyman, M. Brunori, C. Fronticelli, E. Bucci,* and *A. Rossi-Fanelli:* J. Biol. Chem. *240*, 1096 (1965).
118. *Tanford, C.,* and *Y. Nozaki:* J. Biol. Chem. *241*, 2832 (1966).
119. *Kendrew, J. C.:* Angew. Chemie *75*, 595 (1963).
120. *Nobbs, C. L., H. C. Watson,* and *J. C. Kendrew:* Nature *209*, 339 (1966).
121. *Schönborn, B. P., H. C. Watson,* and *J. C. Kendrew:* Nature *207*, 28 (1965).
122. *Nobbs, C. L.:* nach *B. Chance, R. W. Estabrook,* and *R. Yonetani,* Science *152*, 1409 (1966).
123. *Stryer, L., J. C. Kendrew,* and *H. C. Watson:* J. Mol. Biol. *8*, 96 (1964).
124. *Yamazaki, I., K. Yokota,* and *K. Shikama:* J. Biol. Chem. *239*, 4151 (1964).
125. *Hardman, K., E. H. Eylar, D. K. Ray, L. J. Banaszak,* and *F. R. N. Gurd:* J. Biol. Chem. *241*, 432 (1966).
126. *Kendrew, J. C.:* Brookhaven Symp. Biol. *15*, 216 (1962).
127. *Dickerson, R. E.:* In *H. Neurath,* The Proteins, Vol. 2, S. 603. New York: Academic Press 1964.
128. *Rossi-Fanelli, A., E. Antonini,* and *A. Caputo:* J. Biol. Chem. *234*, 2906 (1959).
129. *Gibson, Q. H.:* nach *A. Rossi-Fanelli, E. Antonini,* and *A. Caputo,* Adv. in Protein Chem. *19*, 74 (1964), S. 138.
130. *Rossi-Fanelli, A., E. Antonini,* and *A. Caputo:* Adv. Protein Chem. *19*, 74, 135, 141 (1964).
131. *Gibson, Q. H.,* and *E. Antonini:* J. Biol. Chem. *238*, 1384 (1963).
132. *Cann, J. R.:* Biochemistry *4*, 2368 (1965).
133. *Teale, F. W. J.:* Biochim. Biophys. Acta *35*, 289 (1959).
134. *Kendrew, J. C., R. E. Dickerson, B. E. Strandberg, R. G. Hart, D. R. Davis, D. C. Phillips,* and *V. C. Shore:* Nature *185*, 422 (1960).
135. *König, D. F.:* Acta Cryst *18*, 663 (1965).
136. *Hoard, J. L., M. J. Hamor, T. A. Hamor,* and *W. S. Caughey:* J. Amer. Chem. Soc. *87*, 2312 (1965).
137. *Wang, J. H., A. Nakahara,* and *E. B. Fleischer:* J. Amer. Chem. Soc. *80*, 1109 (1958).
138. *Kao, O. H. W.,* and *J. H. Wang:* Biochemistry *4*, 342 (1965).
139. *Corwin, A. H.,* and *Z. Reyes:* J. Amer. Chem. Soc. *78*, 2437 (1956).

140. *Lemberg, R.,* and *J. W. Legge:* Haematin Compounds, S. 184. New York: Interscience 1949.
141. *Wang, J. H.:* J. Amer. Chem. Soc. *80,* 3168 (1958).
142. — In *J. E. Falk, R. Lemberg,* and *R. K. Morton,* Haematin Enzymes, S. 98. New York: Pergamon Press 1961.
143. — In *O. Hayaishi,* Oxygenases, S. 470. New York: Academic Press 1962.
144. *Pauling, L.:* Nature *203,* 182 (1964).
145. *Alderman, P. R., P. G. Owston,* and *J. M. Rowe:* Acta Cryst. *13,* 149 (1960).
146. *Maggiora, G. M., R. O. Viale,* and *L. L. Ingraham:* In *T. E. King, H. S. Mason,* and *M. Morrison,* Oxidases and related Redox Systems, S. 88. New York: J. Wiley Inc. 1965.
147. *Nelson, S. M., P. Bryan,* and *D. H. Busch:* Chem. Comm. *1966,* 641.
148. *Cohen, G. H.,* and *J. L. Hoard:* J. Amer. Chem. Soc. *88,* 3228 (1966).
149. *Busch, D. H.:* Helv. Chim. Acta Fasciculus extraordinarius "Alfred Werner", 174 (1967).
150. *Viale, R. O., G. M. Maggiora,* and *L. L. Ingraham:* Nature *203,* 183 (1964).
151. *Lang, G.,* and *W. Marshall:* Proc. Phys. Soc. *87,* 3 (1966).
152. *Fritz, H. P., K. E. Schwarzhans,* and *D. Sellmann:* J. Organomet. Chem. *6,* 551 und 558 (1966).
153. *Corwin, A. H.,* and *S. D. Bruck:* J. Amer. Chem. Soc. *80,* 4736 (1958).
154. *Wang, J. H.:* In *J. E. Falk, R. Lemberg,* and *R. K. Morton,* Haematin Enzymes, S. 76. New York: Pergamon Press 1961.
155. *Fabry, L.,* and *H. A. Reich:* Biochem. Biophys. Res. Comm. *22,* 700 (1966).
156. *Smith, J. H. C.:* In Proc. 2nd Intern. Congr. Photobiol., S. 333. Turin: Minerva Medica 1957. Nach *J. E. Falk,* Porphyrins and Metalloporphyrins, S. 47. Amsterdam: Elsevier Publ. Comp. 1964.
157. *Phillips, J. N.:* In *M. Florkin,* and *E. Stotz,* Comprehensive Biochemistry, Vol. 9, S. 34. Amsterdam: Elsevier Publ. Comp. 1963.
158. *Staab, H. A.:* Einführung in die theoretische organische Chemie, S. 630—640. Weinheim: Verlag Chemie 1960.
159. *King, T. E.:* In *T. E. King, H. S. Mason,* and *M. Morrison,* Oxidases and related Redox Systems, S. 35. New York: J. Wiley-Inc. 1965.
160. *Weiss, J. J.:* Nature *202,* 83 (1964).
161. *Vannerberg, N. G.:* In *F. A. Cotton,* Progr. Inorg. Chem., Vol. 4, S. 125. Interscience Publ. 1962.
162. *Peover, M. E.,* and *B. S. White:* Chem. Comm. *1965,* 183.
163. *Slough, W.:* Chem. Comm. *1965,* 184.
164. *Maricle, D. L.,* and *W. G. Hodgeson:* Analyt. Chem. *37,* 1562 (1965).
165. *Briegleb, G.:* Elektronen-Donator-Acceptor-Komplexe, S. 170. Heidelberg: Springer 1961.
166. *Wittenberg, B. A., R. W. Briehl,* and *J. B. Wittenberg:* Biochem. J. *96,* 363 (1965).
167. *Rossi-Fanelli, A., E. Antonini,* and *D. Poveledo:* In *A. Neuberger,* Symposium on Protein Structure, S. 144. London: 1958.
168. *Edmundson, A. B.:* Nature *205,* 883 (1965).
169. *Gratzer, W. B.:* In DMS-UV Atlas organischer Verbindungen, Vol. 1, J 4/1 und J 4/2. Weinheim: Verlag Chemie 1966.
170. *Frick, P. G., W. H. Hitzig,* and *K. Betke:* Blood *XX,* 261 (1962).
171. *Muller, C. J.,* and *S. Kingma:* Biochim. Biophys. Acta *50,* 595 (1961).
172. *Bachmann, F.,* and *H. R. Marti:* Blood *XX,* 272 (1962).
173. *Murawski, K., S. Carta, M. Sorcini, L. Tentori, G. Vivaldi, E. Antonini, M. Brunori, J. Wyman, E. Bucci,* and *A. Rossi-Fanelli:* Arch. Biochem. Biophys. *111,* 197 (1965).

248

174. *Perutz, M. F.:* Proteins and Nucleic Acids, S. 48. Amsterdam: Elsevier Publ. Comp. 1962.
175. *Antonini, E., M. Brunori, A. Caputo, E. Chiancone, A. Rossi-Fanelli,* and *J. Wyman:* Biochim. Biophys. Acta *79,* 284 (1964).
176. — *A. Rossi-Fanelli,* and *A. Caputo:* Arch. Biochem. Biophys. *97,* 336 (1962).
177. *Falk, J. E., J. N. Phillips,* and *E. A. Magnusson:* Nature *212,* 1531 (1966).
178. *Caughey, W. S., J. O. Alben,* and *C. A. Beaudreau:* In *T. E. King, H. S. Mason,* and *M. Morrison,* Oxidases and related Redox Systems, S. 97. New York: J. Wiley Inc. 1965.
179. *Williams, R. J. P.:* In *J. E. Falk, R. Lemberg,* and *R. K. Morton,* Haematin Enzymes, S. 41. New York: Pergamon Press 1961.
180. *Drake, J. F.,* and *R. J. P. Williams:* Nature *182,* 1084 (1958).
181. *Vonderschmitt, D., K. Bernauer,* u. *S. Fallab:* Helv. Chim. Acta *48,* 951 (1965).
182. *Thojo, M.,* and *K. Shibata:* Arch. Biochem. Biophys. *103,* 401 (1963).
183. *Hatano, M.:* Kagaku To Kogyo (Tokyo) *18,* 926 (1965).
184. *Ghiretti, F.:* In *O. Hayaishi,* Oxygenases, S. 517. New York: Academic Press 1962.
185. *Klotz, I. M.,* and *T. A. Klotz:* Science *121,* 477 (1955).
186. — — Arch. Biochem. Biophys. *68,* 284 (1957).
187. *Bayer, E.:* Chimia *16,* 333 (1962).
188. *Klotz, I. M.,* and *S. Keresztes-Nagy:* Biochemistry *2,* 445 (1963).
189. *Roche, A.,* and *J. Roche:* Bull. soc. chim. biol. *17,* 1494 (1935).
190. *Manwell, C.:* Science *139,* 755 (1963).
191. *Keresztes-Nagy, S.,* and *I. M. Klotz:* Biochemistry *4,* 919 (1965).
192. *Groskopf, W. R., J. W. Holleman, E. Margoliash,* and *I. M. Klotz:* Biochemistry *5,* 3779 und 3783 (1966).
193. *Kubo, M.:* Bull. Chem. Soc. Japan *26,* 244 (1953).
194. *Bayer, E.,* u. *W. Parr:* Angew. Chemie *78,* 824 (1966).
195. *Buchanan, B. B.:* In *C. K. Jørgensen, J. S. Neilands, R. S. Nyholm, D. Reinen,* and *R. J. P. Williams,* Structure and Bonding, Vol. 1, S. 109. Heidelberg: Springer 1966.
196. *Bayer, E.,* u. *A. Röder:* Angew. Chemie *79,* 274 (1967).
197. *Keresztes-Nagy, S.,* and *E. Margoliash:* J. Biol. Chem. *241,* 5955 (1966).
198. *Bayer, E., H. Hagenmaier, P. Krauss* u. *D. Josef:* unveröffentlicht.
199. *Manwell, C.:* Science *132,* 550 (1960).
200. *Boeri, E.,* and *A. Ghiretti-Magaldi:* Biochim. Biophys. Acta *23,* 489 (1957).
201. *Manwell, C.:* Ann. Rev. Physiol. *22,* 191 (1960).
202. *Klotz, I. M.,* and *S. Keresztes-Nagy:* Biochemistry *2,* 923 (1963).
203. *Florkin, M.:* Arch. Intern. Physiol. *36,* 247 (1933).
204. *Müller, J.,* and *K. Dehnicke:* J. Organometallic Chem. *7,* P1 (1967).
205. *George, P., J. Beetlestone,* and *J. S. Griffith:* In *J. E. Falk, R. Lemberg,* and *R. K. Morton,* Haematin Enzymes, S. 105. New York: Pergamon Press 1961.
206. *Scheler, W., G. Schoffa,* u. *F. Jung:* Biochem. Z. *329,* 232 (1957).
207. *Pauling, L.:* In *F. J. W. Roughton,* and *J. C. Kendrew,* Haemoglobin, S. 57. London: Butterworths 1949.
208. *Orgel, L.:* In *K. Bloch,* and *O. Hayaishi,* Biological and Chemical Aspects of Oxygenases. Tokyo: Maruzen Comp. 1967; nach *P. Feigelson,* Science *155,* 615 (1967).
209. *Knauer, K., P. Hemmerich* u. *J. W. D. van Voorst:* Angew. Chemie *79,* 273 (1967).
210. *Takeda, Y.,* and *S. Senoh:* nach *P. Feigelson,* Science *155,* 611 (1967).
211. *Fernandez-Moran, H., E. F. J. van Bruggen,* and *M. Ohtsuki:* J. Mol. Biol. *16,* 191 (1966).

E. Bayer und P. Schretzmann

212. Zusammenfassung *Klotz, I. M.:* Science *155,* 697 (1967).
213. *Lontie, R.,* and *R. Witters:* In *J. Peisach, P. Aisen,* and *W. E. Blumberg,* The Biochemistry of Copper, S. 455. New York: Academic Press 1966.
214. *Ghiretti-Magaldi, A., C. Nuzzolo,* and *F. Ghiretti:* Biochemistry *5,* 1943 (1966).
215. *Bayer, E.,* u. *H. Fiedler:* Liebigs Ann. Chem. *653,* 149 (1962).
216. *Redfield, A. C.:* Biol. Rev. Cambridge Phil. Soc. *9,* 175 (1934).
217. *Felsenfeld, G.,* and *M. P. Printz:* J. Amer. Chem. Soc. *81,* 6259 (1959).
218. — Arch. Biochem. Biophys. *87,* 247 (1960).
219. *Pickett, S. M., A. F. Riggs,* and *J. L. Larimer:* Science *151,* 1005 (1966).
220. *Peisach, J., P. Aisen,* and *W. E. Blumberg:* The Biochemistry of Copper. New York: Academic Press 1966.
221. *Frieden, E., S. Osaki,* and *H. Kobayashi:* J. Gen. Physiol. *49,* Pt. 2, 213 (1965).
222. *Brill, A. S., R. B. Martin,* and *R. J. P. Williams:* In *B. Pullmann,* Electronic Aspects of Biochemistry, S. 519. New York: Academic Press 1964.
223. *Nakamura, T.,* and *H. S. Mason:* Biochem. Biophys. Res. Comm. *3,* 297 (1960).
224. *Heirwegh, K., V. Blaton,* and *R. Lontie:* Arch. Intern. Physiol. *73,* 149 und 151 (1965).
225. *Jørgensen, C. K.:* In *J. Peisach, P. Aisen,* and *W. E. Blumberg,* The Biochemistry of Copper, S. 1. New York: Academic Press 1966.
226. *Hemmerich, P.:* In *J. Peisach, P. Aisen,* and *W. Blumberg,* The Biochemistry of Copper, S. 15. New York: Academic Press 1966.
227. *Blount, J. F., K. A. Fraser, H. C. Freeman, J. T. Szymanski,* and *C. H. Wang:* Acta Cryst. *22,* 396 (1967).
228. *Zuberbühler, A.:* Helv. Chim. Acta. *50,* 466 (1967).
229. *Diehls, O.,* u. *W. Koll:* Liebigs Ann. Chem. *443,* 262 (1925).
230. *Brown, I. D.,* and *J. D. Dunitz:* Acta Cryst. *13,* 28 (1960).
231. *Nord, H.:* Acta Chem. Scand. *9,* 430 und 438 (1955).
232. *Van Holde, K. E.:* Biochemistry *6,* 93 (1967).
233. *Orgel, L. E.:* Biochem. Soc. Symposium *15,* 8 (1958).
234. *Rawlinson, W. A.:* Austral. J. Exptl. Biol. Med. Sci. *18,* 131 (1940).
235. *Rocca, E.,* and *F. Ghiretti:* Boll. Soc. Ital. Biol., Sper. *39,* 2075 (1963).
236. *Vanneste, W.,* and *H. S. Mason:* In *J. Peisach, P. Aisen,* and *W. E. Blumberg,* The Biochemistry of Copper, S. 465. New York: Academic Press 1966.
237. *Burton, J. D.:* Nature *212,* 976 (1966).
238. *Bielig, H. J.,* u. *E. Bayer:* Liebigs Ann. Chem. *580,* 135 (1953).
239. — — Pubbl. Staz. Zool. Napoli *29,* 109 (1957).
240. — *E. Jost, K. Pfleger* u. *W. Rummel:* Z. Physiol. Chem. *325,* 132 (1961).
241. *Kovalskii, V. V.,* and *L. T. Rezaeva:* Zh. Obsch. Biol. *25,* 339 (1964).
242. *Kaden, T.:* Helv. Chim. Acta *49,* 1915 (1966).
243. *Shetna, Y. I., P. W. Wilson, R. E. Hansen,* and *H. Beinert:* Proc. Natl. Acad. Sci. US *52,* 1263 (1964).
244. *Hollocher, T. C., F. Solomon,* and *T. E. Ragland:* J. Biol. Chem. *241,* 3452 (1966).
245. *Appleby, C. A.:* Biochim. Biophys. Acta *60,* 226 (1962).
246. *Okazaki, T.,* and *J. B. Wittenberg:* Biochem. Biophys. Acta *111,* 503 (1965).
247. *Wittenberg, J. B., R. W. Noble, B. A. Wittenberg, E. Antonini, M. Brunori,* and *J. Wyman:* J. Biol. Chem. *242,* 626 (1967).

(Eingegangen am 28. März 1967)

Structure and Bonding: Contents Vol. 1 and 2